Genders and Sexualities in History

Palgrave Macmillan's series, Genders and Sexualities in History, accommodates and fosters new approaches to historical research in the fields of genders and sexualities. The series promotes world-class scholarship, which concentrates upon the interconnected themes of genders, sexualities, religions/religiosity, civil society, politics and war.

Historical studies of gender and sexuality have, until recently, been more or less disconnected fields. In recent years, historical analyses of genders and sexualities have synthesised, creating new departures in historiography. The additional connectedness of genders and sexualities with questions of religion, religiosity, development of civil societies, politics and the contexts of war and conflict is reflective of the movements in scholarship away from narrow history of science and scientific thought, and history of legal processes approaches, that have dominated these paradigms until recently. The series brings together scholarship from Contemporary, Modern, Early Modern, Medieval, Classical and Non-Western History. The series provides a diachronic forum for scholarship that incorporates new approaches to genders and sexualities in history.

More information about this series at
http://www.palgrave.com/gp/series/15000

Yuki Terazawa

Knowledge, Power, and Women's Reproductive Health in Japan, 1690–1945

palgrave
macmillan

Yuki Terazawa
Hofstra University
Hempstead, NY, USA

Genders and Sexualities in History
ISBN 978-3-030-10312-5 ISBN 978-3-319-73084-4 (eBook)
https://doi.org/10.1007/978-3-319-73084-4

Cover illustration: "The Birth of the First Child (Uizan)", Suzuki Harunobu, 1725–1770 (Detail). A scene of a childbirth at a presumably wealthy merchant's household in Edo (Tokyo). The mother sits on a birth chair to the left. The woman who is wiping the child is most likely a hired midwife. Photograph © 1769. Museum of Fine Arts, Boston.

Printed on acid-free paper

This Palgrave Macmillan imprint is published by the registered company Springer International Publishing AG part of Springer Nature
The registered company address is: Gewerbestrasse 11, 6330 Cham, Switzerland

Series Editor' Preface

One approach to the study of human bodies and sexuality involves uncovering the diverse interpretations of physicality in other cultures and during different periods of history. Yuki Terazawa's *Knowledge, Power, and Women's Reproductive Health in Japan, 1690–1945* explores how sexual difference was understood in Japan from the early modern era to the mid-twentieth century. She traces the gradual introduction into Japanese culture of the "modern" sexed body, unpicking the political, cultural, institutional, and medical forces that led to the acceptance of the anatomical body as revealed in dissections. She is interested in unlicensed healers as well as organized medicine. Crucially, she argues that shifting interpretations of the body had major repercussions for gender relations. In common with all the volumes in the "Gender and Sexualities in History" series, *Knowledge, Power, and Women's Reproductive Health in Japan* is a multifaceted and meticulously researched scholarly study. It is an exciting contribution to our understanding of gender and sexuality in the past.

John H. Arnold
Joanna Bourke
Sean Brady

Acknowledgments

This book is a culmination of scholarly endeavors over 20 years, including my years as a graduate student and as a Hofstra University faculty member. Without the support of mentors and colleagues, I could not have worked on this extensive and, at times, overwhelming project.

I would first like to express my gratitude to my mentors at the University of California, Los Angeles (UCLA), including, professors in Japanese history, Fred Notehelfer, Herman Ooms, and the late Miriam Silverberg. Professors Benjamin Elman and Francesca Bray in Chinese studies helped me acquire a background in the history of East Asian science and medicine. Professors Janet Goodwin at UCLA and Anne Walthall at the University of California, Irvine (UCI), generously gave me professional advice, and they continued to support me after I graduated from UCLA. Lastly, special thanks to Professor Sharon Traweek, who works in the cultural studies of science, technology, and medicine, for her crucial theoretical insights into this project.

Many colleagues supported me through the long and rocky road of graduate studies. The following are only a fraction of the friends who helped me during my years in LA: the late David Case, Yüan-ling Chao, Elyssa Faison, Karen Flint, Elisabeth Frost, Betsy Jacobson, Susie Kim, Jin-kyung Lee, Elizabeth Leicester, Yue Meng, Bettina Oka, Toshiko Scott, and Rumi Yasutake.

Colleagues who have given me important support in the history of science, medicine, and technology in Japan and East Asia include Juliette Chung, Yoko Matsubara, Keiko Ogawa, and Sumiko Otsubo.

I am also greatly indebted to two copy editors, Dan Harwig and Robert Sauté, and the staff at Hofstra's Educational and Research Technology Services, Joseph Zona, in particular.

Special thanks to Professor Shizu Sakai in the Department of Medical History at Juntendō University for her research advice, and to the staff both in the department and at Juntendō Library for their generous help.

Although I did not incorporate it into this book, my interview with Sumi Nagasawa, who practiced the profession as a well-known, highly respected midwife, gave me invaluable insights. I would also like to thank independent scholar Sachiko Harashima who informed me about the work of Shihoko Muramatsu, who pioneered the midwife profession in the Meiji period.

Professor Noriyo Hayakawa provided continuous and warm support all these years, and professors Yuki Sakurai and Shōko Ishizaki gave me essential assistance for the Tokugawa and modern sections, respectively.

I would also like to thank my colleagues in the departments of History, Asian Studies, and Women's Studies at Hofstra University for their support and encouragement. In particular, I would like to note the late John Walsh whose warm and thoughtful advice was of invaluable support, especially at an early stage of my career there. Hofstra University's School of Arts and Science provided funds that enabled me to carry out research in Japan and to write the later chapters of this book.

Prior to my graduate study at UCLA, I pursued a Master's degree at New York University (NYU). I would like to thank professors Molly Nolan and the late Marilyn Young for directing me to the study of history. I also greatly benefited from studying Marxism with Professor Bertell Ollman. The greatest support I received while I was at NYU was from Professor Timothy Mitchell, who was my advisor in the Department of Politics. It is no exaggeration to say that without his kindness and intellectual guidance at this initial stage of my journey as a young scholar, I would not have developed my academic career as I did.

Finally, I would like to thank my family in Japan. My sister Tomoko Nishioka shared anxieties about my uncertain future for many years, and both of us also felt great pain when I could not fly to Japan to attend her wedding because of a visa issue. My nephew and niece, Yoshinori and Saki Nishioka, have provided great joy in my life. My greatest gratitude goes to my mother, Itsuko Terazawa, and my late father, Hatsuyoshi Terazawa, for their unwavering support and encouragement without which I could neither have completed this book nor successfully pursued a professional career.

Contents

Notes to the Reader

The Periodization of Japanese History

The commonly accepted periodization of Japanese history is listed below. The 1868 Meiji Restoration bridges the pre-modern and modern eras, and the early modern period is equivalent to the Tokugawa period. The modern period is divided into emperors' reigns.

Tokugawa: 1603–1868
Meiji: 1868–1912
Taishō: 1912–1926
Shōwa: 1926–1989
Heisei: 1989–

The Tokugawa System

The feudal military class—that is, samurai (which constituted about 6% of the total population)—dominated Tokugawa society. The samurai class was at the top of the Tokugawa class system, followed in rank order by the farming, artisanal, and merchant classes.

The Tokugawa government, the *bakufu*, was led by the shogun, and thus is conventionally called the Tokugawa shogunate. The Tokugawa house monopolized the hereditary position of the shogun. While the shogunate nominally presided over 250 feudal lords (Jp *daimyō*) who led the military houses, these lords held far-reaching administrative and financial autonomy and power, controlling their fiefs and the samurai retainers who served them. The fiefs and local governments that the lords maintained are referred to as their "domains (Jp *han*)." In this book, "the Tokugawa state" connotes the combination of the shogunate and domain governments, while the Tokugawa government is the shogunate.

Names and Romanization

Throughout the text (except in Acknowledgments), I place the family name first, followed by the given name for Japanese, Chinese, and Korean names. Also note that pre-modern figures who were active prior to the Meiji era are usually referred to by their first names or pseudonyms.

Macrons are used to denote long vowels; however, they are omitted in cases where Japanese words have been adopted into English, such as Tokyo or shogun. Chinese words are romanized according to the pinyin system.

Japanese nouns do not distinguish between singular and plural, and thus words such as "samurai" can be either, depending on the context.

List of Figures

List of Tables

CHAPTER 1

Introduction

For those of us who have grown up in the latter half of the twentieth century, it is difficult to imagine an alternative picture of the body presented in college anatomy textbooks. From our earliest education in science, health, and hygiene, we encounter images of a dissected body with cardiovascular, respiratory, and digestive systems laid out as carefully delineated component parts. Within this modern medical schema, it is the reproductive system, isolated as a set of functional parts, that marks the body's sex. The reproductive organs show the body to be "male" or "female," each charged with a specific reproductive function. These images, combined with the authority of science as the ultimate arbiter of reality, have made these sexual categories seem "natural."

If we look to history, however, we find that this two-sex model is merely one of a variety of ways of envisioning the sexed body. Thomas Laqueur shows that from antiquity through the early modern period, European medicine made no fundamental distinction between male and female bodies. There was only one model: the male. The female body was conceived as a weakened or inferior version of this master template, with the vagina categorized as an inverted penis.[1] Within this frame of reference, it was difficult for women to articulate a feminine experience of the body.

Barbara Duden provides another model of the female body, using a doctor's notes from eighteenth-century Germany, which reveal how women perceived their own bodies. The doctor who was educated in

Y. Terazawa, *Knowledge, Power, and Women's Reproductive Health in Japan, 1690–1945*, Genders and Sexualities in History,
https://doi.org/10.1007/978-3-319-73084-4_1

modern medicine found the bodily experiences of his female patients "strange." In his mind, women tended to rely on "feelings" rather than visual perception to describe their bodily processes. They tended to talk about their corporeal experiences by relating them to outside occurrences in the form of stories. The body they described was porous and amorphous. Airstreams and fluids passed through the skin, and fluids, such as milk, naturally transmuted themselves into other fluids, such as menstrual blood.[2]

People in Tokugawa Japan (1603–1868) also understood bodies in ways that seem as strange to modern eyes as those studied by Laqueur and Duden. Before dissection became popular in the mid-eighteenth century, scholars and physicians described a body without the reproductive organs, which distinguish the body's sex in the modern medical system. Terajima Ryōan, an Osaka-based physician, addressed the question of how to distinguish between a man and a woman in his widely circulated 1712 multi-volume encyclopedia, *Wa-kan sansai zue*. He listed the "outwardly manifested" characteristics commonly used to distinguish the sexes: yin yang, face up and face down, concave, convex, whether a beard was present, and the length of the hair.[3] Ryōan adhered to the dichotomy of yin and yang, roughly female and male qualities derived from long-established Chinese medical traditions. He noted, however, the widespread contemporary belief that a male child was born face up and a female face down. The same principle applied in cases of drowning: a man's drowned body would be found face up, a woman's face down. Ryōan's distinction between concave and convex most likely referred to the genitals, which he saw as one among many signs of sexual difference. He maintained, however, that one could not reliably determine sexual difference from these traits. Rather, the definitive markers of sexual difference were found in skeletal anatomy. In his description, a man's skull consisted of eight bones, while a woman's possessed only six, and the female skull lacked the horizontal seam found at the back of the male skull.[4]

These pre-modern depictions of sexed bodies present modern researchers with a series of questions. When and how were modern anatomical depictions of the body introduced into Japan—as scholarly descriptions and as imagery circulating in popular culture? What kinds of cultural, political, and institutional force were behind the ascendancy of the modern anatomical body in Japan? Which individuals and groups appropriated and disseminated this new conception of the body? How did the shift reshape power relations between the state, organized medicine,[5] unlicensed healers, and medical patients? And, finally, how did this shift in understanding male and female bodies affect gender relations?

Even though this book addresses women's reproductive bodies from the early modern era to 1945, my starting point for the project is women's bodies today, where reproductive and maternal health have significance far beyond individual women's concerns and are vitally important to the interest of the state and medical professionals. As long as we live in this contemporary world, we cannot escape the dominance of modern bio-medical knowledge and government health policies. That is, our bodies have become captive of or colonized by bio-medical knowledge and practices, and government public health measures. While modern bio-medicine is of European origin, its use spread to other parts of the world, and eventually people residing in Japan became subject to the modern power exercised in the field of health and medicine. I clarify how this happened by focusing on the area of midwifery and maternal health.

My approach invokes historical processes of the seventeenth to the mid-twentieth century in which the state and organized medicine exerted an ever-increasing influence over women's bodies and reproductive lives in Japan. Drawing on insights from Foucault's writings, I reject theoretical frameworks of the state as a distinct, coherent, oppressive agent, on one hand, and "women" as an undifferentiated, oppressed group, on the other. Instead, I want to focus on midwifery and maternal health policies, discourses, and practices, and to analyze the ways that the engagement of certain groups of women in such practices led to the generation and spread of bio-medical power over those women.

The extensive research into the history of obstetrics and midwifery I have carried out points to two important historical shifts that reshaped the field of health and medicine in Japanese history, radically changing the ideas and practices of obstetrics, midwifery, and maternal health. First, the medical field in mid-eighteenth-century Japan saw the rise of new approaches to analyzing the body, finding causes of diseases, and developing medical remedies, concomitant with what Foucault would call an "epistemic shift" or an overarching change in the criteria for determining what constituted legitimate medical knowledge. These new approaches prepared Japanese physicians to accept Western medical methods and the medical body that developed within this tradition in the mid- to late nineteenth century. In this eighteenth-century field of obstetrics and midwifery, physicians belonging to the new obstetrics school, called the Kagawa School, invented new methods that took advantage of newly introduced perceptions of the human body as a distinct and tangible object. Their school was opposed to the Chinese medical tradition's age-old perception that the body and larger universe were systemically connected. While Kagawa School obstetricians expanded their practices and client base by developing new methods, the central Tokugawa government and regional (domanial) governments presided over by local feudal lords (the *daimyō*) rejected collaboration with the Kagawa School or other medical factions. Instead, the central and local government sought to control the biological reproduction of those they governed.

The second shift arose after the 1868 Meiji Restoration with the advent of the modern nation state along with new institutions and disciplinary practices. This institutional restructuring profoundly reshaped the politics of the state, organized medicine, and individual citizens whose bodies became the object of medical surveillance and intervention. The restructuring led to the colonizing of bodies by what Foucault calls "bio-power"—that is, the force that works in conjunction with the circulation of knowledge and the implementation of the practices relating to modern science and medicine. In midwifery and maternal health, much of the efforts of organized medicine and the state toward reforming childbirth and maternal health practices were channeled into the nationwide spread of trained midwives and public health nurses. Modern power also transformed women from being merely objects on which power works into subjects with a will that voluntarily participate in the "improvement" of their own bodies through their exposure to modern health education. I maintain that although there have been changes in government policies on reproductive and maternal health throughout the history of modern Japan, the new institutional landscape that emerged in the late nineteenth century became the basis for the development of modern relations of power, involving the state and individuals, through the twentieth century and beyond. Those relations of power have created oppressive conditions for women, who are divided into different groups according to categories such as class, ethnicity, nationality, occupation, and fitness for reproduction, with each group being subject to a particular type of oppression.

My interest in the emergence of modern power and medical institutions initially led me to single out the 1868 Meiji Restoration as the historical point of rupture. After nearly 700 years of military rule, during which Japan's interaction with foreign countries—except for countries in East and Southeast Asia—was limited, the internal crisis encompassing the economic, social, and political spheres along with the pressures of Western imperialism led to the breakdown of the Tokugawa shogunate and the establishment of a modern nation state in Japan. To survive the age of imperialism the Meiji ruling elites embarked on a modernization policy that involved a fully fledged Westernization of medicine, among many other fields.

Further research into this topic suggested problems with this initial working hypothesis, however. Although modern medical institutions and practices arose in conjunction with the establishment of the Meiji state, a crucial shift in the system of scholarly and medical knowledge had occurred during the mid-eighteenth century. What prompted this sweeping change in the academic topography were the movements of progressive scholars and physicians who questioned the then more established scholarly and medical systems—namely, the neo-Confucian, Zhu Xi School tradition [Jp Shushi-gaku] whose medical counterpart was called the Goseihō system. Challenging these systems of thought and medicine, progressive and dissident scholars and physicians prioritized empirical and practical knowledge, corroborated by their own experience of seeing and touching the objects of their investigations. It was the eighteenth-century move toward empiricism within Chinese studies and Chinese medical traditions in Tokugawa Japan, not specific medical models imported from Europe, that led to the development of innovations in medicine and a new understanding of the human body.

To understand this overarching shift in the scholarly and scientific method that took place in eighteenth-century Japan, I draw on Foucault's theory of the "discursive shift." Basic to understanding this concept is our recognition of the importance of general principles governing a particular cultural and historical moment, which collected together form an "episteme." This overarching epistemic framework structures the formation of knowledge within various scholarly fields and determines what kinds of knowledge are legitimate. In his important work earlier in his life, *The Archaeology of Knowledge*, Foucault discusses a decisive epistemic shift during the European Enlightenment that led to the emergence of an academic landscape governed by a new set of conceptual norms.[6] A similar shift occurred in eighteenth-century Japan in Chinese studies as well as in the fields of science and medicine. The resulting new intellectual framework challenged the neo-Confucian conception of nature, society, the state, and morality that had been widely accepted as the mainstream knowledge of late-seventeenth-century scholars and physicians.

As early as the late seventeenth century and continuing through the eighteenth century, prominent physicians and scholars vigorously criticized neo-Confucian methods, originally developed in Song China (960–1279). To undermine neo-Confucianism, and the scientific and medical models associated with it, they endorsed the study of the most ancient Chinese texts available at the time. They deemed these to be more authoritative because they pre-dated neo-Confucian texts and were presumably uncontaminated by the thinking of later, and thus lesser, scholars, including the advocates of neo-Confucianism. The dissidents' emphasis on the importance of ancient texts led them to scrutinize their original meanings, which in the medical field prompted physicians to question the efficacy of therapies and medications recommended in more recent Chinese and Japanese texts. It is not that they used remedies discussed in these ancient texts. Rather, they developed drastically new and original therapeutics, legitimizing them by appealing to more ancient texts than those written after the rise of neo-Confucianism in twelfth-century Song China.

These dissident physicians, whom historians later grouped into the Ancient Practice Method factions (Ko-ihō and Ko-hō), developed a new anatomical model of the human body, emphasizing the importance of its visibility, tangibility, and materiality. While the mainstream Goseihō method viewed the human body as a microcosm inseparable from the larger macrocosm, or the universe in which it was situated, innovative physicians perceived the body as a discrete and self-contained entity that lacked a systemic connection with its environment. The growing enthusiasm for the study of tangible body parts and internal organs prompted many physicians to seek a way to perform dissections despite their continuing prohibition for religious and political reasons.

This new focus on the visible and tangible anatomy of the body led to the emergence of new types of sexed bodies, bodies marked with reproductive organs. Instead of defining the body's sex by grounding it in yin yang theory, and the system of circulation and preservation of *ki* [Ch *qi*] and other vital bodily substances, such as the so-called "blood" and "essence," the body's physicality emerged as an important criterion for determining its sex. That is, one's sexual identity became rather confined, circumscribed by the reproductive anatomy of the body as delineated by the new medical theory.

This new approach to the medical body reshaped the field of reproductive medicine. Physician Kagawa Gen'etsu, for instance, established a new set of obstetric methods based on the newly emerging anatomical model of the female body. Competing with established obstetricians versed in Chinese medical classics in an ever more commercialized medical market in the mid-Tokugawa period, Gen'etsu developed procedures using his hands, arms, and eventually a hook to deliver a live child or to extract a dead fetus and afterbirth. Such techniques, based on new representations of the female body, empowered predominantly male obstetricians to the detriment of local midwives.

The growing intervention of medical professionals during the Tokugawa era into women's reproductive lives, however, was limited compared with the more intrusive methods introduced by organized medicine during the modern period. The principal difference between the two periods is the extent to which the state, in collaboration with organized medicine, was involved in regulating and guiding women's reproductive lives. Although many obstetricians during the Tokugawa period sought to bolster their influence through an ongoing search for advanced anatomical knowledge and new obstetric techniques, the state gave these efforts limited support. The Tokugawa shogunate and the governments of individual feudal domains employed and favored some highly regarded obstetricians because they deemed these physicians capable of delivering healthy children in the family of the Tokugawa shogun or domanial lords. The early modern state, however, did not mobilize physicians, midwives, and pharmacists to improve the reproductive health of the general populace. Nor did it attempt to prevent abortions and infanticide by seeking the collaboration of obstetricians and midwives. When the shogunate and domanial governments sought to crack down on the practice of abortion and infanticide, they relied on existing policing systems, such as the "five-people group" [Jp *gonin-gumi*]—village neighborhood groups—to enforce laws they had promulgated. Efforts by the state and the medical establishment to control women's bodies remained separate until after the 1868 Meiji Restoration.

The Meiji Restoration marked Japan's entry into the global modern nation-state system and its adoption of modern institutions and practices, which led to fundamental changes in medical institutions and the politics of medicalized bodies. The establishment of modern institutions encompassing government, military, economy, education, and health created a framework where individuals were subject to modern power derived from a scientific understanding of human beings and newly introduced science-based institutional practices.

Foucault elaborated his thoughts about the expanding networks of power through the workings of modern discourses and techniques.[7] The production and accumulation of scientifically authorized knowledge about individual bodies, he argues, are central to the operation of this modern bio-power. He suggests that power in modernity is created and circulated locally within and beyond discursive fields and pays particular attention to such novel practices as military training, schooling, medical treatment, and factory work routines. Through these practices, the human body is broken down into parts, analyzed, and reconstructed through the application of specific methods developed in each field. Individual bodies are also reorganized to constitute aggregated masses through a deliberate management of time and space—resulting in collections of bodies useful for production. The creation of a web of knowledge that operates on individual bodies and the development of mechanisms for programming bodies on a society-wide scale are part of this second historical shift in Japanese history.

As Japan entered the modern era, the state and organized medicine, equipped with the expertise of modern Western medical science, gained a renewed interest in gendered bodies and the female reproductive body. The female reproductive body was one of a variety of bodies (e.g., adult male bodies to be used for production and military purposes or children's bodies to be properly developed) that were identified and defined by experts in public health.[8] The newly formed alliance between the state and organized medicine was instrumental in establishing an increasingly close surveillance and management of women's bodies. The attempts made by the state and organized medicine to enhance the well-being of women in the modern era drew on a shared understanding that women's reproductive health was important for national purposes. In the mind of many state officials and civic-minded physicians, the primary goal for improving Japanese women's health was the production of a healthy population in general: serving the interests of women themselves was secondary, if it was ever an issue at all.

Thus, women's bodies became an important object of investigation. Modern European science and medicine provided specific ideas and techniques to form authorized knowledge about bodies. Amassing information about individual bodies was facilitated by newly introduced institutional practices such as schools, factories, the military, hospitals, and the practices of midwives and physicians, all of which gave researchers systematic access to bodies in the general population. Statistical data organized through taxonomies of sex, race, nationality, geographical regions, and age made it possible for government officials and medical researchers to process information into meaningful and manageable forms.

This production of knowledge created new power relationships between ordinary women, on the one hand, and organized medicine and the state, on the other. The medical establishment and the state used the knowledge they had gathered about women's bodies for public debates and policy-making. Even though ordinary women became crucial objects of these surveys and medical studies, they had virtually no control over how their bodies were investigated and how these data were used.

The introduction of modern discourses and practices in the late nineteenth century inserted women's bodies into new webs of power. The model of the female body developed by the modern Western scientific and medical tradition gradually replaced local understandings of the female body, including those held by women themselves. This laid a foundation for women to become active in "improving" their own bodies and to use them for a presumably greater cause such as the well-being of the nation. At the same time, the emergence of modern institutional practices and new forms of material culture transformed women's health practices without women themselves being aware of it. As women were incorporated into schools, factories, and a modern medical care system, their bodies were subject to new disciplinary techniques. Women's physical movements and even the shape of their bodies were reformed by newly introduced spatial and temporal arrangements in their daily lives. Changes in material culture—including modes of transportation, the physical structure of public spaces and home, furniture, clothes, hairstyle, and beauty products—reshaped how women used and treated their bodies. The results were not always beneficial. Industrialization subjected many factory workers and poor people to terribly unhealthy environments. Whether they liked it or not, women came to be incorporated into the new health practices mandated by changing institutional systems and cultural customs.

Although women increasingly experienced forms of invasive power that investigated and disciplined their bodies, this did not necessarily lead to a one-sided and uncomplicated disempowerment of women. Modern power working through discursive systems is often not really repressive but productive and constitutive. Foucault's analysis, if read carefully, leaves open a crucial space for individual agency and personal choice. Within webs of discourses, individuals are consigned to a particular position; discursive practices constitute individuals as speakers or listeners in the narratives they construct. As some poststructuralists argue, however, these "subject positions" can be engaged strategically, allowing individuals to reposition themselves within a given discursive environment.[9] The incorporation of women into discursive systems does not mean that women are simply passive objects; they frequently become active, vocal players situating themselves in strategic subject positions available within a given discursive environment.[10] This leaves a possibility for women to subvert and resist oppression even though women's active engagement in particular social and cultural practices may result in power that oppresses them.

For example, women's responses to newly introduced health practices and beauty norms cannot be described simply as resistance or complicity. Rather, a complex combination of collusion, accommodation, and resistance integrates women into dominant discursive structures. In Chap. 5, for instance, I present examples of women's reactions to the health reform campaign to improve the bodies of young women led by health officials, prominent physicians, and progressive educators at the turn of the twentieth century. The range of their responses shows women engaged in a simultaneous process of resistance and collaboration that made them co-participants in the evolution of their own subjectivity. Women both adopted and adapted health practices that were endorsed by patriotic health reformers occupying influential positions, and some women became vocal activists, taking up unique positions vis-à-vis elite and predominantly male reformers.

It is worth being cautious about using Foucault's theories to understand transformations taking place in Japanese history because Foucault developed his ideas based on his examination of European history and the evolution of specific intellectual trends within it. A good place to start in determining just how far Foucault's ideas can be applied to the Japanese case is recent scholarship on colonialism. Japan is usually considered a "non-Western" nation, whose culture and history are presumably dissimilar from those of the West. Scholars such as Timothy Mitchell, Gyan Prakash, and Ann Stoler, criticize Foucault for overlooking how European colonialism maintained and extended European knowledge/power structures in Europe and its colonial empires.[11] For example, Prakash argues that in India, British colonial administration and scientific scholarship worked closely with the colonial state to introduce networks of power that reworked the spatial arrangement of colonial lands and captured Indian people within a newly organized European-style grid. Although Indian nationalists fought against British occupation and aspired to establish an independent nation, they inherited the institutions of the colonial state—the government bureaucracy, the school system, industrial and urban infrastructure—that were the basis of national development after the British departure from India. This led them to espouse nationalist ideas and sentiments aimed at deflecting the influence of colonial policy, but they nonetheless had to work within the modern institutions and practices that had been laid out by British colonialists. Prakash names this condition "colonial modernity" and contrasts it with the "Western modernity" that developed in the West.

If we look closely at how modern power developed in the epistemic universe of Japan, we see that the Japanese case fits neither the model of Western nor colonial modernity. Although Japan and the Japanese have been classified as "non-Western"—frequently subject to Orientalist representations, for instance—Japan was never the colony of any European or US power. Rather, the shogunate and the governments of some domains under Tokugawa rule already pursued expansionist and colonialist policy toward non-Japanese peoples. For example, the Matsumae and Satsuma domains pursued colonization projects directed towards ethnic minorities living in the vicinity of their territories: respectively, the Ainu in Hokkaidō and the Ryūkyūans in today's Okinawan islands.[12] The Meiji government undertook an even more unambiguously colonialist policy toward the Ainu and the Ryūkyūans, and later toward Taiwan and Korea. Colonial modernity as it was experienced in India and other Western colonies was not a part of Japanese experience.

During the late Tokugawa and Meiji periods, the forceful incorporation of Japan into the international system—particularly the imposition of unequal treaties—left no choice but for Japanese political leaders, bureaucrats, military, and medical experts to introduce modern institutions and practices. Although neither Europe nor the USA blatantly imposed modern institutions and industrializing projects onto the Japanese, there were internal pressures in the Meiji leadership that pushed them toward building a modern nation state and adopting Western science and technology. In this way they were in a different position from former colonies, which had to rely on the institutional and technological infrastructure set up by the colonial state to build new, independent nations.[13] Japanese leaders never had to face the dilemma of drawing on the legacy of a former colonial administration while simultaneously repudiating the legitimacy of colonial rule.

The rise of modernity in Japan was thus different from both the experience of the nation states created in the aftermath of colonialism and from the modernity of the European Enlightenment. These differences are especially apparent in the scholarly fields of medicine and science. To begin with, the institutional mechanisms for certifying the truth claims of scientific knowledge in Tokugawa Japan were quite dissimilar from their European counterparts. Moreover, in Foucault's description of the European Enlightenment, scientific knowledge developed in tandem with its deployment as a mechanism for enhancing institutional and governmental power. During the latter half of the Tokugawa period, on the other hand, there was a shift toward empirical and practical methods for investigating—and "constructing"—natural objects, including the medical body, but knowledge of human subjects procured through a systematic application of scientific methods was not yet available, nor was this knowledge deliberately used by the government to organize and control the people.

Intellectual innovations in scholarly and medical fields by those affiliated with the Ancient Learning and Ancient Practice factions during the Tokugawa period were important precursors to the eventual ascendancy of Western scientific and medical models in Japan. But it was not until the late nineteenth century that scientific and medical knowledge was used in conjunction with various modern institutional practices to reconfigure the power relations between individuals and the state. It is only at this point that the theory of bio-power that Foucault developed later in his academic life could be applied to the Japanese scenario. If Japan's case was unique—a non-Western nation whose own governing elites took the initiative for advancing the nation's modernization—the major consequences of such endeavors were similar to developments in Western nations and their colonies, the colonization of the population at large by invasive bio-power.

In addition to the theoretical works referred to above, this book is indebted to a growing collection of scholarly works on body, gender, and sexuality, which has been accumulating since the 1980s. In Japan and elsewhere, feminist scholars in a number of fields were already working on issues involving women's bodies prior to the late 1980s, and one of the important topics they focused on concerned women's reproductive rights. An urgent practical issue that feminist activists in Japan have dealt with since the 1970s was ordinary women's unrestricted access to the contraceptive pill, which at that time was only available for medical use and with a physician's authorization. Foucault's writings on the body and sexuality opened up new intellectual horizons. Studies on the body and reproduction, and debates on women's reproductive rights developed in new ways, providing insights into the issues of language and identity. Whether historians and feminist scholars embrace or reject Foucauldian post-modern theory, those who work on the history and theory of the body after the late 1980s, in either Japanese- or English-speaking academia, have been in dialogue with his theoretical perspective. Keeping in mind this basic understanding about the evolution of this field, I locate this book within the existing literature on the history of bodies, reproduction, and gender in early modern and modern Japan below.

The growing interest, from the late 1980s and 1990s, in the history of the body prompted Japanese feminist sociologists and historians to pay close attention to questions of gendered bodies in both pre-modern and modern Japan. Scholars of the Tokugawa period who have worked on sex and reproduction during the past 30 years or so have been dispersed across multiple disciplines. Researchers working on issues of biological reproduction, infanticide, and maternal health, including Sawayama Mikako and Ōta Motoko, described discourses and practices of infanticide, using diaries of educated male farmers as well as records left by samurai administrators who staffed domanial governments.[14] Sakurai Yuki has looked at abortion practices and the Tokugawa shogunate's policy to suppress them. To overcome the serious problem of the scarcity of sources, she made creative use of evidence from popular culture, such as *senryū*, or satiric verses, and illustrated flyers to advertise abortion medicine.[15] Fabian Drixler's English-language study on infanticide focused on the history of fertility control and domanial policies for curbing infanticide in northeastern Japan from the Tokugawa period onward.[16] Drixler's research is noteworthy for his incorporation of the insights and relatively recent findings of Japanese demographic historians and his use of statistical data in combination with written records (e.g., farmers' diaries and domanial edicts) and graphic representations (e.g., votive pictures displayed at temples). Drixler makes important efforts to pin down the fluctuations in the numbers and rates of infanticides and abortions as well as the degree to which specific reproductive policies were effective.

This book does not focus on the topics of abortion and infanticide. I leave to other studies, including the aforementioned works, discussions of infanticide and the effectiveness of government fertility control during the Tokugawa period. What I would like to show is how women's bodies became an important object of study and treatment in the medical field in increasingly commercialized Tokugawa society. I discuss the topic of abortion and infanticide to show established obstetricians' attitudes toward them and to demonstrate that the early modern state at neither the shogunate nor domanial (local government) level collaborated with professional medical doctors in its attempt for fertility control.

One of the main topics I address in the early modern section of this book— namely, the rise of the new obstetric methods of the Kagawa School in the eighteenth century—had already been discussed by feminist scholars in the late 1980s and early 1990s. They reworked the narratives of the revolution in the field of obstetrics of such medical historians as Ogata Masakiyo and Sakai Shizu.[17] Ochiai Emiko in her pioneering articles, appropriating theory discussed in Michel Foucault's *The History of Sexuality*, suggests examining the seminal revolution of the body and the making of modernity in Tokugawa Japan, focusing not on sexuality but on the transformation of the field of obstetrics.[18] Susan Burns, on the other hand, emphasizes the continued significance of Confucian discourse in obstetric texts written between the late seventeenth and early nineteenth centuries. The strong influence of Confucianism helped make pregnancy and childbirth become increasingly valorized from the late seventeenth century onward, shaping a new ideal of womanhood and empowering male obstetricians. Although her analysis does not necessarily contradict the approaches taken by Ochiai and me, she places much less emphasis on the revolutionary nature of Kagawa School obstetrics.[19]

Compared to these existing works, my analysis of the eighteenth-century revolution in the field of obstetrics is based on a close reading of several key medical texts, which I locate in the context of the shift in the medical field as a whole and Foucault's theory of "epistemic shift." These approaches, I believe, are useful to meet the following goals. The first is to show a systemic understanding of more established Chinese medicine to which dissident physicians, including Kagawa School obstetricians, were reacting. What was iconoclastic about the changes that they brought about in eighteenth-century Japan was the breaking up of the existing system of Chinese medicine, which well-educated physicians generally thought essential to medical study and practice. The Foucauldian concept of the episteme locates this shift in the medical field in the larger rupture that also involved other scholarly and scientific fields. By discussing the nature of this systemic change, I point to the importance of the eighteenth-century revolution in obstetrics in its much broader historical context, which encompasses early modern and modern Japanese history. In other words, although Chap. 3, which covers the history of the Kagawa School of Obstetrics, could be read separately, it has a greater value when read with the modern section of this book.

More important, the value of this study rests in its coverage of both the early modern and modern periods and the way it discusses the rise of the modern nation state and the modern medical system, accompanied by the proliferation of new "bio-power," and how it changed power relationships among the state, organized medicine, and individual citizens. Drixler's work also covers both the early modern and modern periods, but the important difference between his scholarship and mine is that I focus on the drastic change in the institutional structure that occurred during the transition from the Tokugawa to Meiji that allowed the state and organized medicine to exert a new type of power over women's reproductive lives. Drixler draws on the tradition of demographic history and evaluates the effectiveness of government fertility control measures from the Tokugawa and modern period. His analysis of statistics is beyond the scope of my study. What I would like to show in this study, instead, is how the range of modern institutional practices in public health, medicine and education introduced in the late nineteenth century through to the early 1940s led to recurrent reconstructions of systems of reproductive control.

For the chapters covering the post-Meiji Restoration periods, I am indebted to Japanese women's historians who have investigated the significance of the Meiji Restoration and the emergence of the modern state, especially Fujime Yuki and Hayakawa Noriyo.[20] Central to the new state system were the administrative and legal systems that placed the emperor above the law and constitution. The system left ample possibility for a radical shift toward authoritarianism. The patriarchal family system that posited the emperor as the father of all Japanese citizens became one of its backbones. Ideas that derived from Shinto and Confucianism ideologically buttressed this state system along with the patriarchal family structure.

Fujime Yuki examines Meiji government policies on sex work, reproduction, midwifery, and the ways that the state took advantage of women's bodies and reproductive lives for its own purposes.[21] While Fujime's work has become one of the classics in Japanese women's history, other scholars who work on state reproductive policy such as Ishizaki Shōko and Fabian Drixler have recently challenged Fujime's thesis that the government of modern Japan has consistently pursued pronatalist policies to increase the population.[22] My volume incorporates Fujime and others' invaluable insights and research; however, rather than postulating the state as a monolithic entity that pursued logical and consistent policies, I focus on specific state-endorsed policies and practices, tracing their development in generating bio-power. Further, I identify diverse agents, interests, and ideological forces other than the state and the undifferentiated group "women" that supported or resisted government reproductive policies.[23]

Important research on midwifery practices from the Taishō and Shōwa periods has largely been carried out in anthropology and sociology rather than in history. Two important anthropological works detailing childbirth practices before World War II are Nishikawa Mugiko and Yoshimura Noriko's monographs, both of which draw on oral interviews with retired midwives and their former clients.[24] Ochiai Emiko documented a detailed oral interview with a midwife who began her practice in 1916 in a village in the northeastern part of Japan. These works provide an excellent window on practices in rural Japan that is largely invisible to traditional historical sourcing. Also, although historians tend to be highly critical of publications in folklore studies for their tendency to fit research data into pre-conceived images of the past or to romanticize it, we should not dismiss them categorically, considering the scarcity of sources that reveal day-to-day sex and reproductive practices among ordinary people from the pre-World War II period. There has been much useful information documented in this scholarly tradition that is often absent in the data sets of other disciplines. For example, research that excludes folklore studies and anthropological sources is likely to overlook the significance of ritual pollution and the ways the idea of pollution shaped everyday practices, such as childbirth and midwifery, as late as the mid-twentieth century.[25] I use these works in anthropology and folklore studies, many of which are case studies, to look at broader historical changes in midwifery and maternal health practices and incorporate a discussion of the roles played by the modern state and organized medicine in shaping such historical shifts.

The decade of 1910 bridges the Meiji and Taishō eras and constitutes another transitional period that saw the rise of new ideas about gender, sex, and sexuality within the context of a new wave of urbanization and commercialization. The period also saw the development of new fields of scholarship in the "human sciences," including psychology, criminology, sexology, and education. Sabine Frühstück, Ogino Miho, Sumiko Otsubo, and Matsubara Yoko have explored the impact of this new scientific discourse in such fields as sexology and eugenics.[26] The rise of these new human sciences often intensified or renewed stereotypes of race, class, and gender, although the period certainly led to the spread of new, "modern" discourses on gender and sexuality that benefited some women wanting to pursue a professional career and escape from the structural and ideological confines of the family system.

The influx of new scholarship in these human sciences did not directly influence the field of midwifery; it reshaped, however, the ideas and practices of abortion and birth control. The Meiji state's renewed emphasis on Confucianism and the importance of honoring the emperor and the imperial institution by the expansion of the Japanese population led to the development of anti-abortion ideology and the increasing criminalization of abortions as well as infanticide. The state's stance on the emerging birth-control movement in the early twentieth century, however, was complex because of competing ideas about whether the population should decrease or increase, a eugenicist urge to encourage the reproduction of "desirable" over "undesirable" people, and the aspiration to preserve Confucian morality toward sexual and marital practices. Although the national government attempted to suppress the birth-control movement, the rise of left-wing and feminist activism along with the expansion of print media through which birth-control products were advertised helped spread ideas of birth control in the 1920s and 1930s. So far, scholars, including Ishizaki, Ogino, Burns, and Drixler, have published important works on abortion and birth control in pre-World War II Japan.[27] What is to be remembered is that although we have access to some medical and legal records on abortion, books written on birth control, and advertisements published in various magazines, we have few sources, except for rare oral history interviews, that describe *actual practices* of abortion and birth control.[28] This area may remain a blank page in the history of women and reproduction in modern Japan.

Another important development during the early twentieth century that reshaped the way women related to expert knowledge of maternal health was the remarkable growth of print media and a massive increase in the number of literate women. Barbara Sato and Sarah Frederick show that younger and more educated women along with middle-class housewives became avid readers of women's magazines.[29] They read experts' advice on fashion and beauty, tips for efficient housekeeping, and ways to manage family finances, along with popularized scientific ideas about how to handle pregnancy, childbirth, childrearing, and birth-control methods. Female readers of the women's press were not just passive subjects; instead, they actively read and obtained expert knowledge *of their own accord*. Instead of being reluctant followers, they became thinking and acting players with a stake in the system, a collective force that could promote specific causes—women's rights and issues involving their own identities, but also national objectives such as supporting the war effort—once they were convinced to pursue them.

Fabian Drixler argues that the changes in reproductive patterns in twentieth-century Japan, by and large, occurred owing to the modernization of lifestyles. But an important element of this emerging modernity was that women became active modern subjects displaying their own will and desire in the new networks of power. As I mentioned above, historians covering pre-World War II Japan have no choice but to rely on extremely scant sources to reveal ordinary women's thoughts and feelings about their bodies and reproductive health. But it is crucial to identify and recognize women's agency for bringing out modernity in the area of reproduction and maternal health.

The works I have discussed so far constitute an abridged list of the publications directly relevant to my discussions in this book. I have, indeed, drawn on sources and ideas from a wider range of fields than indicated above. My ultimate goal is not simply to document the changes in midwifery and maternal health practices, but to examine the nature of power that began to proliferate with the rise of modern medical science, the dominance of the nation-state system, and the ways this power affected different groups of women.

This book is roughly divided into early modern and modern sections, and chapters are arranged in a loosely chronological order. In Chap. 2, I begin with an overview of the field of medicine in the early Tokugawa period, followed by an exposition of the body, reproduction, and obstetric methods described in a late seventeenth-century obstetrics manual, Katsuki Gyūzan's *Fujin kotobukigusa* [*A Guide for a Woman's Celebratory Event*, 1692].[30] Gyūzan followed the Chinese medical tradition known in Japan as the Goseihō School. Leading Japanese physicians imported the founding ideas of the Goseihō School from China beginning in the sixteenth century, and this faction gained orthodoxy by the seventeenth century. The perception of the body presented in Gyūzan's manual, therefore, was familiar to students of mainstream Chinese medicine at the time. Gyūzan analyzed the body's physiology using the Yin Yang Five Phases theory and argued for assuring adequate circulation of *ki* (i.e., a vitality vested with a power to form material objects), "blood," and other bodily fluids as conceptualized in traditional Chinese medicine as a way of promoting the body's health. Japanese physicians were also able to take advantage of the fact that orthodox Chinese medical theories aligned with neo-Confucian concepts in their formulation. By conflating medical advice with the moral authority deriving from prestigious neo-Confucian texts, they countered the widespread view that medicine was a lowly and mercenary profession. The medicalization of the female reproductive body thus provided an opportunity for educated and ambitious physicians—whose status was generally far below warrior retainers (i.e., the samurai)—to expand their power and secure higher status and wealth.

In Chap. 3, I discuss the major epistemological shift in medicine that occurred in the mid-eighteenth century and the emergence of the Kagawa School of Obstetrics that occurred in conjunction with this shift. I ask why physicians from relatively disadvantaged social positions were at the forefront in developing new medical methods and how their innovative methods brought about an overarching transformation of the medical field. Their work led to the establishment of the Ancient Practice tradition [Kohō and Koihō], which, contrary to its name, introduced a new set of ideas and methods. The rise of the Kagawa School of Obstetrics was a part of this movement. Kagawa School physicians rejected the theoretical underpinnings of the Chinese medical tradition, including yin yang and the Five Phases. Instead, they projected the body as an object demarcated and isolated from the surrounding environment by its skin and marked it with tangible bone structures and reproductive organs. The body thus lost its organic connections with the outside world and became a freestanding, self-contained object. The flourishing of the Kagawa School occurred concurrently with the substantial growth of wealthy clients who sought out reputable, first-rate midwifery services offered by renowned obstetricians.

Chapter 4 explores the expansion of modern bio-power in biological reproduction and maternal health between the late 1880s and through the early twentieth century. The key to this process was the spread of "modern" and "hygienic" childbirth practice. This transition involved not only changes in government reproductive policy but also the restructuring and introduction of medical and other types of institutions and practices after the 1868 Meiji Restoration, setting in motion and promoting the modern medicalization of bodies. Building upon the newly created modern health bureaucracy and medical system, the government attempted to work on maternal and infant health by producing trained midwives, the so-called "new midwives" [*shin sanba*] or "modern midwives" [*kindai sanba*], educated in modern Western medicine. The number of trained midwives did not increase until the 1910s, and, not infrequently, expectant mothers and their families, particularly in rural areas, resisted using them; by the 1920s, however, the use of trained midwives who worked under licensed obstetricians had become a common practice except in isolated areas. The government and medical experts intensified their supervision of mothers and reproduction by way of trained and licensed midwives. I show how the spread of modern midwifery and the new system of surveilling reproduction involved multiple players, including government health officials, local administrators, physicians, trained midwives, expectant mothers, and the mothers' families.

By focusing on the women's health reform movement from the turn of the twentieth century, Chap. 5 discusses the ways that women's bodies were colonized by the encroachment of modern bio-power. As Chap. 4 also illustrates, Japan sought in the late nineteenth and early twentieth centuries to bolster its national strength through the health of the women who would bear the next generation of Japanese. In this context, women's bodies became an object of scientific study and reform. Medical researchers heavily scrutinized Japanese women's bodies through the taxonomies of sex and race. They recreated women's bodies by introducing mandatory physical education courses to make them fit for a healthy "modern" life. They attempted to make women amenable to urban living and transportation and disciplined them in repetitive and co-ordinated movements required in factory work or militaristic drills. The adoption of health science into women's education reshaped their understanding of their own bodies—colonizing them from the inside out—but there were instances where women resisted the imposition of "reformed" clothes and hairstyles as well as new health regimens by predominantly male physicians, educators, and government officials.

In the final chapter, I explore how the networks of bio-power expanded in the area of maternal health and became more dense and intense between the 1920s and 1945, responding to the growth of government and non-governmental services and programs. The maternal health of the urban poor and rural residents emerged as a national issue in the 1920s, and the bodies of women from less privileged backgrounds were increasingly "colonized" by becoming objects of research and reform. During the war years between 1937 and 1945, government surveillance and reform of maternal health practices became even more intense with the passage of several key pieces of legislation and the establishment of the Ministry of Health and Welfare (MHW). The results produced by new measures, such as the maternity passbook system and the district nurse programs of pregnancy and maternity counseling, were far from what was desired; however, the institutions and laws that were established during this period constituted the foundation for maternal and infant health policies in the second half of the twentieth century.

Taken together, these five chapters encompass the late seventeenth century through to 1945 and present a broad historical perspective on the evolution of midwifery and maternal health. They highlight the distinct discursive and institutional shifts that redefined women's bodies and relocated them within a newly formed network of knowledge and practices in the field. Central to this study is an analysis of the types of power involved in the politics of reproduction and how these power networks were structured through successive historical periods. With these larger goals in mind, the next chapter discusses the development of obstetrics and the growing importance of women's bodies as an object of medical discourse in the early Tokugawa period.

Notes

1. Thomas Laqueur, *Making Sex: Body and Gender from the Greeks to Freud* (Cambridge, MA: Harvard University Press, 1990).
2. Barbara Duden, *The Woman Beneath the Skin: A Doctor's Patients in Eighteenth-Century Germany* (Cambridge, MA: Harvard University Press, 1991).
3. Terajima Ryōan, *Wa-kan sansai zue*, Vol. 2 (Tokyo: Heibonsha, 1986): 108–109.
4. Ibid.
5. Organized medicine connotes a collectivity of physicians and other medical experts who act together to advocate for the medical profession. Historically, medical associations pursued higher status and control in the medical field by defining other types of health care as deviant or "quackery." While organized medicine has sought professional autonomy for medical practitioners, it has worked closely with the state because laws, licensing, and other government regulations are central to its domination of the field. For the purpose of this book, modern organized medicine emerged in Japan after the 1868 Meiji Restoration.
6. Michel Foucault, *The Order of Things: An Archaeology of the Human Sciences* (New York: Vintage/Random House, 1973); and *The Archeology of Knowledge* (New York: Harper Colophon, 1972).
7. Michel Foucault, *The History of Sexuality, Vol. I* (New York: Pantheon, 1979); and *Discipline and Punish* (New York: Vintage, 1979).
8. Sabine Frühstück presents illuminating discussions on this issue in the Introduction and Chap. 1 of *Colonizing Sex: Sexology and Social Control in Modern Japan* (Berkeley: University of California Press, 2003).
9. Bronwyn Davies and Rom Harre, "Positioning: The Discursive Production of Selves," *Journal for the Theory of Social Behavior* (20:1): 46.

10. Ian Hacking discusses how both individuals and categories into which individuals are classified are constructed within discursive structures. See Ian Hacking, *The Social Construction of What?* (Cambridge, MA: Harvard University Press, 1999).
11. See Timothy Mitchell, ed., *Questions of Modernity* (Minneapolis: University of Minnesota Press, 2000). See also Timothy Mitchell, *Colonising Egypt* (Cambridge, UK: Cambridge University Press, 1988) and *Rules of Experts: Egypt, Techno-politics, Modernity* (Berkeley: University of California Press, 2002); Gyan Prakash, *Another Reason: Science and the Imagination of Modern India* (Princeton, NJ: Princeton University Press, 1999); Ann Laura Stoler, *Race and the Education of Desire: Foucault's History of Sexuality and the Colonial Order of Things* (Durham, NC: Duke University Press, 1995) and *Carnal Knowledge and Imperial Power: Race and the Intimate in Colonial Rule* (Berkeley: University of California Press, 2002).
12. See, for example, Brett L. Walker, *The Conquest of Ainu Lands: Ecology and Culture in Japanese Expansion, 1590–1800* (Berkeley: University of California Press, 2001); David L. Howell, *Geographies of Identity in Nineteenth-Century Japan* (Berkeley: University of California Press, 2005).
13. The Japanese experience is similar to that found in Thailand, which also never became a colony of European or US powers, but became instead what Tamara Loos calls a "semi-colonial" state subject to pressures exerted by imperialist powers. See Tamara Loos, *Subject Siam: Family, Law, and Colonial Modernity in Thailand* (Ithaca, NY: Cornell University Press, 2006).
14. Sawayama Mikako, *Shussan to shintai no rekishi* (Tokyo: Keisō Shobō, 1998); Ōta Motoko, *Kodakara to kogaeshi: kinsei nōson no kazoku seikatsu to kosodate* (Kyoto: Fujiwara Shoten, 2007); and *Kinsei no "ie" to kazoku: kosodate wo meguru shakaishi* (Tokyo: Kadokawa Gakugei Shuppan, 2011).
15. Sakurai Yuki, "Mabiki to datai," in Hayashi Reiko, ed., *Josei no kinsei* (Tokyo: Chūō Kōronsha, 1993): 97–128; and "Kinsei no ninshin shussan gensetsu," *Rekishi hyōron*, No. 600 (April, 2000): 27–38.
16. Fabian Drixler, *Mabiki: Infanticide and Population Growth in Eastern Japan, 1660–1950* (Berkeley: University of California Press, 2013).
17. Ogata Masakiyo, *Nihon fujinka gaku shi* (Tokyo: Maruzan, 1918, reprinted by Kagaku Shoin, 1980); and *Nihon sanka gaku shi* (Tokyo: Maruzan, 1918, reprinted by Kagaku Shoin, 1980); Sakai Shizu, *Nihon no iryōshi* (Tokyo: Tokyo Shoseki, 1982).
18. Ochiai Emiko, "Edo jidai no shussan kakumei," in *Kindai kazoku to feminizumu* (Tokyo: Keisō Shobō, 1989): 56–78; "Kinsei matsu ni okeru mabiki to shussan," in Wakita Haruko and Susan Hanley, eds., *Jendā no nihonshi* (Tokyo: Tokyo Daigaku Shuppankai, 1994): 425–459; "The

Reproductive Revolution at the End of the Tokugawa period," in Hitomi Tonomura, Anne Walthall, and Wakita Haruko, eds., *Women and Class in Japanese History* (Ann Arbor: Center for Japanese Studies, The University of Michigan, 1999): 187–215.

19. Susan Burns, "The Body as Text: Confucianism, Reproduction, and Gender in Tokugawa Japan," in Benjamin Elman, John Duncan, Herman Ooms, eds., *Rethinking Confucianism: Past and Present in China, Japan, Korea, and Vietnam* (Los Angeles: University of California, UCLA Asian Pacific monograph series, 2002): 178–219.
20. Fujime Yuki, *Sei no rekishi gaku* (Tokyo: Fuji Shuppan, 1997); Hayakawa Noriyo, *Kindai ten'nōsei kokka to jendā: Seiritsu ki no hitotsu no rojikku* (Tokyo: Aoki Shoten, 1998).
21. Fujime, *Sei no rekishi gaku.*
22. Ishizaki Shōko, "Meiji-ki no seishoku wo meguru kokka seisaku," *Rekishi hyōron,* No. 600 (April 2000); Drixler, *Mabiki.*
23. In her sociological study, Hiroko Takeda also uses Foucault's theory of "governmentality" to discuss the politics of biological reproduction in modern Japan. Her major focus is the post-World War II period. Hiroko Takeda, *The Political Economy of Reproduction in Japan: Between Nation-State and Everyday Life* (London: RoutledgeCurzon, 2005).
24. Nishikawa Mugiko, *Aru kindai sanba no monogatari: Noto, Takejima Mii no katariyori* (Toyama: Katsura Shobō, 1997), and for its abridged English version, see Mugiko Nishikawa, "The Modern Midwife (sanba) and the Transformation of Childbirth at the Local Level in Japan," *U.S. Japan Women's Journal,* No. 24 (2003): 82–101. Other major works include Ochiai Emiko, "Aru sanba no nihon kindai" in Ogino Miho et al., eds., *Seido to shite no "on'na": Sei, san, kazoku no hikaku shakai shi* (Tokyo: Heibon sha, 1990): 257–322; and "Modern Japan through the Eyes of an Old Midwife: From an Oral Life History to Social History," in Wakita Haruko, Anne Bouchy, and Ueno Chizuko, eds., *Gender and Japanese History, Vol. 1* (Osaka: Osaka University Press, 1999): 235–296; Yoshimura Noriko, *Osan to deau* (Tokyo: Keisō Shobō, 1985); and *Kodomo wo umu* (Tokyo: Iwanami Shoten, 1992); Fujime Yuki, "Aru sanba no kiseki: Shibahara Urako to sanji chōsetsu," *Nihonshi kenkyū* 366 (February 1993): 90–112, and its English translation, Yuki Fujime, "One Midwife's Life: Shibahara Urako, Birth Control, and Early Shōwa Reproductive Activism," Wakita et al., eds., *Gender and Japanese History, Vol. 1*: 297–325; Shirai Chiaki, ed., *Umi sodate to josan no rekishi* (Tokyo: Igaku Shoin, 2016).
25. See, for example, Onshi Zaidan Boshi Aiikukai, ed., *Nihon san'iku shūzoku shiryō shūsei* (Tokyo: Daīchi Hōki Shuppan, 1975); Akamatsu Keisuke,

Hijōmin no minzoku bunka (Kyoto: Akashi Shoten, 1986); and *Hijōmin no sei minzoku* (Kyoto: Akashi Shoten, 1991); and Chiba Tokuji and Ōtsu Tadao, *Mabiki to mizuko: kosodate no fōkuroa* (Tokyo: Nōsangyoson Bunka Kyōkai, 1983). For an important recent study on the birth house custom in twentieth century Japan, see Fushimi Yūko, *Kindai nihon ni okeru shussan to ubuya* (Tokyo: Keisō Shobō, 2016).

26. Frühstück, *Colonizing Sex*; Matsubara Yoko, "Minzoku yūsei hogo hōan to nihon no yūseihō no keifu," *Kagakushi kenkyū, dai ni-ki*, No. 36 (April 1997): 42–50; and "Dai go-shō: Sengo no yūsei hogo hō to iu na no danshu hō" in Yonemoto Shōhei, Nudeshima Jirō, Matsubara Yoko, and Ichinokawa Yasutaka, eds., *Yūsei gaku to ningen shakai* (Tokyo: Kōdansha, 2000); Ogino Miho, *"Kazoku keikaku" e no michi: kindai nihon no seishoku wo meguru seiji* (Tokyo: Iwanami Shoten, 2008); Sumiko Otsubo and James R. Bartholomew, "Eugenics in Japan: Some Ironies of Modernity, 1883–1945," *Science in Context* 11.3 and 4 (1998): 545–565; "Between Two Worlds: Yamanouchi Shigeo and Eugenics in Early Twentieth Century Japan," *Annals of Science*, 62.2 (April 2005): 205–231; Sumiko Otsubo, "The Female Body and Eugenic Thought in Meiji Japan," in Morris Low, ed., *Building a Modern Nation: Science, Technology and Medicine in the Meiji Era and Beyond*, (New York: Palgrave Macmillan, 2005): 61–81; and "Engendering Eugenics: Feminists and Marriage Restriction Legislation in the 1920s," in Barbara Molony and Kathleen S. Uno, eds., *Gendering Modern Japanese History*, (Cambridge, MA: Asia Center, Harvard University Press, 2005): 225–256.
27. Susan Burns, "When Abortion Became a Crime: Abortion, Infanticide, and the Law in Early Meiji Japan," in David Howell and James C. Baxter, eds., *History and Folklore Studies in Japan* (Kyoto: International Research Center for Japanese Studies, 2006): 37–55; Drixler, *Mabiki*; Ishizaki Shōko, "Nihon kindai no kazoku to seishoku: 1910-nen dai-1950-nen dai," *Sōgō joseishi kenkyū*, No. 24 (March, 2007): 1–18; and "Principles of Procreation and the Family in Modern Japan: Factors behind Decisions on Family Size," in Hiroko Tomida and Gordon Daniels, eds., *Japanese Women: Emerging from Subservience, 1868–1945* (Folkestone, Kent: Global Oriental, 2005): 278–306; Ogino, *"Kazoku keikaku" e no michi.*
28. Working on the local history of Shizuoka prefecture, historian Owada Michiko published an invaluable article that provides a window into the practices of birth control and abortion during the pre-World War II period. Owada Michiko, "Sanji seigen kara tasan shōrei e: bosei ni miru shizuoka-ken kindai josei-shi no tenkai," *Shizuoka-ken kindai-shi kenkyū* 14 (October 1988): 61–63.
29. Barbara Sato, *The New Japanese Woman: Modernity, Media, and Women in Interwar Japan* (Durham, NC: Duke University Press, 2003); Sarah

Frederick, *Turning Pages: Reading and Writing Women's Magazines in Interwar Japan* (Honolulu: University of Hawai'i Press, 2006). For a useful summary on women and modernity in early twentieth-century Japan, see Vera Mackie, "Modern Selves and Modern Spaces: An Overview," in Elise K. Tipton and John Clark, eds., *Being Modern in Japan: Culture and Society from the 1910s to the 1930s* (Honolulu: University of Hawai'i Press, 2000).

30. Fujikawa Yū, Kure Shūzō, and Masuda Tomomasa, eds., *Nihon sanka sōsho* (Kyoto: Shibunkaku, first published in 1895, reprinted in 1971): 53–63.

CHAPTER 2

The Reproductive Body of the Goseihō School

Owing to the brisk development of the commercialized economy and expansion in scholarly and cultural activities, Tokugawa society saw an unprecedented development in medical methods and an increase in the number of physicians and healers. Along with the development of the profession, obstetrics and gynecology became a distinct field where many physicians competed with each other to pursue prestige and financial gain. In this and the next chapter, I discuss the ways two major obstetric methods were established based on an understanding of the female body and its reproductive system. Obstetricians along with physicians in other medical fields relied largely on Confucian philosophy, fundamental to the rule of the Tokugawa regime, to legitimize and develop their authority. The rise of Tokugawa obstetrics meant that learned obstetricians expanded their influence over expectant mothers and competing birth attendants, such as midwives, taking advantage of the fact that midwives and mothers lacked access to medical knowledge of the female body and reproduction that physicians possessed.

In this chapter, I first briefly introduce the institution of *ryū*, or "school," the system through which masters transmitted knowledge and skills to their heirs and disciples in the Japanese art and craft traditions, including the field of medicine. The rest of the chapter is devoted to an analysis of the text of obstetrician Katsuki Gyzūzan, who adhered to the mainstream Chinese medical tradition.

Y. Terazawa, *Knowledge, Power, and Women's Reproductive Health in Japan, 1690–1945*, Genders and Sexualities in History,
https://doi.org/10.1007/978-3-319-73084-4_2

The Tokugawa shogunate created a system in which status and occupation were determined by lineage. At the core of this hierarchy were four status groups (warriors, peasants, artisans, and merchants) and two so-called "non-status" groups of virtual outcasts, the *eta* and *hinin*.[1] Other groups, such as scholars, writers, artists, Buddhist monks, Shintō priests, the nobility, and the imperial family, were not included in this main hierarchy. Physicians constituted one such ill-defined group.[2] Becoming a physician or scholar was one of the few ways a talented person without status could gain reputation and wealth. Many physicians came from déclassé warrior families that had access to Confucian and medical learning but were unemployed. Their drive to prove themselves through ambitious endeavors as physicians eventually led to numerous innovative contributions.

Over the course of the Tokugawa period, the medical profession became increasingly lucrative, even though medicine was often considered a lowly skill.[3] Physicians' lack of status was partly due to the influence of Chinese studies (or Confucianism), which favored purely intellectual activities over practical manual skills such as medicine.[4] Physicians' physical contact with diseased and presumably pollution-ridden dead bodies could also have exacerbated prejudice against the profession. In addition, precisely because practicing medicine was profitable, physicians were often despised as mercenaries who violated the ideals of Confucianism. Such prejudice against the medical profession led many physicians to identify themselves primarily as scholars and not as physicians, even though they earned their living through their medical practices.[5] On the other hand, some leading physicians endeavored to incorporate Confucian ideas and values into their medical theories and practices in an effort to make the profession more respectable.[6]

In the absence of any state regulations on the practice of medicine, physicians were only one of many types of healers—acupuncturists, masseurs, religious practitioners, shamans, midwives, and pharmacists—all vying for the market in treating the sick. There was no licensing or universal examination system for physicians, nor was there a general state-run institution for medical education. The Taki family's private academy, which eventually became the shogunate's medical school, provided medical education only for physicians employed by the shogunate.

The institutional framework for acquiring medical knowledge in the Tokugawa period was the *ryū*. *Ryū* refers to a method founded by a master and transmitted generation after generation in diverse areas such as scholarship, medicine, arts, crafts, and other accomplishments. *Ryū* also suggests an institution or school led by the head master of an artistic, technical, or scholarly tradition, and followed by masters and disciples at various levels.[7] In the late medieval (fourteenth to sixteenth centuries) and early modern (Tokugawa) periods, newly created skills and knowledge in such diverse areas as martial arts, dancing, music, the tea ceremony, flower arrangement, and even arithmetic were often declared a *ryū* by their founders and taught only to disciples who had been initiated into this *ryū*. In the field of medicine, many *ryū* formed during the Ashikaga period (1336–1573). Most of them, however, were composed of a small body of knowledge—a few prescriptions for herbal medicine, for example, or a particular surgery technique of the so-called, "external medicine," or *geka*.[8] Moreover, many *ryū* lasted only during the life of their founder. Knowledge of a *ryū* was conveyed to a small number of disciples and often disappeared within a relatively short time.[9] Accumulation of knowledge was hampered by the tradition of oral transmission, devoid of written texts. The short life span of many medical *ryū* may also have been due to the political instability and incessant wars of this period.

During the Tokugawa period, the exclusive transmission of knowledge from a master to select disciples—a practice central to the older *ryū* institution—was modified. First, many medical *ryū* began to use the institution of *ie* [the house] to transmit specific skills developed by the founder and his heirs over many generations. The family head was considered the grand master of the *ryū*, and his son or adopted son was designated as the heir of the house and next grand master for the *ryū* establishment that was composed of all the disciples. This made possible a more secure transmission and accumulation of medical knowledge over time. Second, many *ryū* established private academies, where the grand master issued diplomas to disciples whose skills were sufficiently advanced to practice independently.[10] Such certified disciples were allowed to declare themselves masters of the *ryū* in which they were trained. Third, beginning with Manase Dōsan (1507–1594) who published numerous medical treatises in the late sixteenth century,[11] written texts began to play a significant role in disseminating medical knowledge and recruiting disciples.[12] Certain fundamental details, however, were often omitted from these texts and conveyed personally only to disciples who went through years of training with the master. Contrary to medieval practice, which was heavily dependent on the secretive transmission of knowledge, the head masters at this time sought an appropriate balance between secrecy and popularization of knowledge in order to bring prosperity to their *ryū*.

The Tokugawa shogunate's policies on foreign affairs also helped shape the way medical knowledge developed. The so-called closed-door policy of 1638 posed significant limitations on Japanese contacts with Europeans and imports of goods, including books.[13] In principle, the measures prevented travel abroad for the Japanese and visits to Japan by foreigners, except for a small number of Chinese, Koreans, and Dutch. To a large extent, this policy was a defense against Christianity, which the Tokugawa authority viewed as an ideological threat. As a result, strict censorship was imposed on books from Europe. Until the 1850s, Dutch technical books were the only Western texts allowed to circulate in Japan.[14] Japanese physicians learned European medicine from Dutch medical books and from the few Dutch and German physicians attending traders in Nagasaki. On the other hand, no such restriction was imposed on the importation and circulation of Chinese and Korean publications. Japanese physicians could also consult Chinese and Korean monks and traders residing in Nagasaki about medical matters. Thus, despite the restrictions, there was a considerable infusion of foreign knowledge into the medical ideas and techniques developed during the Tokugawa period.

Goseihō's Microcosmic Body[15]

The Goseihō medical body is built upon the Yin Yang Five Phases theory. The yin yang system consists of opposites and complements whose constant movement creates energy that generates and transforms everything in the universe, including the human body. In other words, yin yang is a fundamental idea in Chinese medicine that explains that birth, death, growth, aging, and disease result from the mutual interactions of these primal elements. Since health is maintained when yin and yang factors are balanced, much of Chinese medical practice aims at restoring their balance in and around the body.

The concept of the Five Phases (wood, fire, earth, metal and water) also explains the movement of things and the relationships among them. The successive generation and subjugation of the Five Phases constitute principles of movement and change in the natural world. Chinese medical theories focus on correspondences between the Five Phases and parts of the human body, which is divided into the Five Yin Organs (heart, lung, spleen, liver, and kidney) and the Six Yang Viscera. In the first place, we must note that these bodily organs, while they loosely correspond to what we know as internal organs, should be understood symbolically and figuratively, and they are not the anatomical organs we know through the lens of modern medical science. Within this system, liver is connected to wood, as is the gall bladder. Heart and small intestine are similarly linked to fire: spleen and stomach to earth; lung and large intestine to metal; and kidney and bladder to water. The sixth yang viscera, called Triple Burner [Jp *sanshō*; Ch *san jiao*], was a digestive organ and did not correspond to any particular yin organ or phase.

This taxonomic system is usually presented in a table of correspondences that also includes natural phenomena such as seasons, climates, colors, and tastes. Thus, the natural world and the human body are organized into a coherent system by the theory of the Five Phases (see Table 2.1). Fundamental to this scheme is the aforementioned essential elements, wood, fire, earth, metal, and water. When two of these five elements interact, they produce a certain result. For example, they act against each other or an element overwhelms the other. Scientists in ancient China deduced the characteristics of these five elements and applied them to natural and bodily phenomena. Chinese scholars, then, explained changes or the life cycle both in the natural world and the human body using this system.

Here is an explanation of how this system works, focusing on the "wood" phase. During spring, trees and other plants grow, putting forth buds, and thus, the color that is associated with wood and spring is blue. While wood and spring correspond with "germination" in the category of "growth," the climate in this phase is less stable than in other phases, and thus is windy. Among the cardinal six organs of the human body, liver is assumed to have the characteristics of spring and wood because it likes to act freely and energetically. Liver and gall bladder are considered the front and the back. Liver is also linked to eyes and opens up to the outside of the body through them. Liver also presides over muscles. People who have a body in which the element of wood and liver work vigorously are easily irritable and angry. The wood phase is associated with anger in the area of human emotion. The other four elements are linked to natural phenomena or the human body in a similar manner. With this basic understanding in mind, Chinese physicians diagnose the health of the body and figure out appropriate therapeutics.[16]

The following are simple examples of how Chinese medicine physicians diagnose and provide treatments for patients. A headache that is accompanied

Table 2.1 Classification of phenomena in nature according to the Five Phases theory

Phenomena in nature						*Phenomena in human body*				
Five Phases 五行	*Five colors* 五色	*Climatic factors* 五気	*Growth and development* 生成過程	*Seasons* 時令	*Five flavors* 五味	*Zang (yin) organs* 臓	*Fu (yang) viscera* 腑	*Sense organs* 五官	*Tissues* 五体	*Emotion* 五志
Wood 木	Green 緑 or Blue 青	Wind 風	Germination 生	Spring 春	Sour 酸	Liver 肝	Gallbladder 胆	Eye 目	Tendon 筋	Anger 怒
Fire 火	Red 赤	Heat 暑	Growth 長	Summer 夏	Bitter 苦	Heart 心	Small intestine 小腸	Tongue 舌	Vessel 脈	Joy 喜
Earth 土	Yellow 黄	Dampness 湿	Transformation 化	Late summer 長夏	Sweet 甘	Spleen 脾	Stomach 胃	Mouth 口	Muscle 肌肉	Pensiveness 憂
Metal 金	White 白	Dryness 燥	Reaping 収	Autumn 秋	Pungent 辛	Lung 肺	Large intestine 大腸	Nose 鼻	Skin and hair 皮毛	Grief 悲
Water 水	Black 黒	Cold 寒	Storing 蔵	Winter 冬	Salty 鹹	Kidney 腎	Urinarybladder 膀胱	Ear 耳	Bone 骨	Fear 恐

Sources: Liu, *The Essential Book of Traditional Chinese Medicine*, Vol. 1, 50; Nankin chūigakuin, ed., Chūgoku kanpō igaku gairon, 20

by dizziness is considered to be caused by the wind of liver that circulates over the head of a patient. The person afflicted with a headache often becomes emaciated and suffers from fever. Theoretically this problem is caused by a deficiency of water and an excess of wood. Since liver is associated with wood, the blowing wind of liver results in an excess of wood element within the patient's body. A Chinese medical classic asserts that to cure this problem, it is not sufficient to adjust the wind to the optimal level. The best therapy guided by the Five Phases theory is to increase the water of the kidney (water is kidney's corresponding organ).[17]

Another example is when spleen becomes sickly, which negatively affects heart. Since heart belongs to the fire phase and spleen belongs to the earth phase, and fire produces earth, spleen is considered a child of heart, the mother. This problem is thought to be caused by the child (spleen) depriving the mother (heart) of its *ki* [Ch *qi*], or vital energy (see later for a discussion of *ki*). The symptoms that originate in this problem—such as loss of appetite, diarrhea, debility, insomnia caused by palpitations, and amnesia caused by heart acceleration—are tied to irregularity of heart. The physician, however, cannot cure these problems only by remedying the faculty of heart. It is mandatory that the patient supplements the earth element of spleen, so that the *ki* of spleen becomes vigorous, and thus spleen does not deprive *ki* of its mother, heart.[18]

Another example shows how the yin yang theory and Five Phases theory are used simultaneously. The problem of fatigue is often accompanied by fever, sweat, coughing, vomiting blood, and insomnia and it is caused by a deficiency of water of kidney, and the presence of fever that arises owing to a lack of bodily fluids. Insomnia is brought about by a deficiency of water of kidney, where the level of essence (or water) of kidney (i.e., kidney yin) is low and only the fire of heart rises. The cure for this should be focused on amplifying water, which constrains yang. In this way, water of kidney is boosted and the fever, or the fire caused by a lack of fluids, naturally subsides.[19]

As these examples indicate, there are an almost infinite number of ways in which a physician could devise etiological explanations and medical treatments. The system of the Five Movements and Six Qi [Ch *wu yun liu qi*; Jp *goun rokki*], which the Goseihō physicians adopted from China, were the basis for an even more elaborate model that associated changes in the human body with climate, year, and season. The model used the Yin Yang Five Phases theory and the zodiac signs, including ten celestial signs and

12 terrestrial signs. Goseihō School physicians studied basic theories delineated in Chinese medical canons from ancient China and more recent texts written after the Jin period (1126–1234). The shortcomings in their understanding of this complex model, and ensuing failures for providing effective prescriptions, contributed to the rise of the rival faction, the school of Ancient Practice.

In addition to the Yin Yang Five Phases system, *ki*, blood [Jp *chi*; Ch *xue*], other bodily fluids, and the circulation and tract systems (the meridians) are fundamental ideas in Chinese medicine. The operation of *ki*, a vitality vested with the power to form material objects, affects all change in the universe including generation, growth, and death. Therefore, the condition of human life is also dependent on the workings of *ki*, which generates, maintains, and nourishes the entire body by carrying what it needs to each of the body's organs and parts. Different types of *ki* exist. For example, orthopathic *ki* [Jp *seiki*; Ch *zheng qi*] prevents disease by making adjustments within the body and in the environment surrounding the body. Heteropathic *ki* [Jp *jaki*; Ch *xie qi*] causes movements that interrupt the balance necessary to the body's health.

In Chinese medical language, blood is different from its European counterpart. The Chinese version of blood includes a notion of nourishment: by circulating throughout the body, blood strengthens all parts of the body and makes them function. The body's circulation and tract systems are constituted by the meridians, which distribute *ki* and blood. Each meridian belongs to or is controlled by a yin organ, yang organ, or other bodily constituent. Moreover, the meridians are not confined to the interior of the body. In fact, one of the functions of the meridians is to connect the interior of the body to the environment outside. Circulatory substances such as *ki* flow in and out through certain points on the surface of the human body. Acupuncturists use such points to adjust these flows into and out of the body.

The late seventeenth-century obstetric text *Fujin kotobukigusa* [*A Guide for a Woman's Celebratory Event*, 1692] written by Katsuki Gyūzan (1655–1740), a Goseihō School physician, clearly shows that the book adhered to the aforementioned understanding of the human body, describing obstetric methods and advice to pregnant women. In writing the book, Gyūzan relied heavily on Chinese obstetric texts. The appropriation of medical theories and obstetric methods from classical Chinese texts, in the end, reinforced the moral and medical authority that physicians accrued and helped them enter the midwifery profession.

Manase Dōsan and the Goseihō School

Among many medical *ryū*, the Goseihō School, including several offshoots, constituted the mainstream of medical tradition during the Tokugawa period.[20] Its methodology originated in medical discourse developed in China during the Jin (1126–1234), Yuan (1271–1368), and Ming (1368–1644) periods. It organized its medical theories using the systemic metaphysical concepts of neo-Confucianism or Song Confucianism.[21]

Under the influence of Song Confucianism, physicians undertook a systemization of the disparate ideas from the various medical classics inherited from earlier periods. Specifically, they relied on Song Confucian metaphysics and its central concept of *ri* [Ch *li*]. The well-ordered system of Song Confucianism rested on an assumed correspondence between the macrocosm (heaven, earth, and nature) and the microcosm (human beings). Physicians believed the universe and human beings were structured by the same principles. Thus, they conceptualized the body's mechanism through Song Confucian cosmology and a systematic application of older concepts such as yin yang [Jp *inyō*], the Five Phases [Jp *gogyō*; Ch *wu xing*], the theory of Five Yin Organs and Six Yang Viscera [Jp *gozō roppu*; Ch *wu-zang liu-fu*], and the meridian system [Jp *keiraku*; Ch *jingluo*].[22]

This new medical discourse was known in Japan as "Rishu medicine," and constituted the Goseihō School's main method. Although it originated in twelfth- and thirteenth-century China, Japanese physicians did not adopt it until the sixteenth century, and its spread is owed mostly to Manase Dōsan and his descendants and disciples.[23] Dōsan's sons and adopted sons each formed a branch of the Goseihō School, which dominated Japanese medicine until the early eighteenth century. The main branch of the Manase family enjoyed much prestige, providing the Tokugawa shogunate's medical administration chief physicians for six generations.[24]

Furthermore, the medical field was increasingly classified hierarchically according to specializations.[25] Internal medicine, called *hondō* [Major Way], was at the top, followed by sub-fields of lesser importance such as surgery (the aforementioned "external medicine"), ophthalmology, gynecology/obstetrics, and pediatrics. European medical techniques introduced in the sixteenth century by missionaries became known in the first half of the Tokugawa period through several schools of surgery, although these remained marginal until "Dutch medicine" [*Oranda igaku* or *Ranpō*] developed through the study of Dutch medical texts. The

Netherlands, the only European country that had been allowed to trade with Japan at the time, started to make inroads into Japanese medicine in the middle of the eighteenth century.[26] The hierarchical structure of Japanese medical specialties partly explains the marginalization of European medicine until the late Tokugawa period.[27]

Obstetrics and Gynecology Through the Genroku Era[28]

Physicians specializing in obstetrics and gynecology seem to have first appeared in the ancient period as court physicians for women of the imperial family. Their expertise rested on knowledge of the Chinese medical tradition—as embodied, for example, in the Japanese medical classic *Ishinpō* [*Prescriptions at the Heart of Medicine*, 984]. Throughout the medieval period, access to such knowledge was strictly limited to a few hereditary court physicians.[29] A period of incessant warfare in the fifteenth century led to the emergence of barber-surgeons called "physicians for sword wounds" [*kinsōi*], some of whom also developed obstetric skills. By the sixteenth century, a number of physicians specializing in gynecology and obstetrics formed their own *ryū*.[30] One of them was the Chūjō School of obstetrics.[31]

The texts written by Chūjō School physicians, like other medical texts from this period, typically listed symptoms of illness and their corresponding treatment, the majority of which involved prescriptions of herbal medicine.[32] They also included occasional references to the stagnation of blood and the operation of harmful blood as the cause of childbirth-related illnesses; however, they did not present a systematic description of the body's physiology.

The rise of these *ryū* was followed by an overarching methodological transformation in the seventeenth century, centered on the emergence of the Goseihō School and its offshoots. This transformation was evident in the appearance in the early 1690s of a different type of birth manual from the ones written in the style of the Chūjō School text. Inō Shōji's (Kōken's) *Inago gusa* (1691)[33] and *Fujin kotobukigusa* are good examples of the new style. These texts described each stage of pregnancy, childbirth, and puerperium, instead of simply listing medical symptoms and prescriptions. The novelty of these texts also consisted of representing a systemic understanding of the body and its surroundings. Like many texts of the Goseihō School, they often attributed the cause of illness to a malfunction in the circulation of *ki* and blood. This holistic model of the body included moral teachings derived from Song Confucianism. It should also be noted that, unlike the Chūjō School text and its derivatives, these texts were written for a general readership as well as for professional obstetricians.[34]

Thus, in the late seventeenth century, Goseihō School discourse, along with the component of Song Confucian thought incorporated in it, became an integral part of obstetric and gynecological practice. As a result of this shift, the reproductive body came to be understood in terms of a cosmological system that predicated a systematic link between the body and the universe outside of it. Within this worldview, medical affairs and moral matters were often conflated, and taking care of one's health became associated with maintaining moral rectitude.

Katsuki Gyūzan and His Work

In China, Shu Tankei [Ch Zhu Danxi, 1281–1358] and other prominent physicians identified themselves primarily as literati gentlemen and viewed medicine as one of the domestic skills they should cultivate to fulfill their principal duty as the heads of landowning families.[35] In Tokugawa Japan, however, medicine was a career—often followed by men from déclassé warrior families or the families of well-to-do commoners; in addition were those from hereditary medical families who served the imperial court, the Tokugawa shogunate, and domanial governments. Unlike physicians from the Chinese gentry, they often lacked other sources of income. Practicing medicine, either by cultivating clients, serving a domanial government, or both, was how they made a living. The well-known Goseihō School physician Katsuki Gyūzan was among those physicians who earned their living by providing services to domanial lords. Gyūzan was born in 1655 in Chikuzen in northwestern Kyūshū.[36] He received his medical training from the physician Tsuruhara Gen'eki[37] and studied Confucianism with Kaibara Ekken (1630–1714), a well-known Shushi [Zhu Xi] School scholar, also known for his studies of medicinal herbs [Honzō-gaku]. (Both Gen'eki and Ekiken served the Kuroda House.) Gyzūzan's reputation led the daimyō of Nakatsu domain in today's Ōita prefecture to entice him into the lord's service as a physician.

Gyūzan served the daimyō for 14 years, during which time he often accompanied his lord to Edo, where he met prominent scholars. After retiring in Kyoto, Gyūzan spent his time writing books and socializing

Fig. 2.1 A portrait of Katsuki Gyūzan (Source: Ogata, *Nihon sanka gaku shi*, 92)

with noted intellectuals and physicians. His reputation grew even more during this period. He cured several patients with serious ailments, including a retired emperor, and his fame prompted the daimyō of Kokura in northwestern Kyūshū to invite him to reside there as guest physician-scholar. Gyūzan eventually accepted the lord's invitation and stayed in Kokura until his death in 1740. Although Gyūzan remained single throughout his life, he married his niece to a man named Sokkan and adopted him as his son. After Sokkan's early death, Gyūzan adopted one of his disciples, Teian, to carry on the Katsuki House.

Gyūzan's biography shows that he was a highly reputable physician. The prestige attributed to the medical profession, however, did not match that associated with a high-ranking warrior serving a domain or the shogunate.[38] In fact Tokugawa physicians confronted a pervasive view of the profession as lowly. Practicing medicine did not fulfill the Confucian ideal of a gentlemanly lifestyle, since it was done for money and involved manual labor.[39] From this perspective Gyūzan's extensive appropriation of medical theories from Chinese texts could be viewed as an attempt to make the medical profession more respectable. The sophisticated Chinese medical theories and Confucian scholarship to which Gyūzan had access were an effective means to enhance the legitimacy of his practice. One consequence of this was that Gyūzan followed the ideas presented in authoritative Chinese medical texts rather than challenging them. Writing manuals intended for a wider audience were also part of his efforts to extend the influence of learned physicians.[40]

Pronatalist Reproductive Techniques

Gyūzan's recommendations in *Fujin kotobukigusa* were marked with strong pronatalist values originally developed in Chinese society, where maintaining the patrilineage was considered fundamental for the family's prosperity in both material and symbolic terms. In Ming (1368–1644) and Qing (1644–1912) China, the literate medical tradition reinforced the importance of patrilineal continuity by elaborating on the notion of *yuan qi* [Jp *Genki*]. Conceived as essential vitalities that worked to rejuvenate a lineage, *yuan qi* was transmitted generation after generation through male biological heirs.[41]

For the great majority of families in Tokugawa Japan, it was important to preserve a household over generations and to make it more prosperous and respectable. But it was not important to develop a familial genealogy through biological male progeny. This was evident in the widespread practice of adoption, which was not only accepted but also sometimes preferred since it gave the family the advantage of choosing the most capable person possible as heir. Adoption was practiced by all status and non-status groups and took various forms. It was common for families with only daughters to adopt the husband of the eldest daughter. Many childless families adopted a male child, although it was not unusual to adopt a female child whose husband would inherit the household. There were also cases in which families adopted an adult couple.[42] The pervasive custom of adoption in Tokugawa society indicates that having biological male children was not as important in early modern Japan as in late imperial China.

Nonetheless, Gyūzan's discussions of reproductive strategies, borrowed largely from Chinese texts, placed importance on the male child. This emphasis would have helped promote the influence of learned physicians, who presumably possessed special knowledge for the successful conception of a male child and his safe delivery. Moreover, *Fujin kotobukigusa* communicated Confucian familial ethics, which valued a large family and mandated loyalty of all the family members to their house—a practice encouraged through pronatalist sex counseling. Gyūzan's disapproval of abortion was an effort to support such values.

Gyūzan's pronatalist bias is evident in the lengthy advice he gives for a fruitful conception. Using Chinese texts as his source, Gyūzan delineated the body's reproductive mechanism by focusing on the function of circulatory bodily substances such as *ki*, blood, and essence [Jp *sei*; Ch *jing*],[43] and the Yin Yang Five Phase system. A visceral system centered on the kidney played a crucial role in generating and preserving essence. If one takes care of his or her body thoughtfully, following age-old medical advice developed in the Chinese medical tradition, one could accumulate ample amounts of essence in the space called the kidney, and this was vital for a successful conception. Gyūzan claimed that abundant blood in a woman's body and vigorous essence in a man's body were especially necessary for a fruitful conception. Both women and men, however, were thought to possess a fundamentally identical reproductive physiology.

Gyūzan warned against marriage at an early age because young people lacked the strong *ki* and blood that were indispensable for conceiving and nurturing a healthy and intelligent child.[44] Furthermore, the frequency of sexual intercourse should be regulated according to people's age.

> In *Yōjōroku* [Ch *Yang shen lu*],[45] it is recorded that Sojo [Ch Sunü]...said, "People in their 20s should have sex once every four days. People in their 30s should have sex once every eight days. In their 40s, once every sixteen days. In their 50s, once every twenty days. People over sixty should refrain from having sexual intercourse in order to maintain their essence. It will be easy to conceive a child for those who follow these instructions as the blood and ki of those men and women will be strong. Moreover, the child will live long without having any accidents and will be blessed with luck, wisdom and intelligence."[46]

Gyūzan added that there were individual differences in terms of how often one should have sexual relations. People who had weak constitutions should observe a longer period of sexual abstinence to avoid dissipating their essence.

Gyūzan also introduced Chin Jimei's [Ch Chen Jiming, 1190–1270] idea that vigorous *yōki* [Ch *yang qi*] in the man's body and abundant yin blood in the woman's body were particularly important for successful conception.[47] Citing the Chinese physician Zhu Danxi, Gyūzan explained that a failure to conceive resulted from insufficient blood in the woman's body, a condition that prevented it from holding the proper amount of yang essence. This attribution of infertility to a malfunction in the woman's body was challenged by a citation from another Chinese physician, On Inkyo [Ch Wen Yinju],[48] who described the effects of an inadequate amount of yang essence in the man's body: "When there is not sufficient essence or when essence is weak and cold, they do not mix with yin blood."[49]

Gyūzan stressed that women should not be solely to blame for the failure to conceive, since sterility is caused by a great number of problems in both men and women. This point was underscored in an account of conception using a seed metaphor.

> Ryū Sōkō's [Ch Liu Zonghou] theory states that the shortage of yang ki of the father is often the cause of not being able to have a child.[50] Thus, it is wrong to place blame on the shortage of the mother's blood. Plants grow well on fertile land, but if the seeds are bad, they will not grow. Similarly, if the meridians and pulses are not well adjusted, causing hundreds of illnesses to sprout, the woman will not become pregnant. This is similar to plants that do not grow because the seeds have been planted on barren land.

This account contradicted the prevailing view, which placed the blame on women for infertility.[51] It is not certain, however, if Gyūzan's opinion was influenced by sympathy for women. It is possible that he wanted to give an accurate account of conception, drawing on an authoritative Chinese text, and that his defense of infertile women was an unintended by-product. He could also have been attempting to refute the popular view in an effort to establish himself as a learned physician and to distinguish himself from lay people and other medical practitioners who lacked knowledge of Chinese medical canons.

In Goseihō School medical discourse, changes in the body's internal physiology were connected with changes in its external environment. The cycle represented in the Yin Yang Five Phases system and the movement of *ki* affected both the body and the world outside it. Theories of the Five Movements and Six Qi explained how the shifting climate of the outside world influenced the internal mechanism of individual bodies. The theories utilized the calendar based on Yin Yang Five Phases theory along with the systems of the Ten Celestial Stems and the Twelve Terrestrial Branches. Also, the year was divided into four or five seasons, with each season corresponding to one of the Five Phases. Each year and day was labeled by both a celestial stem and a terrestrial branch.[52]

Citing Chinese physician Son Shibaku [Ch Sun Simiao, 581–682], Gyūzan suggested that couples wanting a child should have sexual relations on a date whose celestial stem and terrestrial branch were the same as those of the year in which either the man or the woman was born.[53] Moreover, Gyūzan believed that a person's energy became most vigorous on dates corresponding to the phase representing her/his nature. "For those whose nature is wood, the main corresponding dates are dates marked with *kō-otsu*, or with *tora-u*. Their energy is vigorous on those dates."[54] These were also the recommended dates for sexual intercourse based on this system.

In addition, Gyūzan recommended that couples have sexual relations on yang dates and times. Yang dates were marked by such celestial stems as *kinoe*, *kinoto*, *kanoe*, *mizunoe*, *tsuchinoe*, or with the terrestrial branches *tora* (tiger), *u* (rabbit), *tatsu* (dragon), *mi* (snake), *uma* (horse), *hitsuji* (sheep). The yang period of the day was between the hour of *tora* (tiger, 4 a.m.) and the hour of *hitsuji* (sheep, 2 p.m.).

On the other hand, people should avoid sexual intercourse on dates marked by certain celestial stems. This is because the nature of the phase associated with these celestial signs affected the body's physiology. In the following example, Gyūzan suggested that having sexual intercourse on dates corresponding to the Fire Phases would damage the kidney system, which was indispensable for reproduction.

> The reason why people should avoid sexual intercourse on dates that are marked as *hinoe* or *hinoto* is because *hinoe* and *hinoto* are the celestial stems where fire works vigorously. These dates are characterized as "yang of yang." Yang ki of the human body is its highest [on those dates]. Because of this, if people have intercourse on those days, yin fire[55] proceeds to work inside. Since fire burns things, and it exhausts kidney yin water.[56]

Gyūzan noted that this recommendation originated in Sun Simiao's *Senkinpō* [Ch *Qianjin fang*].[57] However, the way in which he conceptualized the crisis of an overactive yang energy associated with fire has an unmistakable resonance with Zhu Danxi's interpretation of bodily and psychic dynamics through the juxtaposition of heart fire and kidney water. Zhu Danxi defined fire, or the active bodily energy generated by the heart system, as the destabilizing source of human emotions, including sexual desires. When such potent yang fire acted in a disorderly manner—in the form of a powerful sexual impulse, for example—it would influence the kidney system and damage its yin water, a fluid essential for reproduction. Zhu Danxi thus incorporated Song Confucian notions into his medical theories by connecting physiological and psychic aspects of the human self.[58] As a faithful student of Zhu Danxi, Gyūzan also inherited this line of reasoning.

The calendar also indicated dates when *ki* was adequate for a successful conception. *Ki* would become invigorated or dissipated according to the calendar, affecting bodily as well as atmospheric changes. People's sexual practices should be regulated according to this cycle. Gyūzan, for instance, prohibited having sex on the first and the last day of a given month because he considered that *ki* was weak at that time and needed to be vitalized. He also recommended that couples observe sexual abstinence when the full moon appeared on the fifteenth day of each month because people's yin blood was at its fullest and *yinki* was vigorous. Gyūzan proposed that abundant *yinki* should not be undermined by sexual intercourse but should be preserved until other elements necessary for a fruitful conception were also well developed.

Warnings against sexual intercourse were also issued on days when volatile weather indicated an "irregular *ki*": "The reason why there is strong wind, heavy rain, extreme heat or cold, lightning, rainbows, the eclipse of the sun or of the moon on certain days is because irregular *ki* of Heaven and Earth operates on such days…Tankei imposes strong injunctions on sexual intercourse on those days."[59]

In addition to a lack of *ki*, blood, and essence, neglecting to engage in righteous deeds could result in infertility. Gyūzan listed various measures a childless couple could adopt, including the aforementioned reproductive techniques, praying to the gods, or procuring a concubine. According to Gyūzan, if a couple tried these means and still failed to conceive a child, it meant they were engaged in "evil thoughts and conduct." Such couples should mend their ways and cultivate virtue. By practicing virtue through good deeds, acts of kindness, and displays of benevolence, "the charity of the emperor in Heaven will penetrate nature so that the couple will be able to have children and maintain future generations [for the family]."[60]

Although Gyūzan described pronatalist reproductive techniques at length, he refused to discuss abortion methods. His reservation notwithstanding, in large cities such as Edo, Osaka, and Kyoto, methods to induce abortion were widely known among different types of medical practitioners in the late seventeenth century. Starting in the eighteenth century, those who aided in abortion were often called *onna isha* (physicians for women who were abortion specialists)[61] and Chūjō-ryū.[62] Practitioners sold abortion pills containing strong drugs, such as ground cherry roots [*hōzuki*], gallnuts [*gobaishi*], rhubarb roots [*daiō*], dried gadflies [*bōchū*], and liquid mercury. An abortifacient included in a 1668 Chūjō School obstetric text was a combination of mercury and a decoction of betal palm nuts and Japanese mint, which was inserted into the vagina. This formula was believed to putrefy and abort the fetus. A more popular method that caused the same result involved inserting the root of ground cherry into the vagina.[63]

The ready availability of abortion providers in urban areas concerned shogunate authorities as early as the mid-seventeenth century. During the early Tokugawa period, the shogunate prohibited abortions not so much out of concern for a population decline but because it reasoned that abortion impinged on Confucian family ethics. It was commonly believed that women who sought the assistance of abortion specialists had become pregnant as a result of adultery, and restricting abortionists was thought necessary to prevent the decline of sexual mores.[64]

Gyūzan's views on abortion were in line with those of the state. He remarked that eminent physicians from ancient times had referred to certain abortion methods only as a means of saving the mother's life when complications arose during delivery. Noting the activities of abortion doctors, Gyūzan insisted that their businesses should be shut down. He used anecdotes to describe the ways in which these abortionists would receive punishment in the form of various misfortunes, including a dreadful death, for their immoral conduct.

Gyūzan avoided the issue of pollution, which was thought to be generated during the birthing process. The understanding of pollution varied by historical period, geographical region, and social group in China and Japan, but major pollution-producing events, including death, disease, childbirth and menstruation, occur at the boundary of life and death. Things, places and people become polluted by being in contact with defiled matter, such as diseased or dead bodies, or parturition or menstrual blood. In many cases, pollution was customarily addressed by segregation and removed by purification rituals. Physicians and midwives who handled pollution were often relegated to lower status, but contemporaries understood they had a special ability to manage powerfully polluting substances at the border of this and other worlds.

That Gyūzan did not mention the polluting aspects of childbirth was likely to have been deliberate since many of the Chinese texts he cited underscored the defiling aspects of childbirth. Gyūzan also did not mention fetal poison [*taidoku*] in *Fujin kotobukigusa*. According to Charlotte Furth, there was less emphasis on birth pollution in China during the Ming–Qing period than in earlier periods. Instead, obstetricians and pediatricians developed various theories of fetal contamination to highlight the dangerous aspects of childbirth.[65]

In Japan, despite Gyūzan's neglect of the matter, the idea of birth pollution was deeply rooted in Tokugawa childbirth practices. This is evident in the recommendations listed in *Onna chōhōki*. The text suggests that after childbirth, the new mother, the child, and their family should observe appropriate mourning periods and purification rituals.[66] Japanese folklorists and anthropologists have shown that new mothers and their families were subject to childbirth-related pollution taboos in many areas of rural Japan through the nineteenth century, and in some remote regions, this practice remained in place as late as the early twentieth century.[67]

By dismissing pollution taboos, Gyūzan may have intended to shift his readers' attention away from the negative aspects of childbirth. He could also have wanted to disassociate the medical profession from defiling events. This dismissal of Chinese theories about childbirth demonstrates that Gyūzan's appropriation of Chinese medical knowledge was selective, not all-inclusive. At the same time, his strategic use of text-based Chinese medical knowledge was most likely an outgrowth of his desire to make his professional services respectable and attractive to potential clients.

Disciplining Pregnant Women

Fujin kotobukigusa listed various recommendations for pregnant women to assure healthy fetal growth and a safe delivery. Gyūzan introduced a widespread Chinese model that viewed the woman's body and psyche as delicate, vulnerable, and prone to malfunction. Pregnant women should therefore be strictly protected from emotional disturbances and strenuous physical movements. The teachings regarding the educating of the fetus [*taikyō*] emphasized how important it was for women to discipline themselves and cultivate their virtue in order to protect the well-being of the fetus. Such efforts at self-discipline, however, were supposed to be guided by learned physicians, such as Gyūzan, who were well versed in Chinese texts. In this way, Goseihō physicians assumed moral authority as well as medical expertise at the point where medical issues overlapped and converged with ethical conduct.

The pregnant body was considered to be fragile due to the disruptive movements of fetal *ki* [*taiki*], which was believed to emerge with the fetus and nurture it during the period of gestation.[68] Irregular movements of fetal *ki* could also produce "heat" to make the body ill.

> Pregnant women sometimes suffer from an illness, which either causes them to urinate frequently, or makes them unable to urinate due to pain. This illness is called *shikan* in *Fujin ryōhō*. It is caused by volatile fetal *ki* that clogs the lower section of the Triple Burners and generates heat.[69] Fetal *ki* should be made calm to reduce the heat...Women should not urinate excessively. These symptoms often appear five or six months after a woman becomes pregnant.[70]

To treat this illness, Gyūzan provided herbal prescriptions.

Disturbing the movements of fetal *ki* also caused morning sickness. In the second or third month of pregnancy, when fetal *ki* stagnated in and around the throat and the chest, it would interfere with the function of chūshō [Ch *Zhong jiao*, the middle section of the Triple Burner], the digestive organ, and make the woman nauseated. When fetal *ki* became "settled" in and around the uterus or "below the navel" during the fifth or six month of pregnancy, the digestive system usually operated well, and the woman would no longer suffer from morning sickness.

Fetal *ki* was not the only active element during pregnancy. The process of gestation involved the functioning of major visceral organs, which maintained the adequate circulation and preservation of *ki*. For such organs to function well, the pregnant woman should not become emotionally agitated or mistreat her body. Gyūzan believed that failing to observe this lesson would cause complications during childbirth.

> According to *Bensan suchi* [Ch *Bianchan zuzhi*], a pregnant woman should not exhaust the ki of heart by thinking. She should not break the ki of liver by getting angry. She should not damage the ki of lung by exposing herself to the wind and cold of outside. She should not upset her spleen and stomach by excessive eating and drinking. She should not exhaust essence of kidney by having sexual desires. She should not reduce ki of the gods by feeling sad. If she does not follow these instructions, she will experience premature birth or have a difficult delivery.

Unlike the active function of these major organs in the reproductive process, the uterus [*shikyū*] performed a relatively passive role. Gyūzan described it as a container possessing two branches, each for nurturing a fetus of a different sex.

> The place where the fetus dwells is called the uterus. A single system presides over it; however, it has two branches. It looks like two bowls that have

> been conjoined. One is connected from the left [side of the body], and the other is connected from the right. If yang essence overcomes yin blood, yang becomes the main element [to form a fetus]. In this case, the uterus on the left side receives ki to produce a male form. If yang essence does not conquer yin blood, yin becomes the main factor. Then, the uterus on the right side receives ki and creates a female body.[71]

Gyūzan also described the growth of the fetus following Chinese models.

> *Fujin ryōhō* says...in the first month of pregnancy, it [the fetus] looks like a dewdrop. In the second month, it looks like a peach flower. In the third month, the sex of the fetus is determined. In the fourth month, it materializes [its body]. In the fifth month, it develops muscles and bones. In the sixth month, it grows hair. In the seventh month, its soul [Jp *kon*, *tamashii;* Ch *hun*] wanders, and the child repeatedly moves its left hand. In the eighth month, its soul also wanders around, and it frequently moves its right hand. In the ninth month, it rotates its body three times. In the tenth month, it receives ki.

The fetus, by this account, did not take human form until the fourth month; instead, it looked like a dewdrop or a peach flower. Once the sex of the fetus had been fixed and its body began to form, it developed new body parts (e.g., muscle, bones, and hair) with each passing month. In the later months of pregnancy, the fetus became more like a person with a "soul."

Another version of the growth of the fetus illustrated how the meridian systems connected the body of the mother with that of the fetus. Each month different meridians developed specific body parts: blood, veins, sinews, bones, and so on. Moreover, certain body parts developed when the fetus received a particular essence, each of which pertained to one of the Five Phases. For instance, in the fourth month, the "yang minor hand meridian"[72] functioned in conjunction with the essence of the water phase to create the fetus's blood and veins. The lung presided over "meridian of great yin[73] which also provided the *ki* that developed bones during the seventh month, when the fetus received the essence of the wood phase.

The following table shows the months in which the fetus receives each aspect of spirit pertaining to the Five Phases and the body parts that are developed by its reception.

Month	*Phase of spirit received*	*Body parts developed*
4	Water	Blood veins are made. Six viscera are "conditioned."
5	Fire	*Ki* and four limbs
6	Metal	Sinews and muscles, and the mouth and the eyes, and so on
7	Wood	Bones, skin and hair
8	Earth	Skin/hide and nine holes (i.e., nine openings from the body to the outside)
9	Stone	Skin and hair
10	–	Five organs, six visceras and other body parts are complete.

It is evident that Gyūzan's concept of the body classified its parts differently than modern European medicine. The categories shown in this table also correspond to those in Terajima Ryōan's *Wa-kan Sansai Zue.*

Gyūzan included detailed recommendations for pregnant women during this process of fetal growth. In particular, he stressed the importance of regulating the operation of two types of *ki*, including fetal *ki* and cardinal *ki* [Jp *shinki*] because anomalous movements of these *ki* would disturb the fetus. Fetal *ki*, by and large, was affected by a woman's physical movement. Cardinal *ki*, which formed as a result of the function of the heart system, on the other hand, tended to be stirred up by emotional agitation. The movements of either, however, affected the other.

For the two types of *ki* to function in a regular pattern, the expectant mother was counseled to discipline her movements. For example, citing *Retsujoden* [Ch *Lie nü zhuan*; *Biographies of Exemplary Women*], one of the Chinese classics for educating women, Gyūzan remarked that in ancient times when women became pregnant, they slept on their side, sat up straight, and did not get up abruptly.[74] These were the "proper manners" that a woman should follow. Prohibited activities included taking too much medicine, excessive drinking, inappropriate moxibustion and acupuncture, urinating in inappropriate places, carrying heavy things, stretching oneself to grab something out of reach, and walking up a steep path. Failure to follow these recommendations disturbed fetal *ki*, thereby causing a premature birth or a difficult delivery.

Gyūzan recommended that a pregnant woman not eat, see, or listen to what was improper. For example, she should not listen to amorous popular songs that were sung to a *shamisen* (a Japanese three-string instrument). Instead, she should listen to poetry and virtuous lyrics chanted by blind minstrels. If the expectant mother followed these instructions, her child would be handsome, intelligent, and of excellent character.

While fetal *ki* was agitated mainly by the woman's mishandling of her body, cardinal *ki* was disturbed principally by strong emotional reactions. For instance, cardinal *ki* was affected by watching a fire, unexpectedly hearing sad news, or exposure to loud shouting close at hand or noise from construction work in the neighborhood. The disturbance of cardinal *ki* would, in turn, affect fetal *ki*, which could eventually cause illness.

Abnormal functioning of blood along with that of *ki* would also cause problems in the gestating body. In this case as well, emotions were often the source of disorder. For example, Gyūzan remarked that if the expectant mother's desires were stirred up, *ki* and blood would become heated and disturb the fetal *ki*. This made it imperative for the expectant mother to suppress her desires, and particularly to observe sexual abstinence.

The idea of the education of the fetus further underscored the importance of the pregnant woman's proper conduct. In Tokugawa Japan, discourse on fetal education was accompanied by the widespread dissemination of Confucian teachings from the late seventeenth century. Like other scholars and physicians who believed in the effectiveness of fetal education, Gyūzan viewed women's moral conduct during pregnancy as directly affecting the health and intelligence of the child. He suggested that the mother's conduct in specific areas of activity influenced corresponding areas in the developing fetus. For example, to have a handsome and virtuous child, Gyūzan recommended that the expectant mother say only righteous things and not engage in indecent gossip. She should, moreover, discipline herself and not be self-indulgent or engage in shameful conduct. If she wanted a beautiful child, Gyūzan advised the expectant mother to wear pearls and to refrain from looking at ugly women, cripples, and other people of strange appearance. Similarly, if she wanted an intelligent child, she should read *The Book of Documents*, *The Book of Songs*, and other books of the sages.[75] For a child of the preferred sex, Gyūzan remarked: "If the mother wants a boy, she should always carry a bow and arrows. If she wants a girl, she should wear a belt made of beautiful beads." His recommendation was an example of how the outward form affects what develops inside the mother's body. He summed up how a mother's actions influenced the growth of the fetus as follows: "When outward forms are proper, it is felt by what is inside".

Gyūzan's medical representations of a pregnant woman's body were disseminated through major women's household manuals. They may well have influenced how pregnant women were treated. At the mercy of fetal *ki*, the pregnant body required utmost protection. Furthermore, the discourse of fetal education, which emphasized the mother's virtuous conduct and body discipline on the child's physical, intellectual, and spiritual growth, introduced a new order of surveillance. Thus, the pregnant woman should be ideally confined to a quiet family quarter of the house. Not only should she avoid arduous physical labor that would damage her delicate health, but she should also refrain from taking on the management of professional and household affairs that would deprive her of peace of mind. This advice bolstered a patriarchal ideology and practice that assigned wives of childbearing age to the narrowly circumscribed role of mother, and imposed restrictions on their physical mobility, an arrangement that diminished the power of young wives within the family and subjected them to greater control by husbands and parents-in-law.

Gyūzan's recommendations vested with such ideological significance should be situated in the context of late-seventeenth-century Japan where there was an emergence of a class of wealthy merchants and peasants who aspired to learn proper manners authorized either by Confucian or indigenous aristocratic traditions. Adopting such refined customs was a way to demonstrate affluence and underscore a distinction in social standing vis-à-vis other commoners. To sequester women in the family to an inner part of the house was a practice that respectable families followed, according to both Chinese tradition and Japanese aristocratic custom. At the same time, women who excelled in various accomplishments and knew proper manners were assets both for the women's natal families and the households into which they were married. Upwardly mobile commoner families, therefore, had good reasons for confining young wives to the domestic sphere, limiting the range of activities in which they could engage, and scrutinizing their manners.[76]

Newly introduced teachings for expectant mothers generated a new form of oppression for women. It is precipitate, however, to conclude that adopting such guidance authorized by Confucian doctrines resulted in merely producing undue burdens for them. Depending on the family's social standing and domestic setting, there were ways in which women could take advantage of those medical representations and moral instructions. It should be noted that in practice, the majority of commoners' families could not afford to seclude pregnant women within the house because their labor was indispensable for farming or maintaining the family business. In such circumstances, the discourse that dictated providing protection and care for the pregnant woman could have helped some expectant mothers obtain relief from strenuous farm and household labor and thus safeguard their health and well-being during pregnancy.

The discourse on fetal education that underscored the significance of the tie between mother and child had both oppressive and favorable implications for women. In it the mother's moral strength and the level of her cultural refinement were considered the prime determinants of whether a child would grow up to become a talented and virtuous person. Such thinking made the mother particularly liable for the failure to produce qualified progeny.[77] On the other hand, contrary to a line of patriarchal ideology that reduced the value of the wife to her physical ability to procreate, the emergent Confucian teachings for expectant mothers acclaimed the moral and intellectual power of women. They depicted women as possessing a potential for attaining the moral rectitude required to observe maternal responsibilities. Concurrently, women's education was strongly endorsed. Women were not only encouraged to master womanly accomplishments, such as playing musical instruments, dancing, and flower arranging, but also to study Chinese classical texts. Within the confines of motherhood, women could fulfill their desire for learning, engage in character building, and attain self-satisfaction for carrying out suitable maternal responsibilities.

Among women who subverted Confucian norms of motherhood were a significant number of women who took Buddhist orders and relinquished motherhood entirely. Ladies-in-waiting who worked in the Ōoku—the women's quarters in the shogunal castle in Edo—also forwent motherhood for careers that were highly rewarding status-wise and financially. A very small group of delinquents from warrior and merchant families, including some women, called *kabuki-mono* [eccentrics], created a rebellious culture in dressing and lifestyle.[78] Nonetheless, the popularity of Confucian teachings from the late seventeenth century onward was unwavering and growing among people belonging to different status groups and in disparate classes within each group. Within this climate, it often served women's interests to follow Confucian tenets rather than challenge them. There may well have been spaces in which they could maneuver and modify Confucian ethics and medical recommendations to their advantage. As long as women conformed to Confucian norms, however, they ultimately relied on male scholar-physicians, such as Gyūzan, who commanded authority in their knowledge of Chinese texts. In sum, while the proliferation of Confucian-influenced women's education did not invariably entail women's oppression, it certainly reinforced the power of predominantly male educators and medical professionals.[79]

Representing the Birthing Process

Fujin kotobukigusa suggested a model of childbirth that required the attendance of both a physician and a midwife. Although they were to play complementary roles, the physician assumed the authority of supervising the overall process and diagnosing the mother's condition. This portrayal of the birthing process, however, is likely to have been highly fictitious and idealized. For one thing, only a small minority of women from privileged social groups used the assistance of a physician for deliveries during the Tokugawa period. It is also doubtful that physicians, on average, possessed adequate obstetric skills to maintain supervisory authority vis-à-vis midwives, who had much more experience. In particular, the obstetric methods Gyūzan recommended for physicians mostly involved the use of herbal medicine, acupuncture, and moxibustion, rather than active surgical intervention. Moreover, it was the midwife who operated on the woman when the physician decided that the fetus's position should be adjusted during labor. Such an arrangement suggests that midwives, who had direct physical contact with the woman's body, could develop more effective obstetric skills than the book-learned physicians. A midwife could challenge and undermine a physician's authority during delivery. The childbirth scenes that Gyūzan described, thus, concealed possible conflicts between physicians and midwives, or midwives' subversion of physicians' authority.

Little is known about midwives during the Tokugawa period. There is evidence suggesting that, beginning in the eighteenth century, professional midwives, often widows without other means of supporting themselves, appeared on a mass scale.[80] Even earlier in the Genroku period (1688–1703), according to Ihara Saikaku's (1642–1693) novels, professional midwives were customarily employed to assist with childbirth by townspeople in the Osaka area, except for those in dire poverty. Midwifery was described by Saikaku as a profession that yielded profits without capital investment.[81] The historian Sakurai Yuki presents a legal record of a wife from a lower-class warrior in the Yonezawa domain, in today's Yamagata prefecture, who was punished for assisting with abortion as a midwife in 1793.[82] Modern sources derived from anthropology and folklore studies suggest that some midwives came from non-status groups that were targets of discrimination, and that they were often consigned to the work of cleaning up pollution. That is, midwives were supposed to handle defiled matter, such as parturient blood, the afterbirth, the bodies of the mother and the newborn child, and polluting materials left behind after childbirth.[83]

Gyūzan's detailed advice on how to deal with midwives did not involve a discussion of birth pollution and mostly drew on Chinese texts. Inasmuch as Gyūzan presumably intended to give his readers practical help, his recommendations regarding midwives were likely to have, to some extent, reflected Tokugawa practices. Gyūzan suggested that the woman's family should choose a midwife of a quiet nature who would not insist on doing things her way and would not be easily upset during an emergency. He warned against hiring old midwives, who were low in spirit and prone to sleepiness when labor was prolonged. In addition, because these old midwives often told discouraging stories about childbirth-related illnesses, the expectant mother, particularly one who had never given birth before, would become upset, making her delivery difficult.

Gyūzan also advised that the family invite the midwife to their home once a month so that the mother would become well acquainted with her. He suggested that such familiarity would lead the midwife to relay only good news to the mother during labor, preventing her from becoming discouraged. The family should give the midwife generous fees, praise her work, and try to arrange things to her liking during childbirth. The text reveals the midwife's low class origin in the use of the expression, *hodokoshi* for indicating payment, which customarily refers to a charitable gift. Thus, in Gyūzan's portrayal, midwives were significant, able to facilitate or obstruct the childbirth process.

Customarily, an important job of a midwife during pregnancy was to help the woman put on a pregnancy sash. Using *The Tale of Genji*,[84] and an etiquette manual of the Ogasawara ryū[85] as his sources, Gyūzan described the ritual of putting on a pregnancy sash, a *yuhata obi* or *iwata obi*, in the fifth month of pregnancy. The midwife, however, was absent from his account of this ceremony.[86] Gyūzan described the sash as seven or eight inches in width, to be worn from the back to the abdomen. It was to be sewn together, and not untied day or night. As fetal *ki* gradually inflated,[87] and the woman started to have trouble breathing, the seam was undone to lengthen the sash a third or half an inch. The sash, however, should be no more than an inch longer than its original length. Gyūzan explained that applying the sash would prevent fetal *ki* from overinflating and lead to an easy delivery. "[The sash]," he remarked, "helps prevent fetal ki from rising up. Therefore, the gap between the chest and abdomen is relaxed. This helps the woman walk, sit and lie down."

According to Gyūzan, when the mother felt acute pains during labor, the physician would prescribe medication. The midwife, on the other hand, was supposed to attend to the mother during the entire duration of labor. She was also supposed to urge the woman to move from her bed or birth chair to the birthing area covered with straw at the exact moment when she began to feel severe pains in her hips and abdomen. Gyūzan remarked that the timing of this transfer was vital: if such a move took place too early, the expectant mother would exhaust her spirit, which would negatively affect the delivery process. This important judgment was made by the midwife.

The mother was advised to deliver the child kneeling down and holding on to a rope hanging from the ceiling. Gyūzan summed up a "normal delivery" [*seizan*]:

女醫博士が貴婦人を拜診する圖

Fig. 2.2 A court-appointed obstetrician sees a noblewoman (Source: Ogata, *Nihon sanka gaku shi*, 93)

妊婦が陰陽師を招きて方位又は男女
を占はしむる圖

Fig. 2.3 An expectant mother invites a diviner (Onmyōdō master) to reveal auspicious directions or the sex of the fetus (Source: Ogata, *Nihon sanka gaku shi*, 93)

In the case of normal delivery, [the woman] suddenly feels great pains in her abdomen [*hara*] and her lower back [*koshi*] after the full ten months have passed following conception. Subsequently, the ki of the fetus moves abruptly to heighten the pain. She also feels a smarting pain in the area between her hips. Rupture [of the large intestine] occurs. Waters break and leak out. Blood oozes out. The fetus, as if awakened from a dream, separates itself from the afterbirth, and seeks its way out. A child is finally born.

Fig. 2.4 An expectant mother is taking labor-inducing medication and trying to push out the baby (Source: Ogata, *Nihon sanka gakushi*, 94)

The fetus's voluntary movement was the key to a successful delivery. Accordingly, what made this possible was a sufficient supply of *ki* and blood to the fetus.

On the other hand, a difficult labor was often caused by a dispersal of *ki* and blood from the uterus during pregnancy. Gyūzan insisted that one of the major causes of this problem was the arousal of the woman's sexual desires or sexual intercourse during later months of pregnancy. When this happened, "the uterus opens itself up and is no longer firm" to diffuse *ki* and blood. Then, "the breath of the fetus" became weak, resulting in irregular movements of the fetus. In this case, the fetus could not be delivered with amniotic water, because the water had dried up, and "filthy

Fig. 2.5 A midwife takes up a child (Source: Ogata, *Nihon sanka gaku shi*, 94)

blood had blocked up the route." This would cause the fetus to "lose its way," resulting in an abnormal fetal position during delivery.

Gyūzan classified difficult cases of childbirth, explained their causes, and described the treatment by citing Chinese texts on gynecology and obstetrics such as *Fujin ryōhō* and *Nyoka junjō* [Ch *Nüke zhunsheng*].[88] Abnormal cases of childbirth such as "the sideways birth" (*yokozan*, a complicated case with a transverse presentation), in which the child's hand appeared first and "the sitting birth" (*zasan*, a difficult case with a breech presentation), in which the child's buttocks emerged first, often resulted from a mother's hasty pushing as well as the malfunction of *ki* and blood. For dealing with such cases, Gyūzan recommended inserting one's hand into the birth canal to adjust the position of the fetus. The midwife was to conduct this operation under the guidance of the physician. For example, to correct the fetal position in a sideways birth, the mother should first be laid on her back. The midwife should then insert her hand inside the mother's body and "move it up to gradually approach the body of the fetus." Finally, using her middle finger, she should push its shoulder up to adjust it to the correct position. Alternatively, the midwife can "wring its ear to fix its position," upon which the mother should push to deliver the child. The midwife's hand manipulations were the only physical operation that appeared in *Fujin kotobukigusa*. Hooks, levers, forceps, or other instruments were not mentioned.

Thus, unlike the obstetricians of the Kagawa School in the eighteenth century, Gyūzan did not endorse drastic surgical procedures to deal with difficult births. His instructions on labor and delivery involved either prescribing drugs or relaxing the woman, often to restore the adequate function of *ki* and blood. For instance, in the case of an "upside down birth," the midwife should push the child's legs back inside the mother's body, then have her relax and wait until the child emerged naturally. In this model, the body was not something to be actively operated on; rather the physician, the midwife, and the pregnant woman should wait for nature to take its course.

Indeed, Gyūzan repeatedly warns against touching or operating on the vulva, which he called the "birth gate" [*sanmon*]. For example, when the midwife inserted her hand to correct the position of the fetus, Gyūzan recommended that she not tamper with the body but carefully follow the methods described by Yō Shiken [Ch Yang Zijian].[89] Even during a difficult delivery, the mother's body should not be touched unless the midwife possessed adequate skills. If an unskilled midwife tampered with the "birth gate," "the inside" would be damaged and "cause problems afterwards."

As an alternative method to using hands and other instruments for complicated deliveries, Gyūzan suggested using acupuncture to push down the child. The physician should apply acupuncture on specific pressure points on the hands, which were connected to the "large intestine meridian of yang major." This method was effective not only in inducing delivery of the child, but also in removing a dead fetus or the afterbirth.

Gyūzan assumed that physicians possessed specific methods to diagnose the condition of the fetus in the mother's body, which midwives were unable to do. Whether the mother was "hot" or "cold" was an important indication for understanding the condition inside her body.

> Chin Jimei says that the fetus is dead in the woman's body if the mother's body feels heavy [*omoku*], if it is heated [*nesshi*] or cold [*samushi*], if the vein below her tongue is black or blue, and if the surface of the tongue is cool.

産褥婦は産椅に凭れ薬湯を侑る所

Fig. 2.6 A puerperant woman sits on a birth chair while [an attendant] serves a decoction (Source: Ogata, *Nihon sanka gaku shi*, 95)

Certain symptoms appearing on the body's surface, particularly in the area of the face, such as the discharge of phlegm, were important clues that disclosed the body's internal changes. Physicians had to examine the color of the woman's face, lips, and tongue with care. Unusual coloring indicated that meridians, organs, or the circulation of bodily fluids had been impaired. Gyūzan maintained that the physician could find out by such a diagnosis whether the fetus was dead, or if the mother would die.

> In general, if the woman's face is red, it is because *eiki*[90] is circulating. The mother, in this case, will revive.

Fig. 2.7 A lady-in-waiting prepares a decoction for a noblewoman after childbirth (Source: Ogata, *Nihon sanka gaku shi*, 95)

> If the tongue's surface is blue, the conception meridian is shut off. The fetus, therefore, is dead.
>
> If the woman spits a mouthful of phlegm, *ki* of spleen and stomach has expired; therefore, both the child and the mother will die.

The ability of a learned physician to diagnose the body's internal processes and the condition of the fetus could have helped reinforce his authority vis-à-vis the midwife, the mother, and the mother's family.

Fig. 2.8 Author's note: Presumably this illustration is intended to teach a pregnant woman not to bend her head forward and wash her hair by herself. This instruction appears in the text of *Fujin kotobukigusa*. In this illustration, the expectant mother is having an attendant clean her hair (Source: Fujikawa, *Nihon sanka sōsho*, 95)

Fujin kotobukigusa reveals that there were other types of healers whose services were employed for a safe delivery. Gyūzan sought to bolster the legitimacy of his method by discrediting the views of such competitors. For example, he denounced Onmyōdō practitioners (diviners) as spreading deceptive ideas about why women would suffer a difficult labor.

> Even when the woman suffers from such difficult childbirth as "sideways birth" or "upside down birth," she fully recovers [from such an affliction] without any changes [to her body] after she has given birth...Onmyōdō practitioners in Japan as well as in China suggested that the cause [of her

> problem], in this case, is not therefore any illness, or the malfunction of *ki* or blood; but, it is caused by evil spirits of gods...Such a theory is a deceptive one that would confuse the public.

Gyūzan stressed that many of the teachings of Onmyōdō practitioners were vulgar and had no foundation, condemning them as "silly delusions of the Onmyōdō masters."

His model of the need for childbirth to be overseen by a physician is likely to have been instrumental in cultivating new clients for learned obstetricians during the Genroku era, when some *daimyō* and townspeople had become affluent enough to seek such a service. Although the vast majority of commoners still could not afford a learned physician's assistance, an increasing number of literate people had been exposed to the idea of physician-assisted childbirth by reading childbirth and household manuals. *Fujin kotobukigusa* encouraged the expansion of midwifery and the medical profession.

Markers of Sexual Difference

Gyūzan's biography contains an anecdote about "a man giving birth." Gyūzan reportedly "cured" this man, who had hair on his tongue, produced breast milk, and ate only fish.[91] It is not clear if the treatment was targeted at restoring his male reproductive function, correcting his abnormal eating habits, or both. This "man" may have been what we call a hermaphrodite. That the biography talks about this patient as a man points to how the sexual distinction was imbued with ambiguity and how possessing what are today deemed female or male reproductive functions was certainly not the ultimate determinant of one's sexual identity within both text-based and popular medical discourses predominant at the time.[92] I have already shown that within Goseihō School discourse, male and female bodies possessed an identical reproductive mechanism composed of the kidney system and the meridians that worked with it to condition and distribute *ki*, blood, and essence. This does not necessarily mean, however, that Tokugawa physicians were uninterested in the body's sexual distinctions. How, then, did Gyūzan discuss the difference between male and female bodies?

He referred to the question of sexual difference when he presented methods to determine the sex of a fetus. He introduced an old formula that had been handed down from ancient times in China and mentioned by the Chinese physician, Chō Kaihin [Ch Zhang Jiebin].[93] The male fetus

Fig. 2.9 Author's note: This illustration presents an expectant mother viewing a beautiful dance performance. It is accompanied by a recommendation that expectant mothers view beautiful objects and listen to pleasant music to encourage the birth of a good-looking baby (Source: Fujikawa, *Nihon sanka sōsho*, 97)

moves in the third month, while the female fetus moves in the fifth month. Gyūzan explained that this difference emerged because the nature of yang essence was quick and that of yin essence was slow. In other words, the male fetus was more influenced by the yang phase, while the female fetus was more affected by the yin phase. This difference manifested itself when the fetus moved for the first time during pregnancy.

He continued to argue that there was a difference between the direction male and female fetuses faced while inside the mother's body. The female fetus had her back to the mother and looked in the same direction as the mother. The male fetus faced the mother. He felt this explained the mother's "soft" abdomen when she was carrying a female fetus, but "stiff" with a male fetus. To support the theory, he introduced another tradition that drew on a similar idea: "If you look at people who have drowned, men are positioned face down, women, face up." In this way, one's sex was associated with certain movements or positions of the body.[94] Gyūzan did not look for signs of sexual difference on the body itself, thereby revealing that the contours of the body and the sexual organs that modern medicine uses to identify an individual's sex did not have the same significance in Japan in the late seventeenth century.

Terajima Ryōan, another established physician of the Goseihō School, addressed the question of how to make a distinction between a man and a woman in his widely circulated multi-volume encyclopedia, *Wa-kan sansai zue*, compiled in 1712. He listed what he called the "outwardly manifested" characteristics that were commonly used to distinguish the sexes: yin yang, face up and face down, concave, convex, whether one had a beard, and the length of hair.[95] What Ryōan referred to as "concave-convex" most likely indicated genitals, which were merely one among many other signs of sexual difference.

Ryōan maintained, however, that one could not reliably determine sexual difference from these traits. Rather, the definite markers of sexual difference were found in one's skeletal anatomy. Ryōan asserted that a man's skull consisted of eight bones, while a woman possessed only six. The female skull lacked the horizontal seam found at the back of the male skull. Moreover, he suggested that women had additional ribs: "Both men and women have twelve ribs on the right and the left side. Eight are long and four are short. Women, however, have two additional *keifu* bones [*keifu-kotsu*]; thus, they have fourteen on each side." Sexual difference was also manifested in the shape of the coccyx, which Ryōan called the *kamenoo kotsu*, or the bone that looked like a turtle's tail: "Men's *kamenoo* bone is pointed. The same bone in women is flat."

It should be remembered that dissections were not commonly conducted until the mid-eighteenth century in Japan. Since physicians could not examine the skeletal anatomy through dissection, Ryōan's theories lacked any practical applicability. It was significant, however, that he demonstrated uneasiness and dissatisfaction about conventional methods of sexual identification. For Ryōan, such methods allowed too much ambiguity, and therefore became a target of criticism and reconsideration. His search for more reliable markers of sexual difference led him to focus on the body's physique, rather than looking within yin yang dynamics. This move foreshadowed the critical methods as well as the interest in studying the body's anatomy that emerged among physicians of the Ancient Practice School, the Dutch School, and the Kagawa School of Obstetrics from the mid-eighteenth century.

While Ryōan most likely possessed very little knowledge of European medicine, his focus on anatomical features for sexual distinction paralleled the methods of ascertaining the body's sex developed by European physicians during the Renaissance and the early modern period.[96] He and European medical scientists shared a basic desire to draw a clear boundary between the sexes. Both sought to establish legitimate categories to classify human beings, which were made the prime objects of their scholarly investigation. Ascertaining a classificatory system was often a way for medical experts to claim their authority over a lay audience. In this sense, Ryōan's comparison of what he called "the commonly used methods" for establishing sexual distinction and the "definitive" ones he endorsed underscored his scholarly authority. His scholarly disposition, thus, resembled that of European physicians in these maneuvers for bolstering his authority where the issue of sexual distinction played a significant role. Further research will be necessary, however, to pursue a comparative study on why physicians in Europe and Japan demonstrated a renewed interest in the sexual differentiation of bodies by way of anatomical investigation in the early modern period.

The type of medical body and treatments found in *Fujin kotobukigusa* derived from the mainstream Chinese medical tradition. Based on his reading of Chinese texts, Gyūzan delineated the different roles to be performed by the physician, the midwife and the mother during the prenatal period and the delivery process. The appropriation of Chinese medical conventions was accompanied by the introduction of Confucian family values and notions of femininity derived from reproductive medicine in late imperial China. It should be noted, however, that Gyūzan's use of Chinese theories was selective, and worked in the interests of learned physicians in Tokugawa society, who practiced within an open and commercialized medical market.

The Goseihō School was unable to maintain its prestige in the areas of obstetrics and gynecology and in the medical field in general. From the mid-eighteenth century, the expanding market for skilled obstetricians was dominated by the Kagawa School, which developed methods that were radically different from those of the Goseihō School but similar to those of the Ancient Practice School. How did the Kagawa School emerge? What techniques did it develop? In what ways were such methods effective in gaining clients who would reward Kagawa physicians with prestige and generous fees? And, what happened to perceptions of the body?

Notes

1. The *eta* (literary, plenty of defilement) were people belonging to a hereditary outcast group who during the Tokugawa period often engaged in jobs, such as handling leather and funerals, associated with ritual pollution. The polluted nature of the work was linked to the Shinto religious tradition. The *hinin* were also lowly people who did not belong to the four major status groups. Unlike the *eta*, whose membership in most cases was defined by heredity, there were many cases where people could be deprived of their status through demotion to the *hinin*. The *hinin* engaged in a range of occupations, which included begging, entertaining, acting as executioners, and prostitution.
2. It should be noted that in many domains, peasants wanting to become physicians required permission from a local magistrate because it was crucial for local governments to maintain a sufficient number of peasants to pay rice taxes. Nonetheless, during the Tokugawa period, an increasing number of peasants became physicians and provided medical services even in rural areas. Aoyagi Seiichi, *Shinryō hōshū no rekishi* (Kyoto: Shibunkaku Shuppan, 1996): 125–128; Yamazaki Tasuku, *Iji dansō* (Tokyo: Nihon Rinshōsha, 1948): 138–140.

   For physicians' social positions in Tokugawa society, see Fuse Shōichi, *Ishi no rekishi: sono nihonteki tokuchō (Tokyo: Chūōkōron, 1979): 30–72; Mitamura Engyo, Mitamura Engyo zenshū, Vol. 14* (Tokyo: Chūōkoron, 1975): 277–347. Kate Nakai discusses the status of Confucian scholars with a focus on those who served the shogunate. Kate Wildman Nakai, *Shogunal Politics: Arai Hakuseki and the Premises of Tokugawa Rule* (Cambridge, MA: Council on East Asian Studies, Harvard University, 1988): 25–46.
3. See, for example, Tsukamoto Manabu, "Iryō wo meguru hitobito no kōryū," *Shizuoka-ken shi kenkyū*, 4, No. 8 (March 1988): 20–34.
4. Fuse, *Ishi no rekishi*, 30–72.
5. Such examples include the Chinese studies scholar Dazai Shundai (1680–1747) and the Japanese studies scholar Motoori Norinaga (1730–1801). Eminent Ancient Learning scholar Ito Jinsai (1627–1705), when young, was urged by his relatives and friends to practice medicine to make a decent living; however, he rebuffed these suggestions due to his negative view of physicians who he thought used menial skills, flattered the powerful and rich, and were eager to make profits. Hattori Toshirō, *Edo jidai igakushi no kenkyū* (Tokyo: Yoshikawa Kōbunkan, 1978): 40–43, 62–69, 172–232.

6. For example, the renowned Ancient Practice School physician Kagawa Shūtoku declared that the goals of medical studies and Confucian scholarship were the same. Kyōto-fu Ishikai, ed., *Kyōto no igakushi* (Kyoto: Shibunkaku, 1980): 463; Fujikawa Yū, *Nihon igakushi* (Tokyo: Shinrisha, 1941): 351.
7. *Ryū* refers to both the content of teaching and a group of practitioners consisting of the master and her or his disciples. On the other hand, *dō* or *michi* usually indicates the area of learning (e.g., *sadō*, tea ceremony; *kadō* flower arrangement).
8. While "internal medicine (*naika*, or *hondō*, "The Major Way")" used systemic diagnosis and herbal prescriptions, "external medicine [*geka*]" mainly treated sword wounds and tumors by lancing, suturing, and applying ointment. Frequent warfare during the Ashikaga period contributed to the establishment of this specialization. In the modern period, the term, *geka*, indicates surgery as defined in bio-medicine. Nihon gakushi-in, Nihon Kagakushi Kankōkai, ed., *Meiji-zen Nihon igakushi, Vol. 1* (Tokyo: Nihon Gakujutu Shinkōkai, 1955): 19, 23–24; Sōda Hajime, *Zusetsu nihon iryō bunkashi* (Kyoto: Shibunkaku, 1989): 93–94, 111–128.
9. Yakazu Dōmei, "Nihon igaku chūkō no so Manase Dōsan," Tōa Igaku Kyōkai, ed., *Nihon no kanpō wo kizuita hitobito, Kanpō norinshō*, Vol. 9, No. 11–12 (December, 1962): 16–39.
10. About private academies in the Tokugawa period, including some medical academies, see Richard Rubinger, *Private Academies of Tokugawa Japan* (Princeton, NJ: Princeton University Press, 1982).
11. Medical historian Fujikawa Yū lists 29 texts by Dōsan in his *Nihon igakushi*, 270–277.
12. Yakazu, "Nihon igaku chūkō no so Manase Dōsan," 16–39.
13. On the Tokugawa shogunate's foreign policy, see Ronald Toby's classic study where he argues the shogunate, indeed, allowed some foreign contacts as well as an influx of foreign goods and ideas. Ronald P. Toby, *State and Diplomacy in Early Modern Japan: Asia in the Development of the Tokugawa Bakufu* (Stanford, CA: Stanford University Press, 1984).
14. The eighth Tokugawa shogun, Yoshimune (1684–1751), who wanted to study European astronomy, relaxed the ban on the import of technical books of European origin in 1720. This order, combined with his subsequent promotion of the translation of Dutch texts, contributed to the rise of Dutch studies [Rangaku] in the eighteenth century. Masayoshi Sugimoto and David L. Swain, *Science and Culture in Traditional Japan* (Cambridge, MA: MIT Press, 1978): 312–316.
15. This section is based on my reading of Nankin chūigakuin, ed., and Chūigaku gairon hōyaku īnkai, trans., *Chūgoku kanpō igaku gairon* (Tokyo: Chūgoku kanpō, 1965); Fujihara Ken and Ogura Shigenari, *Kanpō gairon* (Osaka: Sōgensha, 1979); Ted J. Kaptchuk, *The Web That*

Has No Weaver (New York: Congdon & Weed, 1983); Nathan Sivin, *Traditional Medicine in Contemporary China* (Ann Arbor: Center for Chinese Studies, Univ. of Michigan, 1987); Manfred Porkert, *The Theoretical Foundations of Chinese Medicine: Systems of Correspondence* (Cambridge, MA: MIT Press, 1974). Also, see the section on the body in Terajima Ryōan, *Wa-kan sansai zue*, an encyclopedic work originally published in 1711–1712. Terajima Ryōan, *Wa-kan sansai zue*, translated into modern Japanese by Shimada Isao, Takeshita Atsuo, Higuchi Motomi (Tokyo: Heibonsha, Tōyō bunko series, 1985–1991).

16. Nankin chūigakuin, ed., *Chūgoku kanpō igaku gairon*, 20.
17. Ibid., 22.
18. Ibid., 21.
19. Ibid., 21–22.
20. "Goseihō" means "the method of a later age," which indicates the Jin-Yuan period (1126–1368) and is used vis-à-vis the period prior to this in Chinese history. The term "Goseihō" School emerged during the first half of the eighteenth century when the Ancient Practice School became a self-conscious movement distinctly separate from what was termed the Goseihō School. This happened when Kagawa Shūtoku (1683–1755) appeared as a master physician of the Ancient Practice School. See Sōda, *Zusetsu nihon iryō bunka shi*, 151.
21. I use the more general term, "neo-Confucianism," interchangeably with "Song Confucianism" in this book.
22. Sakai Shizu, *Nihon no iryōshi* (Tokyo: Tokyo Shoseki, 1982): 157–159.
23. What is called Rishu medicine in Japan consists of the medical theories elaborated by two prominent physicians, Ri Toen [Ch Li Dongyuan, 1180–1251] and Shu Tankei [Ch Zhu Danxi, 1281–1358] from the Jin and Yuan dynasties. For a brief introduction to Rishu medicine, see Norman Ozaki, *Conceptual Changes in Japanese Medicine during the Tokugawa Period* (Ph.D. Dissertation, University of California, San Francisco, 1979): 45–50; Sakai, *Nihon no iryōshi*, 157–159; Fujikawa, *Nihon igakushi*, 185–188.
24. Sakai, *Nihon no iryōshi*, 174; Yakazu, "Nihon igaku chūkō no so Manase Dōsan," 20–22, 29; Achiwa Gorō, *Kindai nihon no igaku: seiyō igaku juyō no kiseki* (Kyoto: Shibunkaku, 1982): 40.
25. Ibid.
26. For various methods of *geka* between the fifteenth and eighteenth centuries, see Ozaki, *Conceptual Changes in Japanese Medicine during the Tokugawa Period*, 70–145; Sōda, *Zusetsu nihon iryō bunkashi*, 116–132.
27. For a recent English language study on Dutch medicine during the late Tokugawa period, see Ellen Gardner Nakamura, *Practical Pursuits: Takano Choei, Takahashi Keisaku, and Western Medicine in Nineteenth-Century Japan* (Cambridge, MA: Harvard University Asia Center, 2006).

28. The Genroku period spanned between 1688 and 1704. The era was marked by a flourishing economy and urban culture.Mary Standlee discusses midwifery and obstetrics in Japan from the ancient period through the modern era. Her impressionistic sketches of Japanese midwifery practices relied on a limited range of secondary sources and for the most part failed to examine original obstetric texts. Mary Walker Standlee, *The Great Pulse: Japanese Midwifery and Obstetrics through the Ages* (Rutland, VT: Charles E. Tuttle Company, 1959). For a more recent study that outlines the field of obstetrics and midwifery in Tokugawa Japan, see Sakurai Yuki, "Kinsei no ninshin shussan gensetsu," *Rekishi hyōron*, No. 600 (April 2000): 27–38.
29. Ogata Masakiyo, *Nihon sanka gaku shi* (Tokyo: Maruzan, 1918, reprinted by Kagaku Shoin, 1980): 1–57.
30. Sōda, *Zusetsu nihon iryō bunkashi*, 94–98.
31. Sakai, *Nihon no iryōshi*, 186–187.
32. *Chūjōryū sanka zensho*, in Fujikawa Yū, Kure Shūzō, and Masuda Tomomasa, eds., *Nihon sanka sōsho* (Kyoto: Shibunkaku, first published in 1895, reprinted in 1971): 1–30. For the Chūjō School methods, see Ogata, *Nihon sanka gaku shi*, 69–90.
33. Fujikawa et al., eds., *Nihon sanka sōsho*, 53–63.
34. These texts were widely circulated in a period when the publishing industry and the book-lending industry were taking off. Demand for books rose because of the increasing number of warriors, townspeople, and wealthy peasants who were learning to read. The obstetric texts of the Chūjō School, on the other hand, were written for use by Chūjō School obstetricians themselves. For the expanding publishing industry during the Tokugawa period, see Peter Kornicki, *The Book of Japan: A Cultural History from the Beginnings in the Nineteenth Century* (Leiden: Brill, 1998): 169–207, 251–276. For more on the book-lending industry during the Tokugawa period, see Nagatomo Chiyoharu, *Kinsei kashihonya no kenkyū* (Tokyo: Tokyodō Shuppan, 1982). For an English language source, see Mary Elizabeth Berry, *Japan in Print: Information and Nation in the Early Modern Period* (Berkeley: University of California Press, 2006).
35. His name was Zhu Zhenheng. Danxi was his pseudonym. For Zhu's life and medical method, see Yoshida Sōjin, *Chūgoku meii retsuden* (Tokyo: Chūōkōron, 1992): 84–92.
36. Identical biographies of Gyūzan are found in Ogata, *Nihon sanka gaku shi*, 92–93; Fujikawa et al., eds., *Nihon sanka sōsho*, 4–5; Fujikawa, *Nihon igakushi*, 366–377. Fujikawa notes *Kinsei sōgo*, Utsuki Kondai's *Nihon ifu* (the Tempo period, 1830–1843), and *Kokuji isō jobun* as his sources for Gyūzan's biography.
37. His biographical data are yet to be researched. The historian Inoue Tadashi, however, mentions him as a domainal physician to the Kuroda House in what is today's Fukuoka prefecture. Inoue Tadashi, *Kaibara Ekiken* (Tokyo: Yoshikawa Kōbunkan, 1963): 301.

38. Fuse, *Ishi no rekishi*, 45–53.
39. Ibid., 30–45, 54–72.
40. Many ideas presented in his *Fujin kotobukigusa* were also presented in other more popular women's manuals. One manual, *Onna Chōhōki*, was compiled in the late seventeenth century and early eighteenth century. Many versions of this work are in existence.
41. Charlotte Furth, "From Birth to Birth: The Growing Body in Chinese Medicine," in Anne Behnke Kinney, ed., *Chinese Views of Childhood* (Honolulu: University of Hawai'i Press, 1995): 162–163.
42. The best English-language source describing adoption practices from the Tokugawa period to the early Meiji period is Kurosu Satomi and Ochiai Emiko, "Adoption as an Heirship Strategy under Demographic Constraints: A Case from Nineteenth-century Japan," *Journal of Family History* 20:3 (July 1995): 261–289. See also Harumi Befu, "Corporate Emphasis and Patterns of Descent in the Japanese Family," in Robert J. Smith and Richard K. Beardsley, eds., *Japanese Culture: Its Development and Characteristics* (Chicago: Aldine Pub. Co., 1962): 34–41. For Japanese sources on this topic, see Wakita Osamu, "Bakuhan taisei to josei," in Joseishi Sōgō Kenkyūkai, ed., *Nihon Joseishi*, Vol. 3 (Tokyo: Tōkyō daigaku shuppankai, 1982): 22–28; Miyashita Michiko, "Nōson ni okeru kazoku to kon'in," 31–62; Hayashi Reiko, "Chōke josei no sonzai keitai," ibid., 95–126; Ōguchi Yūjirō, "Kinsei kōki ni okeru nōson kazoku no keitai," 193–225. For adoption practices within the warrior class, see also Mitamura, *Mitamura Engyo zenshū, Vol. 2* (Tokyo: Chūōkōron, 1975): 374–388. Numerous cases of adoption for physicians' families are indicated in Kyōto-fu Ishikai, ed., *Kyōto no igakushi* (Kyoto: Shibunkaku, 1980). For lineage practices in late imperial China, see Kai-wing Chow, *The Rise of Confucian Ritualism in Late Imperial China: Ethics, Classics, and Lineage Discourse* (Stanford, CA: Stanford University Press, 1994); and David Wakefield, *Fenjia: Household Division and Inheritance in Qing and Republican China* (Honolulu: University of Hawai'i Press, 1998).
43. "Essence" is one of the fundamental bodily substances. There are two types of essence. The first, which is inherited from parents, constitutes a basic element in the reproductive process, and is governed by the kidney system. The second is generated from ingested food to be used as nourishment. The kidney stores this postnatal essence and distributes it throughout the body. Kaptchuk, *The Web That Has No Weaver*, 43–45.
44. Fujikawa et al., eds., *Nihon sanka sōsho*, 65.
45. An unknown source. It is probable that Gyūzan was referring to a section in *The Yellow Emperor's Inner Canon* [Ch Huangdi Neijing], one of the most important Chinese medical classics (reportedly originated in the Qin and former Han periods between the third and first century BCE), where "Sojo" appears. For an English translation, see Ni Maoshing, *The Yellow*

Emperor's Classic of Medicine: A New Translation of the Neijing Suwen with Commentary (Boston: Shambhala, 1995).

46. All passages from *Fujin kotobukigusa* are from a reprint edition in Fujikawa et al., eds., *Nihon sanka sōsho*, 63–113.
47. Chen Jiming was a physician from Southern Song who wrote extensively on medicine, particularly in the fields of surgery and gynecology. His *Fujin daizen ryōhō* [Ch *Furen da quan liang fang*; *All-Inclusive Good Prescriptions for Women*], published in 1237, was one of the earliest comprehensive texts on gynecology. Yoshida, *Chūgoku meī retsuden*, 99–105. See also Charlotte Furth, *A Flourishing Yin: Gender in China's Medical History, 960–1665* (Berkeley: University of California Press, 1999): 85–87.
48. He was also named Wen Daming. He was a physician from the Song dynasty who practiced medicine for more than 40 years starting from 1174. Xie Guan, *Zhongguo yixue dacidian*, Vol. 4 (Taibei: Shangwuyin shuguan, 1988): 4405.
49. For a mechanism of conception as described in *The Yellow Emperor's Inner Canon*, see Furth, *A Flourishing Yin*, 47–48. Although Gyūzan offers a superficial explication of the mechanism of conception, his account does not capture much of the theoretical complexity involved in the process, as described by Furth.
50. Liu Chun, whose pseudonym was Zonghou, was a physician who lived between the Yuan and Ming periods. His father is said to have received medical instructions from Zhu Danxi, and Chun followed his father in studying medicine. Li Yun, *Zhongyi renming cidian* (Beijing: Gouji Wenhua chubanshe, 1988): 207.
51. Childless women were stigmatized and called *umazume* [stone women]. The belief of the Hell of Childlessness [*Umazu no jigoku*] had been developed in Buddhist discourse since the medieval period. Some popular writings from the early Tokugawa period, such as Nakagawa Kiun's *Shikata banashi* (1659) and Kusada Itsuki's *Komori-mimi* (1687), depict traveling nuns describing the story of the Hell of Childlessness to townspeople using a mandala, or a graphic chart that described the worlds of enlightenment and hell through a schematic depiction of various Buddhas, Bodhisattvas, and gods. See Kawamura Kunimitsu, *Jigoku meguri* (Tokyo: Chikuma Shobō, 2000): 177–209; and Nīmura Izuru, *Kōjien* (Tokyo: Iwanami Shoten, first published in 1955, 4th edition, 1991): 1812. Regarding the Hell of the Blood Pool, see Nakano Yūko, "Women and Buddhism: Blood Impurity and Motherhood," Okuda Akiko and Okano Haruko, eds., Alison Watts, trans., *Women and Religion in Japan* (Göttingen: Harrassowitz Verlag, Wiesbaden, 1998): 65–85.
52. This calendrical system had been widely used in the practices of Onmyōdō [The Way of Yin and Yang] in Japan. Onmyōdō developed during the ancient period, and was based on Chinese cosmological beliefs derived

largely from *The Book of Changes*, one of the five Confucian classics. The institution established by the imperial court in the ninth century for Onmyōdō study and ritual performances was maintained throughout the Tokugawa period. Onmyōdō concepts were also incorporated into Japanese religious traditions such as Shinto and Shugendō (the practices of mountain ascetics), and shaped local folk customs in various ways. Gyuzan's discussions of reproductive strategies relied, however, on the use of a calendrical system originating in Chinese texts written by physicians such as Sun Simiao, Chin Jiming, and Zhu Danxi. For the impact of the ideas derived from *The Book of Changes* on Tokugawa medicine, see Wai-ming Ng, *The I Ching in Tokugawa Thought and Culture* (Honolulu: Association for Asian Studies and University of Hawai'i Press, 2000): 150–167. Ng's discussions regarding the influence of the *Yijing* on the Goseihō School, however, do not refer to *Fujin kotobukigusa* or any other works by Gyūzan.

53. Sun Simiao was a physician of vital importance who established the text-based medical tradition between the Tang and Song periods. Charlotte Furth extensively discusses his work in Furth, *A Flourishing Yin*.
54. Gyūzan failed to discuss how people could determine the Phase that their nature pertained to. It is possible that the Phase that represents the nature of a person was discovered by examining the celestial stem and terrestrial branch of the year and date that they were born.
55. It is possible that Gyūzan appropriated this notion of yin fire from Zhu Danxi. Furth, *A Flourishing Yin*, 147.
56. "Yang of yang" and "yin primal" indicate certain phases that are posited by the theory of the Five Movements and Six Qi.
57. It is unclear if Gyūzan indicated *Beiji qianjin yaofang* [*Prescriptions Worth a Thousand for Every Emergency*, 652] or *Qianjin yi fang* [*Supplementary Prescriptions Worth a Thousand*, 682] by noting *Senkinpō*.
58. Furth, *A Flourishing Yin*, 145–151.
59. Zhu Danxi.
60. Gyūzan also introduced a method of predicting the sex of a fetus through *Yijing* divination signs called *ke* [Ch gua]. The age of the parents and the month of conception were used to form a specific divination sign that would indicate the fetus's sex. He also listed charms to turn a female fetus into a male fetus, such as placing an ax with its blade upward under the floor of the woman's bedroom during the third month of pregnancy.One important omission from Gyūzan's pronatalist techniques was the role of orgasm for a successful conception. Female orgasm was often thought as crucial for a successful conception within both the Chinese medical tradition as practiced in China and the European medical tradition. Furth, *A Flourishing Yin*, 200, 204, 211, 213; Thomas Laqueur, *Making Sex: Body and Gender from the Greeks to Freud* (Cambridge, Mass.: Harvard University Press, 1990): 43–52, 64–68, 99–103, 146–148, 161–163.

61. Historian Sakurai Yuki suggests that by the Kyōho era (1716–1735) abortion providers were predominantly women. Sakurai Yuki, "Mabiki to datai," Hayashi Reiko, ed., *Nihon no kinsei, Vol. 15: Josei no kinsei* (Tokyo: Chūōkōron, 1993): 120–121.
62. The name of the *ryū* that specialized in obstetrics, the Chūjō School, came to be used to indicate abortion experts in the eighteenth century. Ibid., 120.
63. Hori Kazuhisa, *Edo fūryū igaku banashi* (Tokyo: Kōdansha, 1996): 146–147.
64. In 1646, the *bakufu* issued a ban on the business of abortion providers in Edo. This was followed by another edict issued in 1667 prohibiting the display of signboards of abortion doctors or discretely conducting abortions. The punishment, however, was minor and involved being banished from the town quarter where the offender resided (*chōnai tsuihō*). Sakurai, "Mabiki to datai," 119–123; Ochiai Emiko, "Kinsei matsu ni okeru mabiki to shussan" in Susan Hanley and Wakita Haruko eds., *Jendā no Nihon-shi, Vol.1* (Tōkyō: Tokyo daigaku shuppan kai, 1994): 445; Takahashi Bonsen, *Datai mabiki no kenkyū* (Tōkyō: Daīchi Shobō, 1936): 40. For abortion practices with a focus on their Buddhist constructions in Tokugawa Japan, see also William LaFleur, *Liquid Life: Abortion and Buddhism in Japan* (Princeton, NJ: Princeton University Press, 1992): 69–118. For a study of infanticides in Tokugawa Japan, see Fabian Drixler, *Mabiki: Infanticide and Population Growth in Eastern Japan, 1660–1950* (Berkeley: University of California Press, 2013).
65. Furth, *A Flourishing Yin*, 178–182; and "Concepts of Pregnancy, Childbirth and Infancy in Qing Dynasty China," *Journal of Asian Studies* 46:1 (February 1987): 20–21.
66. Tōyoko gakuen joshi tanki daigaku josei bunka kenkyūjo, *Eiri onna chōhōki*, 77–78. "The mourning period" refers not only to the period after a person's death, but also the period during which pollution is effective, such as after childbirth.
67. Japanese traditional customs defined death, menstruation, and childbirth as the three major pollution-generating events, which were respectively called "black pollution" [*kuro fujō*], "red pollution" [*aka fujō*], and "white pollution" [*shiro fujō*]. Kizu Yuzuru, *Nyonin kinsei: gendai kegare kiyome kō* (Osaka: Kaihō shuppansha, 1993): 183. Further scholarly research is required to clarify how the concept of birth pollution evolved historically in Japan. On this subject, see anthropologist Emiko Namihira's works, including, "Pollution in the Folk Belief System," *Current Anthropology* 28:4 (1987); and her original work in Japanese, Namihira Emiko, *Kegare* (Tokyo: Tokyōdō Shuppan, 1985). A useful historical discussion on childbirth practices in medieval Japan is also helpful. Īnuma Kenji, "Chūsei zenki no josei no shōgai," in Joseishi Sōgō Kenkyūkai, ed., *Nihon josei sei-*

katsushi, Vol. 2 (Tokyo: Tokyo Daigaku Shuppankai, 1990): 41, 54, 58–59. For historical accounts of birth and pollution in Japan, see also Nakano, "Women and Buddhism"; Taira Masayuki, "Chūsei bukkyō to josei," in Joseishi Sōgō Kenkyūkai, ed., *Nihon josei seikatsushi*, Vol. 2; Matsuoka Hideaki, "Wagakuni ni okeru ketsubonkyō shinkō ni tsuiteno ichi kōsatsu," in Sōgō Joseishi kenkyūkai, ed., *Josei to shūkyō, Nihon joseishi ronshū series, Vol. 5* (Tokyo: Yoshikawa Kōbunkan, 1999): 257–279.

68. As Furth demonstrates, this portrayal of the pregnant woman's body accords with childbirth texts in late imperial China. Charlotte Furth, "Concepts of Pregnancy, Childbirth, and Infancy in Qing Dynasty China."
69. The Triple Burners [Jp *Sanshō*; Ch *San jiao*] is one of the Five Organs and Six Viscera. Its main function is the digestion of food.
70. By *Fujin ryōhō*, Gyūzan most likely was referring to the aforementioned *All-Inclusive Good Prescriptions for Women* by Chin Jiming, or its annotated version, *Jiaozhu furen liang fang* [*Revised Good Prescriptions for Women*, 1547] by Xue Ji.
71. Gyūzan cited Zhu Danxi as his source. For Zhu's description of the uterus, see also Furth, *A Flourishing Yin*, 211. Miscarriages were often described without any particular emphasis on the uterus's role, but were viewed as a consequence of the failure of adequate functioning of *ki* and blood: "When *ki* and blood are not healthy enough to nurture the fetus, the fetus naturally 'falls down' from the woman's body, just as leaves and flowers do from dead plants."
72. This meridian is presided over by the Triple Burner. For its pictorial depiction, see Kaptchuk, *The Web That Has No Weaver*, 100–101.
73. For its graphic depiction, see ibid., 84–85.
74. This seven-volume Chinese classic comprising lessons for women transmitted through biographies of exemplary women was written by Ryūkō [Ch. Liu Xiang] in the former Han period (202 BC–8 AD).
75. *The Book of Documents* and *The Book of Songs* are among the Five Classics of Confucianism.
76. Historical sources show that, during the Tokugawa period, it became more and more pervasive for the upper echelons of peasant families to sequester young unmarried women within the house, and thereby, prevent them from having a sexual relationship with young men in the village. See Mega Atsuko, *Hankachō no nakano onnatachi* (Tokyo: Heibonsha, 1995): 49–96.
77. See, for example, the Confucian moralist Nakamura Tekisai's (1629–1702) teachings for mothers in his widely distributed didactic text for women, *Himekagami* (1709). Mashita Michiko, "Shussan ikuji ni okeru kinsei," in Joseishi Sōgō Kenkyūkai, ed., *Nihon josei seikatsu shi*, Vol. 3, 146–147. Through her analysis of *Kosodate hanjō tebikigusa*, a popular Confucian text on pronatalist ethics, Kanazu Hidemi points out the tendency in the

Tempo era (1830–1843) and after to place blame on the mother for abortion and infanticide. Such propaganda popularized the idea that women who were prone to commit such a crime should be guided by men. Kanazu Hidemi, "'Datai mabiki' kō," in Kinugasa Yasunobu, ed., *Kinsei shisō shi no genzai* (Kyoto: Shibunkaku, 1995): 401–421.

78. The word *kabuki-mono* was used to describe young men and women from warrior and merchant families who formed a gang of delinquents. The strange fashion they created for themselves helped demarcate their identity. The degree to which their presence implied a political threat to the *bakufu* changed during the Tokugawa period, as their activities gradually lost political significance. *Kokushi daijiten*, Vol. 3 (Tokyo: Yoshikawa Kōbunkan, 1982): 512.
79. Gyūzan's recommendations for pregnant women to engage in self-cultivation should also be reviewed in conjunction with the rise of the discourse on health care (*yōjō ron*). *Fujin kotobukigusa* and other childbirth manuals intended for a wider readership represented a specific type of health care manual within the genre.
80. Sawayama Mikako, *Shussan to shintai no kinsei* (Tokyo: Keisō Shobō, 1998): 236.
81. All the following citations from Saikaku's works are from Asō Isoji and Fuji Akio, eds., *Taiyaku Saikaku zenshū*, Vols. 7–16 (Tokyo: Meiji Shoin, 1974–1978). For references on midwives, see *Shoen ōkagami* (1684), Vol. 2: 4, 192, 278; *Saikaku shokoku banashi* (1685), Vol. 5: 23; *Danshoku taikan* (1687), Vol. 6: 150; *Budō denraiki* (1687), Vol. 7: 56, 233; *Honcho ōin hiji* (1689), Vol. 11: 28; *Nihon eitaikura* (1688), Vol. 12: 76; *Seken Munesanyō* (1692), Vol. 13: 47, 52, 105; *Saikaku oridome* (1694), Vol. 14: 165; *Saikaku okimiyage* (1693), Vol. 15: 81; *Saikaku nagori no tomo* (1699), Vol. 16, 204.
82. Sakurai, "Mabiki to datai," 109, 111.
83. For example, see Nishikawa Mugiko, *Aru kindai sanba no monogatari: Noto, Takejima Mii no katariyori* (Toyama: Katsura Shobō, 1997).
84. *The Tale of Genji* [Jp *Genji monogatari*] was written by Murasaki Shikibu, a lady-in-waiting for the Japanese imperial court in Kyoto, around the turn of the eleventh century.
85. Ogasawara ryū was one of the schools that taught manners for warrior households. The Ogasawara family is one of the families descended from the Seiwa Genji, one of the major warrior families in the Heian period, which had been established by the emperor Seiwa's posterity. Ogasawara families in Kyoto and Shinano (today's Nagano prefecture) handed down old customs and rites, which the ruling Tokugawa family and daimyō houses followed in the Tokugawa period.
86. The sash was commonly put on by a midwife. Gyūzan may have omitted mention of the midwife because he took her presence for granted.

87. In this context, "fetal *ki*" is used almost interchangeably with "the fetus."
88. *Nüke zhengzhi zhunsheng* [*Guide for the Treatment of Female Disorders*] was first published in 1608 as part of Wang Kentang's (1549–1613) collected works. Furth, *A Flourishing Yin*, 159.
89. Furth notes that Yang Zijian's name was associated with a text-based medical tradition in the field of midwifery [Ch *chanke*] from around the twelfth century in China. Furth, *A Flourishing Yin*, 117.
90. *Eiki* is commonly rendered as *ying qi* in Chinese. Kaptchuk translates it as Nutritive *qi*. This substance, a purified form of digested food and drink, circulates within the body and nourishes it, working closely with blood. Nankin chūi gakuin, ed., *Chūgoku kanpō igaku gairon*, 75–80; Kaptchuk, *The Web That Has No Weaver*, 39.
91. Ogata, *Nihon sanka gaku shi*, 92.
92. We could also speculate that in the absence of modern state institutions (e.g., household registration, schools, the military, and medical, welfare, and penal institutions) to confine individuals to specific categories (sex, race, nationality, age, and quality of health and bodily capacity) through the use of discursive knowledge, Tokugawa society may have been more lax in binding individuals to a specific sex.
93. Zhang Jiebin wrote various works of medicine around the mid-seventeenth century. Paul U. Unschuld, *Medicine in China: A History of Ideas* (Berkeley, Los Angeles, and London: University of California Press, 1985): 199–200, 220–222.
94. Sakai Shizu points out that until they began to conduct dissections, many physicians believed that the location of major organs in the male body and those in the female body were reversed because maleness corresponded to yang and femaleness to yin. For example, it was believed that the heart was located on the left side in a man's body, while it was found on the right side in a woman's body. Sakai, *Nihon no iryōshi*, 271–272. No such concept of sexual difference in the position of organs seems to have been developed in late imperial China.
95. See Terajima Ryōan, *Wa-kan sansai zue*, Vol. 2 (Tokyo: Heibonsha, 1986): 108–109, for all citations from *Wa-kan sansai zue*.
96. For debates by physicians in early modern Europe on sexual difference, see Laqueur, *Making Sex*, 149–243; Londa Schiebinger, "Skeletons in the Closet: The First Illustrations of the Female Skeleton in Eighteenth-Century Anatomy," in Catherine Gallagher and Thomas Laqueur, eds., *The Making of the Modern Body: Sexuality and Society in the Nineteenth Century* (Berkeley: University of California Press, 1987): 42–82; and *The Mind Has No Sex?* (Cambridge, MA: Harvard University Press, 1989): 189–213.

CHAPTER 3

Changing Perceptions of the Female Body: The Rise of the Kagawa School of Obstetrics

The field of medicine underwent a major transformation during the mid-Tokugawa period. During the late seventeenth and eighteenth centuries, a group of physicians belonging to the Ancient Practice School (Kohō and Koihō[1]) established a new medical tradition based on their critique of the mainstream Goseihō School. The Ancient Practice School moved away from a systemic understanding of the body based on cosmological concepts and reconceived it as a self-contained, tangible object of hands-on investigation. Its methodology focused on visible and tactile features on the surface and interior of the body; it rejected locating the body within the traditional Chinese cosmology that systematically connected the body and the world outside it.

The rise of the Ancient Practice School paralleled the growing influence of the Ancient Learning School in the scholarly field. The Ancient Learning School challenged the neo-Confucian or Song Confucian scheme that originated in China during the Song period (960–1279) and that aimed to make sense of natural and moral affairs by situating them in its cosmology. Within the neo-Confucian worldview, the dual cosmic concepts of *ri* [Ch *li*] and *ki* [Ch *qi*] explain both natural and moral phenomena, the former being the universal principle governing all existence and the latter vital, cosmic energy that shapes and transforms things in the universe. One way Ancient Learning School scholars departed from this neo-Confucian method was to rebuff its theoretical

Y. Terazawa, *Knowledge, Power, and Women's Reproductive Health in Japan, 1690–1945*, Genders and Sexualities in History,
https://doi.org/10.1007/978-3-319-73084-4_3

foundation in which natural and human affairs were conjoined and assert that they should be studied independently from each other.

The growing influence of the Ancient Learning School helped develop new medical theories for the Ancient Practice School. To legitimize their methods, ambitious scholars and physicians who pursued these new lines of inquiry brought to light older classical texts that preceded texts from the Song period and used the authority of those ancient texts to challenge Song Confucian and medical ideas. This is why these newly established schools of scholarship and medicine are called the schools of *Ancient* Learning and Medicine.

The changes that occurred both in the scholarly and medical fields by the rise of the Ancient Learning and Ancient Practice schools did not simply involve cumulative improvements on the existing methods. Rather, they were the results of intellectual and medical explorations from a decidedly different angle to the existing neo-Confucian theoretical framework. Thomas Kuhn suggested such a drastic shift in scientific method with his concept of "paradigm shift." Kuhn's "paradigm shift," however, seems more suited to analyzing a methodological shift in a particular scientific or medical field. That is, he never attempted to explain the connections between a methodological change in one field to that in another or in the larger intellectual terrain.[2] Considering the overarching methodological shift encompassing philosophy, medicine and various natural sciences, we might want to apply Michel Foucault's concept of "epistemic change."[3] In other words, scholars and physicians who pursued new lines of investigation used a different criterion for ascertaining and legitimating the truthfulness of the knowledge they developed. With this break from the earlier Song Confucian and medical traditions, scholars and medical practitioners began to accumulate knowledge while moving into a different intellectual terrain, a new "episteme."

In this chapter, I first discuss the ways physicians who formed the Ancient Practice School developed their medical theories and perceptions of the body, paying particular attention to the influence of its counterpart in the field of Confucian Studies, the aforementioned Ancient Learning School. The rest of this chapter is devoted to an investigation of the Kagawa School of Obstetrics that emerged parallel to the movement of the Ancient Practice School physicians. Combined with other endeavors and achievements of Ancient Practice School physicians, Kagawa School obstetricians helped shape a new medical landscape where innovative

Ancient Practice School methods arose and even overshadowed the Goseihō School that, with the endorsement of the Tokugawa regime, had widely been accepted as the mainstream method.

The Rise of the School of Ancient Practice

The Goseihō School dominated the field of medicine until the late seventeenth century when independent-minded physicians began to develop a competing medical method called the Ancient Practice Method. Physicians who formed the newly emerged Ancient Practice School of medicine departed from the Goseihō School by severing the ties between the internal process of the human body and that of the larger universe, rebuffing such key concepts as the Yin Yang Five Phases theory, *ki*, and the meridians I describe in Chap. 2. In other words, Ancient Practice School physicians developed a new and revolutionary concept of the body viewing it as a self-contained entity.

Such a drastic shift in perceiving the body was possible owing to the larger transformation in the field of Tokugawa scholarship with the emergence of the Ancient Learning School. Prominent Ancient Learning School scholars such as Itō Jinsai (1627–1705) and Ogyū Sorai (1666–1728) suggested that natural phenomena (which they called "heavenly affairs") and human affairs were governed by different rules, and should therefore be studied independently from each other. The resulting epistemological break was the basis for a new form of inquiry. Instead of explaining objects and their attributes through neo-Confucian metaphysics that tied the occurrences of natural phenomena with moral consequences, scholars and physicians began to conduct empirical investigations of natural objects, focusing on their tangible features. Pursuing this line of thinking, Sorai conceived of the human body as an entity that functioned according to its own rules and therefore suggested that it should be investigated and treated separately from all the other phenomena of the universe.[4]

In rejecting the legitimacy of the Song Confucian metaphysical framework that was central for both mainstream Confucian scholarship, called the Shushi [Ch Zhu Xi] Learning, and the equally dominant Goseihō School of medicine, Sorai along with other scholars and physicians challenging these traditions used pre-Song-era Chinese texts. For example, the scholar Itō Jinsai used Confucius's *Analects* and Mencius's *Mencius* from ancient China to criticize Zhu Xi's and other Song Confucian ideas. Likewise, the physician Nagoya Gen'i also emphasized the importance of reading pre-Song medical classics such as *Shōkanrōn* [Ch *Shanghanlun*; *On Cold Damage Disorders*][5] from Han China (206 BC–220 AD) to legitimize his critique of the Goseihō School.[6]

Yoshimasu Tōdō (1702–1773) furthered radical medical methods based on the new epistemological foundation established by pioneering Ancient Practice School physicians like Gen'i. He focused solely on tangible symptoms appearing on the surface of the body. Not only did he reject the Goseihō etiological system I introduce in Chap. 2, he also repudiated the notion of *ki* because, in his view, it was something intangible. Disease was not caused by the stagnation of *ki* but resulted from "poison" [*doku*], whose existence Tōdō believed he could prove by touching the patient's body.[7]

Another physician of the Ancient Practice School, Kagawa Shūtoku (1683–1755), developed his predecessors' work, emphasizing the importance of empirical verification—testing theories through experimentation. That is, he refused to accept the authority of any of the therapeutic methods presented in medical texts until he verified their effectiveness through his own experiments and experience.[8] His critical and empirical attitude drove him as far as to cast doubt on cardinal Chinese medical classics such as *the Yellow Emperor's Inner Canon* [Jp *Kōtei Daikei*; Ch *Huangdi Neijing*].[9]

Having pursued his medical studies in the Ancient Practice School, the physician, Yamawaki Tōyō, was eager to test through dissection the accuracy of the Theory of Five Yin Organs and Six Yang Viscera, central to the mainstream Chinese medical tradition. His exposure to Dutch anatomical texts was also likely to have provoked his curiosity for viewing the insides of the body.

Dissection, however, had actually been strictly if tacitly prohibited in Japan since the ancient period for religious and political reasons. Despite that the Taihō law codes (702) and the Yōrō codes (718) were virtually the only laws prohibiting dissection, those who conducted dissection were severely punished. Behind this taboo were religious and moral doctrines developed under the influence of Buddhism and Confucianism, which viewed dissection as a desecration of the dead. Physicians also refrained from dissection because they wished to avoid physical contact with a dead body that was considered polluted.[10]

Overcoming these taboos through prestige and connections he had developed with government authorities, Tōyō carried out the first publicly approved dissection in Japanese history in 1754.[11] Tōyō publicized his findings by publishing a text titled *Record of Internal Organs* [*Zōshi* 1759], which contained four leaves of illustrations, produced by an artist who attended the dissection, as well as Tōyō's commentaries. In it Tōyō argued that the Theory of Five Organs and Six Viscera was erroneous.

Medical historian Achiwa Gorō investigates how Tōyō's interests in human anatomy were a precursor to the mode of inquiry adopted by Japanese physicians trained in European medicine (what was then known as "Dutch medicine"). He argues that Tōyō's exploration of human anatomy was another turning point in the discursive development of medicine in Tokugawa Japan and that Tōyō's mode of inquiry went beyond the methods of the Ancient Practice School. That is, Ogyū Sorai's line of thinking, which strongly influenced the Ancient Practice School, was based on his opposition to one of the main tenets of the Zhu Xi School—the assumption that human beings had unlimited intellectual capacity for acquiring knowledge about the world. For Sorai the intellectual capacity of humans was limited, a conviction which led him to distinguish the realm of heaven from the realm of humans. Matter that belonged in the realm of heaven was, according to Sorai, unknowable.

For some major Ancient Practice School physicians, such as Yoshimasu Tōdō, the internal anatomy of the human body belonged to the realm of heaven, and was thus not to be examined. Yamawaki Tōyō, however, wanted to explore this prohibited realm through dissection. In other words, he attempted to procure knowledge not only by observing the body's surface, but also by investigating the mechanisms of its inner organs. In this sense, Achiwa argues, Tōyō departed from the conceptual framework of the Ancient Practice School.[12]

His departure from the modes of investigation advocated by the Ancient Practice School and the Ancient Learning School foreshadowed new approaches seeking to procure knowledge through a combination of rational guesswork and empirical observation. Such new lines of studies were developed by some Dutch School physicians.[13] Together with Tōyō, they ignored the limitations Sorai placed on the human capacity for acquiring indefinite knowledge about all the phenomena in the universe, including those that occurred in the human body. Seeking a systematic understanding of the body's internal mechanism, they reintroduced the notion of universal principle, or *ri*, which had once been discarded by the

Ancient Practice School. Dutch School physicians, thus, revived theoretical conjectures for their study of the human body, but this time, they used this method in conjunction with the observation of tangible, material features inside and outside the body.[14]

It was, however, not until the late 1850s and the 1860s that the Tokugawa regime gave Dutch medicine significant official standing and support. Until then, the Ancient Practice School continued to thrive, overshadowing the mainstream Goseihō School. The recognition and popularity of the Kagawa School of Obstetrics rose within this medical landscape.

The Kagawa School of Obstetrics

The founder of the Kagawa School of Obstetrics, Kagawa Gen'etsu (1700–1777), developed his medical methods in the mid-eighteenth century in a similar way to Ancient Practice School physicians. Seeking more effective ways to address the complications of childbirth, Gen'etsu and the Kagawa School physicians developed surgical techniques involving the use of instruments, techniques radically different from those appearing in Chinese medical texts. These developments led to a new conception of the female reproductive body. For the Kagawa School, the body was a self-contained entity that functioned as an object of empirical observation and surgical intervention.

One factor behind Kagawa School physicians' continual search for improved obstetric methods was a competitive and profitable market for skilled obstetricians. While physicians' practices were thus increasingly subject to commercializing forces, the production and dissemination of obstetric knowledge were mediated by the Kagawa's *ryū* establishments.[15] The shogunate and domanial authorities, on the other hand, played limited roles in directing or supporting the efforts of Kagawa School physicians in developing new techniques and educating neophyte disciples.

The Kagawa School's approach demonstrates the influence of that variety of empiricism advocated by the Confucian scholar Ogyū Sorai. Gen'etsu believed in a body that was visible, tangible, and subject to empirical investigation, much like the physicians of the Ancient Practice School who also respected Sorai's views. Gen'etsu renounced what he considered the highly speculative theory of the Yin Yang Five Phases, which could be proven neither by seeing nor touching. In *Sanron* [*Treatise of Childbirth* 1765], he refuted conventional views widely disseminated by the Goseihō School by adopting the confrontational style typical of Sorai and the Ancient Practice School physicians. He and his followers also employed aspects of European anatomical knowledge; the experiential methods developed by the Ancient Practice School were a more fundamental influence, however, than those of Europeans in shaping Gen'etsu's approach.[16]

Kagawa Gen'etsu

Kagawa Gen'etsu, the son of a low-ranking warrior family, served the lord of the Hikone domain located north of Kyoto.[17] He shared social position, personality, and medical methods with other pioneering physicians of the Ancient Practice School. He had almost no access, for instance, to formal medical training and Confucian studies. As the son of a concubine, unable to inherit his father's family name, he was adopted by his mother's family, a family of lower status than his father's (Fig. 3.1).[18]

From early in life, Gen'etsu disliked the idea of taking up farming, the occupation of his maternal family. He wanted to become a physician instead. His passion for studying medicine led him to Kyoto, the center of

Fig. 3.1 A portrait of Kagawa Gen'etsu (Source: Fujikawa et al., *Nihon sanka sōsho*, 5)

medical learning at the time. There, he studied medicine on his own while making a living by selling old copper tools and practicing acupuncture and massage. Gen'etsu's biographies suggest that his career as an obstetrician was prompted by an incident in which he helped a neighborhood woman who was experiencing a difficult birth. The child's hand had emerged, but the mother was unable to push the child out. It occurred to Gen'etsu that he could pull the child out by using a hook that people used for weighing things. He performed this operation and saved the mother although he could not salvage the child with this technique. After this experience, Gen'etsu began experimenting with various techniques by volunteering to attend poor women giving birth.

As Gen'etsu developed more effective procedures, drawing in part on his skills as a masseur, he successfully dealt with many complicated births. His reputation gradually rose and numerous domanial lords, aristocrats, and even the imperial family sought his assistance. Eventually, he established the Kagawa School of Obstetrics (Kagawa-ryū sanka). With the help of the noted Confucian scholar Minagawa Kien (1734–1807), who took up the task of writing the text for Gen'etsu (who reportedly had very limited ability in reading and writing), he published *Sanron* [*Treatises of Childbirth*] in 1765.[19] In 1769, he was granted warrior status by the daimyō of Awa on Shikoku island and became his retainer, enjoying a yearly stipend of 100 *koku* of rice. His stipend, and similar stipends received by subsequent head masters of the Awa branch of the Kagawa House, indicated the relatively low status of physicians and other scholars as non-military personnel within the domanial bureaucracy.[20] Kagawa head masters in Awa most likely earned enough from commoner clients to supplement their low stipend and make a decent living.

Gen'etsu's career path stood in marked contrast to that of Katsuki Gyūzan. While Gyūzan was trained by a notable physician and a Confucian scholar, Gen'etsu had no such support. The difference in the medical training these two men received was reflected in the clinical methods they developed. Gyūzan's training in the tradition of the Goseihō School and Song Confucianism confined him to the conventions of established Chinese medical methods. In contrast, Gen'etsu's lack of traditional medical training led him to develop entirely new obstetric methods. Because he did not possess the prestige associated with Chinese learning, developing effective treatments was almost the only way to gain recognition and cultivate his clientele.

THE RECONCEPTUALIZATION OF THE BODY IN *SANRON* AND *SANRON YOKU*

Gen'etsu radically revised Goseihō School methods, which drew on visual and pulse diagnosis, and herbal medicine. That is, he relied on knowledge from his own observations and experience and rejected ideas derived from Chinese texts or Japanese customs. He asserted that obstetricians should diagnose and operate on the body through physical contact, including the use of surgical instruments when necessary. His efforts culminated in establishing a set of obstetric procedures that required the physician's active intervention for all types of childbirth, not just difficult or critical cases. Establishing a new system of obstetrics entailed new ways of studying and handling the body.

In *Sanron*, Gen'etsu made numerous references to the body's anatomy and physiology based on his experiences in obstetric practice. This information was often specific, with time, size, and weight enumerated in detail. A good example was his observation of the standard term for pregnancy.

> It takes about 300 full days between the time of conception and the time of delivery if the woman is pregnant for the first time. For a woman who has already experienced childbirth, it takes 275 days. People can take these numbers as a guide and calculate the due date by making adjustments in response to each situation.[21]

He also detailed the period of severe morning sickness.

> [Women] suffer from morning sickness 30 days after conception. They probably do not have the next menstruation [after conception], and they have morning sickness around the time when menstruation should occur [unless they were pregnant]. Some women suffer from minor headaches. Others become irritated. Or, they feel lazy and do not want to do anything. These [symptoms] become increasingly severe 45 to 50 days [after conception].[22]

These passages demonstrate the way Gen'etsu limited himself to describing observable symptoms. He refrained from speculating about the causes of illness using the etiological reasoning typical of mainstream Chinese medicine, which explained illness as malfunctions of *ki* and blood.

Gen'etsu's practice of privileging his own observations and presenting them in quantitative terms is also apparent in his description of the female reproductive anatomy.

> Beside the hip bones beneath the navel, there are "*yokobone*" [side bones, pubic bones]. The extensions of the side bones are curved diagonally inward and extended from both sides between the legs. These are called "*kōkotsu*" [cross bones].[23] Between the "cross bones," there is a gap, which is 3 *bu* for men, 5 *bu* for women, and over 1 *sun* for women who have given birth. This is called the "*ein*" [perineum]. In front of the cross bones, there is the "*in*" [vulva]. Behind it is the "*kōmon*" [anus]. The "*shikyū*" [uterus] is located above the anus, 4 *sun* above the *in*. The length of the uterus is 8 *sun*. Its opening always faces the back.[24]

In this way, not only did Gen'etsu identify specific bones and organs that constituted the reproductive anatomy, but he also defined their size and location (see Fig. 3.2 for our contemporary understanding of pelvic anatomy).

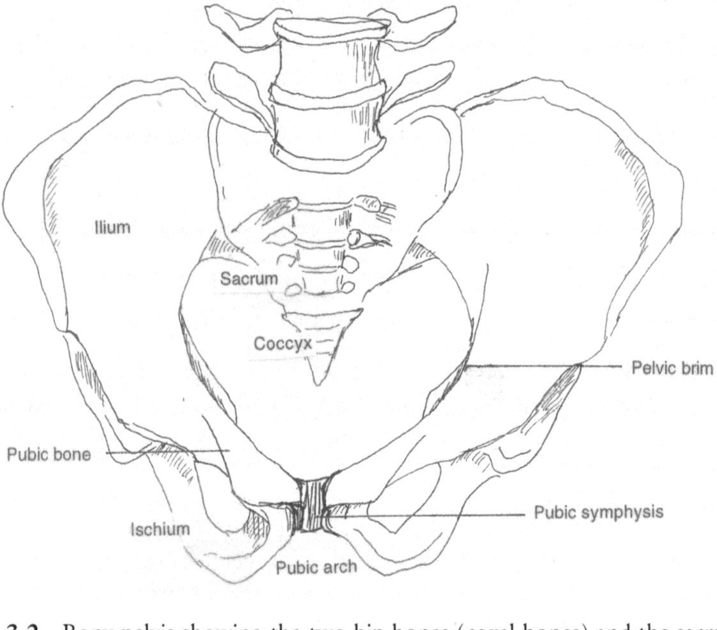

Fig. 3.2 Bony pelvis showing the two hip bones (coxal bones) and the sacrum. Illustrated by the author

Gen'etsu also critically reviewed commonly accepted ideas and customs. For instance, he challenged the widespread idea that a pregnant woman should not lie down with her legs stretched out, which was believed to cause complications during birth. He suggested that a woman should only avoid lying down with her legs folded. If she lay in this position, her body would be curved and its insides contracted, which would cause "the fetus to have a difficult time staying within [her abdomen]." As this example shows, it was typical of Gen'etsu to focus on the body's anatomical and mechanical features.

This empirical perception of the body also characterized his discussion of the detrimental effects of using a pregnancy sash. He noted that beginning in the fifth month of pregnancy, women commonly wore a sash around the chest in order to calm down fetal ki and prevent it from moving upward. Gen'etsu asserted that, since the fetus had already been positioned upside down by the fifth month, its ki could not move up to pose any such danger. By assuming that the sash was commonly bound around the woman's chest just where the fetus's buttocks were located, Gen'etsu became convinced that the practice would cause the fetus's buttocks to be covered with harmful "amnion blood." He thought this would make it difficult for the afterbirth to emerge and cause serious problems with the uterus after delivery. Gen'etsu also reasoned that the use of the sash could prevent the fetus from shifting in accordance with the mother's movement, causing it to become positioned wrongly.[25]

The perception of the fetus in *Sanron* was radically different from that given in *Fujin kotobukigusa.*

> In general, the shape of the four limbs [of the fetus] is formed 120 days after a woman becomes pregnant. Before this date, there is just a round *hō* [*taiban*; placenta]. It looks like an acorn. The white membrane which wraps it is quite thick.[26]

Thus, Gen'etsu dismissed the theory that the fetus's body parts were formed by receiving essence pertaining to each of the Five Phases each month. Instead, he developed a new perception of female reproductive anatomy focusing on the tangible features of fetal development.

Gen'etsu also attempted to ascertain the fetus's position inside the mother's body by means of his own observations rather than by reading medical texts. He remarked that Chinese medical books and Japanese traditions described the fetus positioned with its head-up, then turning upside down just before emerging from the mother's body. He even noted that anatomical illustrations in Western medical texts also showed the fetus with its head upright.[27] But Gen'etsu contested this theory by pointing out that it would be impossible for the fetus to turn itself upside down immediately before delivery due to its size and the size of the uterus.

> When months have passed [after conception], how large is the fetus? And, how large is the space inside the uterus? If the fetus actually turns around, logically, [the uterus] will burst. What a preposterous and confusing idea it is![28]

Gen'etsu remarked that an obstetrician would naturally notice that the fetus was positioned with the head down if he or she[29] massaged the mother's abdomen using proper guidelines. He described its position inside the woman's body as follows.

> When five months have passed [after conception], the size of the fetus inside [the mother's] abdomen is similar to that of a melon [*uri*]. Its back faces [the front of the abdomen], and the position of the neck is upside down. [The fetus] remains [inside the mother's abdomen], its head touching the upper side of the side bones [*yokobone*; pubic bones]. Its afterbirth covers its buttock.[30]

Gen'etsu thus developed his anatomical knowledge through somatic diagnosis that involved active touching, a method shared by some Ancient Practice School physicians. Gen'etsu's empirical approach is also evident in his response to questions regarding the fetus's sex. Within the Goseihō system, the application of yin yang theory was central to guessing the fetus's sex. Gen'etsu dismissed the theory, claiming it could not be supported by tangible evidence. Thus, when asked how to determine the fetus's sex, he simply replied that he did not know any such method. He also contested the theory that the male fetus remained on the left side of the mother's abdomen while the female one stayed on the right side. Gen'etsu suggested that the fetus was generally positioned in the center along the conception meridian [*nin*].[31]

Operating outside the framework of yin yang theory, Gen'etsu began to find sexual differences in the body's morphology and anatomy, particularly in relation to its reproductive functions. For example, in the passage discussing pelvic structure, Gen'etsu noted that the gap between the "cross bones" cited above was different for men and women.[32] In another passage, he emphasized the idea that male and female bodies were formed differently by linking the shape of the hips to the reproductive functions of each sex.

> The shape of the woman's hip is always curvy and its inside is spacious. This is the place of conception as given by Heaven. Thus, the shape of the man's hip is straight, and its inside is unable to accommodate [the fetus]. The bodies of women and men are not the same.[33]

Thus, Gen'etsu reconceptualized the body based on his experiential anatomical knowledge. This led him to posit sexed bodies rather than the fundamentally androgynous body found in traditional Chinese medical discourse.

The Kagawa School's development of new anatomical descriptions was part of a general trend at this time. The study of human anatomy had become remarkably popular among many physicians of the Ancient Practice School and the Dutch School in the late eighteenth and early nineteenth centuries. Inspired by the publication of Yamawaki Tōyō's *Record of Internal Organs* and Sugita Genpaku's *New Book of Anatomy* [*Kaitai shinsho,* 1772],[34] many physicians sought to conduct dissections, despite the enormous difficulty of obtaining permission from the shogunate or domanial authorities. But historian Kuriyama Shigehisa argues that dissections alone did not advance Japanese physicians' knowledge of human anatomy. They also needed to correlate their observations with illustrated anatomical texts, without which they could not identify and classify the internal organs in dissected bodies.[35] Japanese physicians sought out Dutch anatomical texts, which were praised for their precise and realistic representations of the body's interior. The Kagawa School obstetrician Hara Masakatsu (1752–1820) realized the value of combining these two approaches and advised physicians to conduct dissections and refer to illustrations in Dutch texts.[36]

In the late eighteenth and early nineteenth centuries, few Japanese obstetricians could read Dutch. Nonetheless, they looked at anatomical illustrations in Dutch medical texts. Since he did not know the Dutch language, Gen'etsu was likely to be among those who obtained anatomical knowledge simply by observing illustrations.[37] Although *Sanron* did not include drawings of bodies, *Sanron yoku* [*A Supplement to Treatise of Childbirth,* 1775], a text written by Gen'etsu's foremost disciple and adopted heir Kagawa Genteki (1739–1779), represented the newly introduced anatomical body in pictorial form.

In *Sanron yoku,* the mother's body was anatomized and reduced to a few select sections. As shown in Figs. 3.3, 3.4, 3.5, 3.6, 3.7, 3.8, and 3.9, the head, arms, shoulders and most of the legs have been removed from

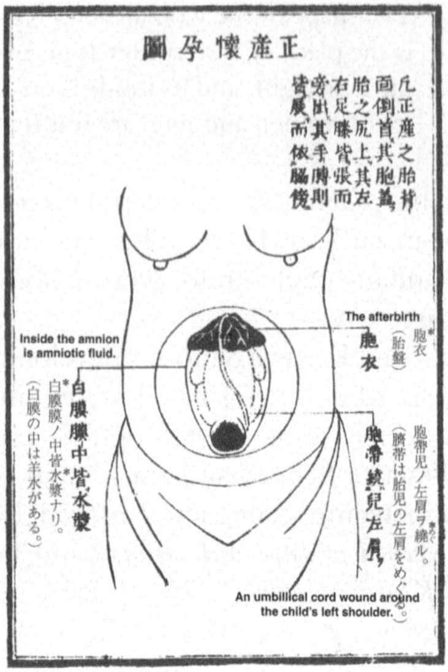

Fig. 3.3 An illustration of gestation for a normal birth (Source: Sanka Bunken Dokushokai, ed., *Heiseiban Sanron, sanron yoku*, 276)

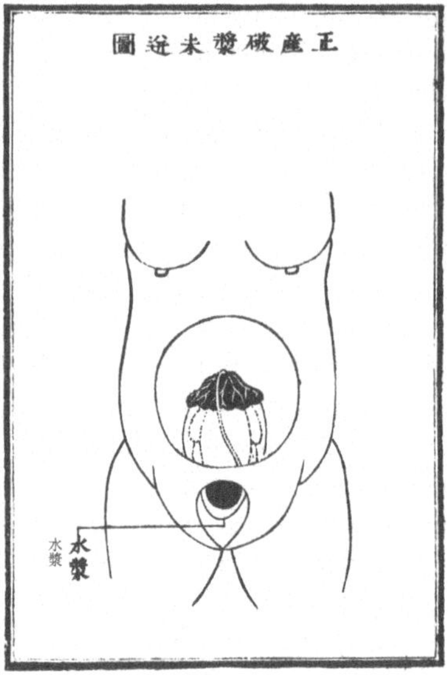

Fig. 3.4 Water has not yet broken (normal birth) (Source: Sanka Bunken Dokushokai, ed., *Heiseiban Sanron, sanron yoku*, 277)

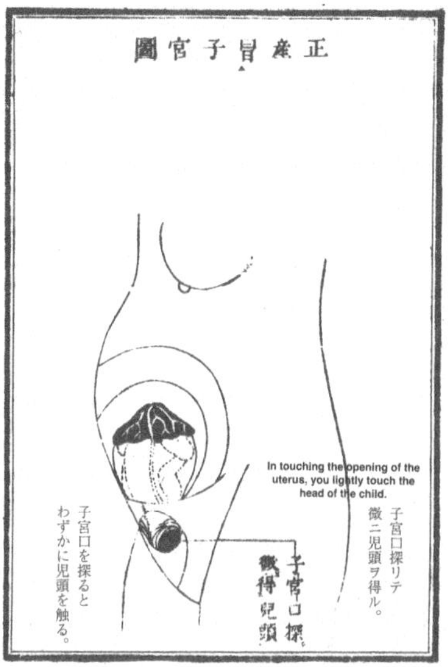

Fig. 3.5 Feeling out the uterus (normal birth) (Source: Sanka Bunken Dokushokai, ed., *Heiseiban Sanron, sanron yoku*, 278)

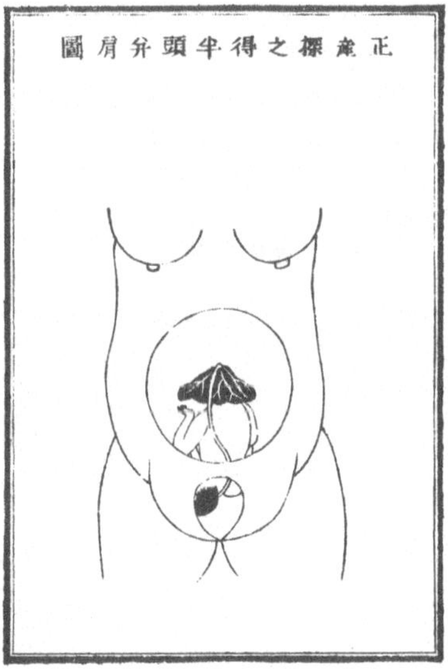

Fig. 3.6 Half of a head and a shoulder are shown in the case of normal birth (Source: Sanka Bunken Dokushokai, ed., *Heiseiban Sanron, sanron yoku*, 279)

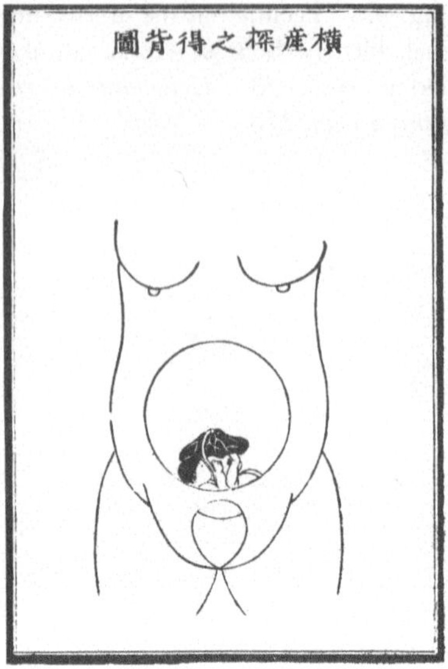

Fig. 3.7 The child's back is detected in the case of a transverse birth (Source: Sanka Bunken Dokushokai, ed., *Heiseiban Sanron, sanron yoku*, 297)

Fig. 3.8 An elbow and an arm come out in the case of a transverse birth (Source: Sanka Bunken Dokushokai, ed., *Heiseiban sanron, Sanron yoku*, 298)

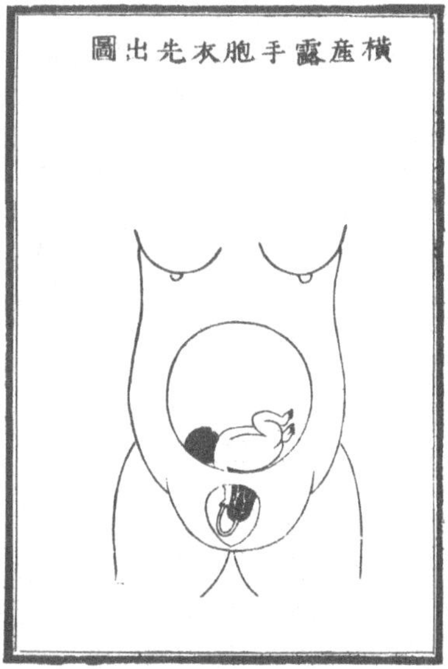

Fig. 3.9 A hand and the afterbirth emerge first in the case of a transverse birth (Source: Sanka Bunken Dokushokai, ed., *Heiseiban sanron, Sanron yoku*, 299)

the mother's body, and only frontal views of her breasts, abdomen, and the upper part of the legs are shown. A large opening in the center of the abdomen allows one to view the inside. The illustration shows the back of the fetus in an upside down position as Gen'etsu claimed in *Sanron*. The afterbirth is located on top of the fetus's buttocks, and an umbilical cord hangs from the fetus's back. In Figs. 3.3 and 3.4, the amniotic fluid and the amnion are explicitly labeled, as well as the umbilical cord and the afterbirth. In Fig. 3.5 the text within the figure states, "In touching the opening of the uterus, you lightly touch the head of the child."

These illustrations show that the Kagawa School rejected the holistic system of yin yang dynamics and circulatory functions. Instead, the body was stripped of its diverse attributes and functions, and reduced to a simple mechanism for holding and delivering the child. This method of partitioning the body and isolating its parts was fundamental to modern European medicine but would have made no sense in conventional Chinese medicine. Thus, these illustrations provide further evidence of how the Kagawa School was moving in an entirely new direction.

The illustrations in *Sanron yoku* represent the mother's body in stasis. The dynamic movements and traces of severe pain characteristic of the birthing process are absent. This representation of the woman's body as a passive object resembles the view of the body found in modern European obstetric practice. Recent scholarship on the history of obstetrics in Europe and North America suggests that obstetricians wanted quiescent patients with bodies suitable for their operations and for overall management of the birthing process. This view of the maternal body is reflected in the overuse of anesthesia by obstetricians in Europe and the USA during the nineteenth century.[38] Kagawa School obstetricians did not go as far as to use anesthesia, but they seemed to want their patients' bodies to be docile objects amenable to elaborate obstetric techniques without any obstruction. It is not far-fetched to conclude that the pictorial depictions in *Sanron yoku* were partly a product of the Kagawa School's trivialization of the mother's active and spontaneous movements during labor.

A New System of Obstetric Procedures

Gen'etsu accompanied reconceptualization of the body with a redefinition of the physician's role during delivery. He declared that a physician's assistance was necessary for both normal and complicated births and that he or she should be able to provide a surgical operation, which he called a "rescue operation."

> In terms of treating a pregnant woman, the most important question is how to treat her during labor. In eight out of ten cases, it is necessary to use a hand-operated or surgical technique and provide a proper post-delivery care [*kyūgo-jutsu*]. Herbal medicine is needed in two or three cases out of ten. If one could not provide an operation, herbal medicine is of no use.[39]

Gen'etsu severely criticized the incompetence of physicians whom he claimed did nothing but indulge in futile discussions about various prescriptions of herbal medicine.

> However, physicians nowadays only vainly discuss the characteristics of herbal medicine and do not discuss medical techniques. They do little to treat a mother who is in labor nor do they look for signs on a child to see whether or not it is dead, leaving everything to the midwife instead. Facing a dangerous situation, they are simply perplexed and sit by and watch both the mother and the child die. Is this the deed of a physician who is supposed to cure illness and save people's lives [*kyūkan saisei sha*]?[40]

These statements should be understood in light of Gen'etsu's status as an upstart social climber competing with physicians who were well versed in Chinese and Japanese medical texts, and familiar with elaborate methods of prescribing herbal medicine. His excoriation of existing methods was most likely prompted by the need to underscore the legitimacy of his own methods. He also attempted to support his approach by overstating the risks of childbirth. The claim that "eight out of ten cases" of childbirth are difficult is certainly an exaggerated figure.[41]

Gen'etsu also disapproved of the common practice of midwives handling childbirth instead of physicians.

> Those who are called midwives generally are widows or old women who do not have families to rely on, and who practice midwifery out of necessity. They are just ignorant women. All they know how to do is to bathe or wipe the child. How could they make judgements on the matters of life and death, and help make the rescue operation successful, collaborating with a physician?[42]

The underlying assumption here is that childbirth was no longer an event that should be left to nature; it had become instead an occasion for active intervention and management by physicians. In Gen'etsu's view, a properly trained obstetrician possessed the capacity to change the fate of the mother and the child, a power that untrained midwives lacked.

To provide adequate treatment for each case of childbirth, Kagawa School physicians considered it a top priority to identify whether or not a birth was normal and, if it was complicated, the exact position of the fetus. To accomplish this task, Gen'etsu developed a system of pelvic examination called the *tankyū-jutsu*. It would help a physician determine the type of delivery and assign a particular technique to be used for each case: normal delivery [*zasō*] (see Figs. 3.3, 3.4, 3.5, and 3.6); breech delivery [*jotō*], correcting sideways birth [*seiō*] (see Figs. 3.7, 3.8, and 3.9); dealing with twins [*kyoren*]; and the "restoration technique" [*kaisei-jutsu*].

Gen'etsu detailed the procedures to be followed for a pelvic examination and delivery assistance in each case. For example, a normal delivery should be treated in the following way. Once the mother began to feel pain and the need for a bowel movement, the physician should sit beside her and support her by holding her back with his left arm. The mother's head should be supported by the physician's shoulder (see Fig. 3.10).

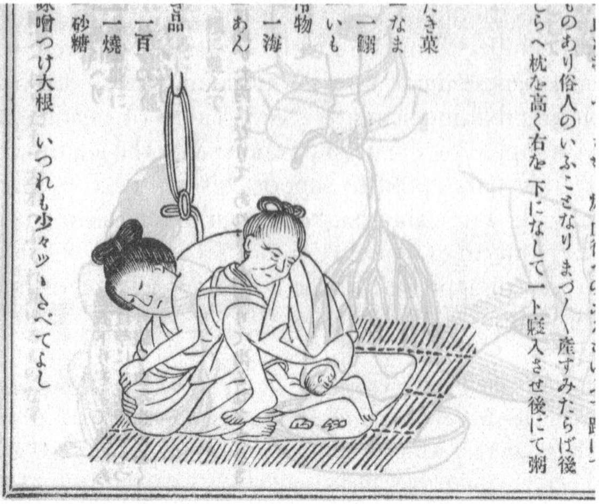

Fig. 3.10 The position to handle a normal birth in the Kagawa School tradition. An illustration presented in the obstetric manual, *Yōka hatsumō zukai* (1852) by Yamada Hisao (Source: Fujikawa et al., *Nihon sanka sōsho*, 741)

The physician then should wrap his right hand with a cotton cloth, move it along between her legs, feel for her coccyx [*bitei-kotsu*], and have her sit (while holding her by the groin). After the physician and the mother took this position the physician was supposed to press and protect her anus with his right hand, while holding her body up with his left arm each time the mother felt pain.[43]

After the mother had felt labor pains for a sufficient period of time, the physician was supposed to insert his index and middle fingers into her vulva [*in*] to identify the position of the fetus. Gen'etsu specified that "the fingers should be first pointed towards the anus and then they should be moved upward" to determine the size of the opening of the uterus. If the opening of the uterus was small, the child remained in the uterus. The child's head was covered by fetal membrane [*maku*]. In a pelvic examination, the fetal membrane would feel like a wet silk cloth. The physician might find that the child's head had already emerged from the uterus into the birth canal. If the child had not yet emerged, however, amniotic fluid would be full and soon burst. The physician should then wait until the mother felt pain in her hips and groin area. This meant the amnion was about to burst, at which point he should tear it with his fingernail. This procedure allowed the amniotic fluid to leak out and yang *ki* to reach its peak, resulting in pain being reduced by more than half. Gen'etsu believed that when the mother's *ki* was vigorous, the delivery would be easy.[44]

During the entire time of labor, the physician was to massage the mother's abdomen with his right hand while holding her up with his left arm whenever pain was felt. Once the child emerged from the uterus, it would proceed downward with its head held by the mother's perineum in the upside down position. Hernia of the large intestine might develop as well. The mother would then typically experience extreme pain, and the child would finally emerge. Gen'etsu also noted that this procedure required the assistance of a midwife. The midwife did not disappear from the birthing scene; she assumed a supporting role. To prevent rupture of the perineum, the midwife was to hold the area behind the mother's vulva at the time of delivery, while a physician supported and massaged the mother.

In Gen'etsu's depiction of childbirth, the woman was treated as an inanimate object that the physician should touch and operate on. Although the procedures he described did not require any instruments, the physician's whole body, including arms, hands, fingers and fingernails, functioned in a machine-like fashion, supporting, massaging, and operating on the mother's body. The obstetrician's entire body served as an instrument. The mother, on the other hand, was not viewed as an active agent. Her actions and input were not relevant to the success or failure of the birthing process.

In the treatments Gen'etsu devised for difficult deliveries, the passivity and infirmity of the woman's body is further highlighted by its supine position, while the use of the physician's body as an instrument is also evident. A good example is the technique called the *jotō*, which was used when a child's leg or buttock emerged first. If the physician felt the child's foot in his pre-delivery examination, he was told to push it back into the mother's body. The mother should then be laid down with her upper body supported by high pillows. Before applying this procedure, Gen'etsu considered it of foremost importance for the physician to find out which foot, the right or the left, had emerged by examining and feeling for the big toe. Once the physician identified which foot it was, he or (she) should take hold of one of the big toes tightly between his forefinger and middle finger, and the other big toe between his middle finger and third finger. While assuming a squatting position, the mother would put her weight on the physician's knees as he swiftly pulled out the child.

In cases where the physician detected the child's arm and hand during the pre-delivery diagnosis, another specific set of procedures was to be applied. In this case, too, the mother was laid on her back with her legs apart. The physician then inserted his left hand into the mother's body and grabbed the child's hand or elbow. He should place his right hand on the left side of the mother's lower abdomen, and push it up to put the child into the correct position. This technique was called "correcting sideways [birth]" [*seiō*].

Among the procedures that Gen'etsu devised, the only technique employing an actual surgical instrument was the "restoration technique" [*kaisei-jutsu*], which involved removing the dead fetus from the mother's body by using a special iron hook. Gen'etsu did not describe this procedure in detail, remarking that it required subtle skills and could therefore only to be taught to one or two of his disciples. Gen'etsu's concern about the possible misuses of this method could have been genuine. Such secrecy, however, could also be understood as a conventional practice of *ryū* establishments, which transmitted certain knowledge only to senior disciples. Despite Gen'etsu's intentions, this "restoration technique" seems to have become widely known among physicians and midwives by the early nineteenth century. Some Kagawa School texts described the procedure in detail, sometimes accompanied by an illustration of the hook, with a specific description of its size, shape, and weight (see Fig. 3.11).[45]

This indicates that the secret transmission of medical knowledge, which was the norm prior to the Tokugawa period, gradually gave way to a more open circulation of medical ideas through publications.

Existing scholarship on the Kagawa School often highlights a drastic change in obstetric techniques by focusing on the introduction of surgical instruments to the process of childbirth.[46] But procedures other than the "restoration technique" were equally physical and mechanical in nature. The hook merely supplemented the physician's body, which also functioned as an instrument. Another significant characteristic of Gen'etsu's method was the systematic use of diagnosis and classification. A correct judgment of the fetal position in each case increased the chances of a successful obstetric intervention. Under Gen'etsu's system, the birthing process was rigorously controlled and managed by the obstetrician, who would physically manipulate and operate on the mother's body. The introduction of a particular surgical instrument was perhaps less important than the creation of a system of obstetric procedures that underestimated the woman's ability to deliver a child and imposed regimented postures and

Fig. 3.11 An illustration of a hook used by Kagawa Gen'etsu, presented in Oku Ressai's obstetrics manual, *Kaisei kōhō hikketsu*. The size and weight of the hook are specified: length, nine-and-a-half *sun*; and weight, 35 *sen*. The sizes of particular parts of the hook are meticulously described in the attached caption (1 *sun* = 3.03 centimeters = 1.193 inch; 10 *bu* = 1 *sun*; 1 *sen* = 3.75 gram = 0.13 ounce). From Fujikawa et al. (1971: 764)

movements. This was the major shift that Gen'etsu's innovation brought about.

Notions of childbirth and the body developed by the Kagawa School reached physicians and the general public through *Sanron* and *Sanron yoku*, both of which were widely hand-copied and read. Not all physicians who read Kagawa School texts, however, accepted the ideas and techniques presented in them without question. Physicians criticized the Kagawa School's techniques and suggested alternative treatments. The publication of *Sanron* and *Sanron yoku* thus generated intense debates over obstetric techniques and customs related to childbirth, issues such as the pros and cons of using a pregnancy sash. In particular, the use of a hook to remove a fetus was a controversial procedure. For example, Hiruta Katsumei (1746–1817), a principal rival of the Kagawa School, emphasized the cruelty of cutting a fetus with a hook. He also believed that physicians would do more harm than good by using a hook, since they had to handle the instrument inside a mother's body, which was "as if they were operating in the dark."[47]

Physicians who challenged Gen'etsu, however, shared his attitudes toward practicing medicine: being critical, thinking independently, and refusing to accept existing theories without good reason. They also privileged knowledge gained through their own experience and observations. This attitude can be seen as a correlate of their adoption of the new anatomical body. Although they sometimes mentioned *ki* and blood, most obstetricians after Gen'etsu abandoned the holistic understanding of the body represented in *Fujin kotobukigusa*. Discussions about the Yin Yang Five Phases theory and the meridian system disappeared from their writings. Instead, physicians were eager to learn about the body's reproductive anatomy by following the empirical tradition to which Gen'etsu belonged. Obstetricians in general were also convinced of the benefits of active physical intervention during delivery. Although Hiruta Katsumei refused to use surgical instruments, for instance, he considered it crucial to use pertinent hand operations in all cases of childbirth. As such attitudes gained ground, prescribing herbal medicine became a minor part of most obstetricians' practice.

It should also be noted that, even though many physicians challenged Gen'etsu's recommendations, they often framed their discussions by applying the same categories and terms used in *Sanron* and *Sanron yoku*. Most later obstetric texts were organized with subject headings similar to those of *Sanron*. *Sanron* and *Sanron yoku* played an important role in establishing a shared understanding of obstetric terms and issues that were indispensable for meaningful debates among obstetricians.[48] In this way, the emergence of the Kagawa School led to the valorization of a new taxonomy of obstetrics and to the formation of a new order of uniformity in the field in the latter half of the eighteenth century.

Obstetric Practice in Commercialized Society

The lucrative market for skilled obstetricians helped the Kagawa School flourish in the late eighteenth and early nineteenth centuries.[49] Kagawa School physicians after Gen'etsu were unhappy with the hook method because the procedure did not allow them to deliver a living fetus. To overcome this problem, obstetricians continued their search for methods and instruments that would allow them to save both the mother and the child during a difficult labor.

For practicing obstetricians in Tokugawa Japan who desired approval and recognition from their clients, the most important measure of success was the mother's well-being. This led them to prioritize the mother's life over the child's when technological limitations forced them to choose between the two. Unfortunately, the decision to use the hook to save the mother often meant killing the child. The instrument was supposed to be used only when the fetus was already dead, but it seems to have been common practice to use it on living fetuses as well. Evidence for this can be found in the vociferous criticisms of the hook method by obstetricians outside the Kagawa faction, as well as by Kagawa School masters' repeated injunctions against the hook's misuse. To prevent a child's death by way of the hook, Kagawa Mantei (1772–1833), head of the Kyoto branch of the Kagawa House, stipulated a list of "eight rules" to identify whether a fetus was alive or dead (*Shitai hassoku*).[50]

The practice of sacrificing the child's life was tolerated partly because physicians could often hide that the fetus had been alive during the operation from the mother and her family. Gen'etsu and other Kagawa School masters stressed that the operation should be performed beneath a cover to prevent others from observing. In addition, the physician was instructed to prepare a pot to dispose of the fractured fetus, which should be hidden from the eyes of the mother and her family. Gen'etsu's remark in *Sanron* that a dead fetus sometimes utters a cry during an operation is further evidence that in many cases physicians killed a live fetus with the hook. Historical research and folklore studies have shown that very young children were often not considered full human beings, and this made the act of infanticide morally acceptable. This perception may have partly shaped the Kagawa School's prioritizing of the mother's health over the child's.[51]

Although they prioritized the life of the mother over that of the child, physicians still wanted to save the child if possible. Indeed, some Kagawa School physicians worked towards achieving this new goal. For example, Kagawa Mantei and Oku Ressai (1780–1835) invented special procedures to turn the child's body and to pull it out by using only the hands.[52] Mantei also invented a set of instruments called *tangan-ki* ("instrument to find a jaw") to replace Gen'etsu's hook (see Figs. 3.12 and 3.13).

The main part of this instrument was a strong but flexible wire-like material made of a whale's barbel. This instrument was inserted into the pregnant woman's body, with the help of a flat stick, to be hooked on the child's jaw. Once the child was caught, the physician would pull the special handle attached to the barbel to remove the child from the mother's body.[53] A similar method was devised by Mizuhara Gihaku (1782–1864) and was promoted through his extensive publications. Gihaku also devised a net with long flexible handles that could remove a child and the afterbirth in certain situations.[54]

Kagawa Mansō (1796–1864) produced another obstetric tool called *tentō-ken* ("silk cloth to wrap the head") (see Figs. 3.14 and 3.15) to replace the *tangan-ki*, which had the disadvantage of leaving marks on the child's body. The *tentō-ken* consisted of two slender sticks, each having a long hole, and a silk cloth.

The cloth was inserted into the holes and rolled up from both sides so that the sticks came together. The physician inserted the sticks into the mother's vagina and unrolled the silk cloth to wrap the child's head from both sides. When this procedure was complete, the sticks were removed and replaced with a sturdy stick with a hole on the top, into which the cloth was inserted. The stick was then pushed in until it almost touched the child, and the cloth was tightly wrapped around the child's head. Finally, the physician held the stick in the left hand and both ends of the silk cloth in the right hand and gradually pulled the child out. Mansō reportedly assisted in the successful birth of a prince of the imperial family in 1833 by using this instrument—an accomplishment that significantly advanced his rank in the imperial court.[55]

As this case shows, Kagawa School physicians' continued efforts to improve obstetric tools were prompted, to a large extent, by a clientele that was able to grant substantial prestige and remuneration for perform-

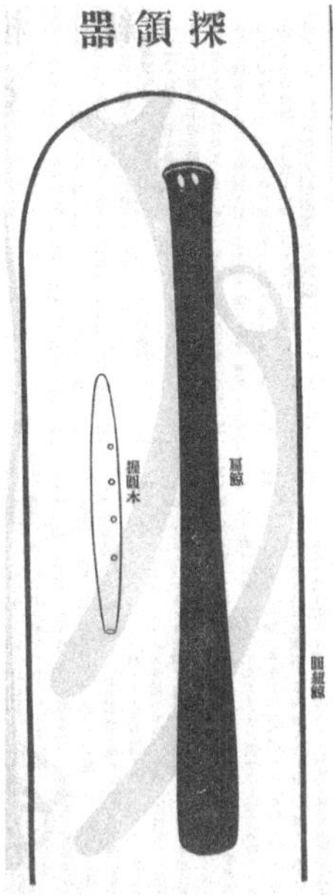

Fig. 3.12 An illustration of *tangan-ki* in a Kagawa obstetric text from the Meiji period, entitled, *Sanka kikai yōhō sho* (1891) by Kagawa Mansai (Source: Fujikawa et al., *Nihon sanka sōsho*, 318–319)

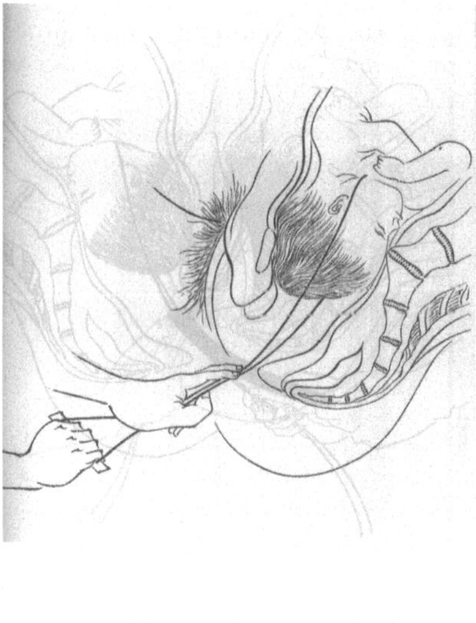

Fig. 3.13 An illustration of a procedure using *tangan-ki* (Source: Fujikawa et al., *Nihon sanka sōsho*, 318–319)

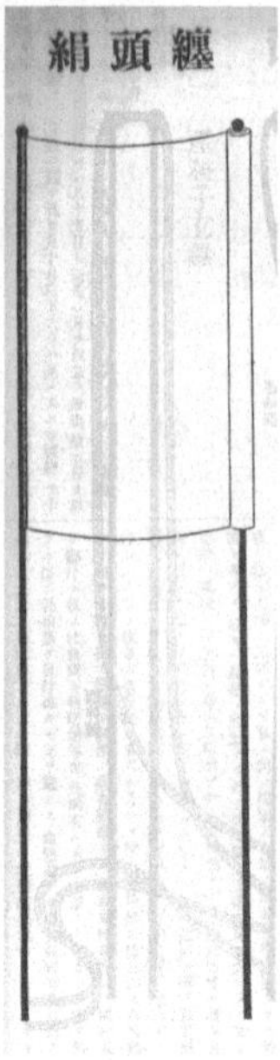

Fig. 3.14 An illustration of *tentō-ken* in *Sanka kikai yōhō sho* (1891) by Kagawa Mansai (Source: Fujikawa et al., *Nihon sanka sōsho*, 318–319)

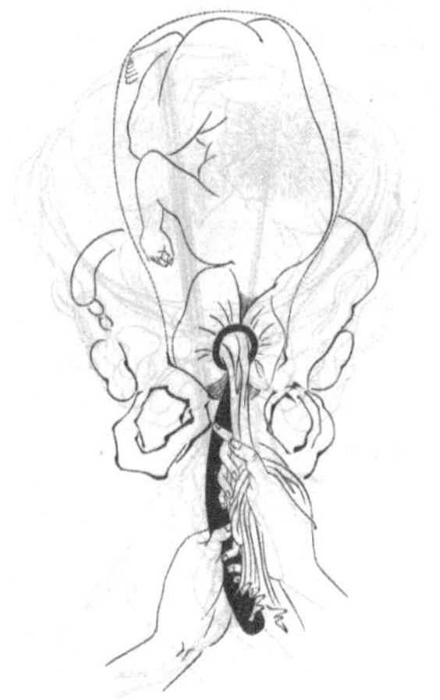

Fig. 3.15 An illustration of a procedure using a *tentō-ken* (Source: Fujikawa et al., *Nihon sanka sōsho*, 318–319)

ing successful deliveries.[56] The prestige endowed by the imperial court was instrumental in attracting large numbers of disciples, which was indispensable for the prosperous operation of a *ryū* establishment. The granting of court rank, however, did not mean government support for the new inventions of the Kagawa House. During the Tokugawa period, the imperial court possessed no governing power. The shogunate and *daimyō* exercised real authority, supported by their economic and military forces. Thus, the imperial court was in no position to promote Kagawa physicians' techniques for general use by the public; it simply recognized their achievements when they managed a successful birth within the imperial family. Since the yearly stipend they received from the court was small, presumably a few dozen *koku* of rice, the Kagawa head masters relied on other clients and their disciples for financial support.

The above examples show that Japanese obstetricians invented adequate tools on their own, even though European obstetric instruments were known to them through Dutch obstetric texts. In some cases, European forceps, levers, and hooks were introduced in obstetric texts written by Japanese physicians in the late eighteenth and early nineteenth centuries. For example, in his book *San'iku hen*,[57] Yamabe Bunhaku described levers and hooks, which were made widely known by the English physician William Smellie (1697–1763).[58] Another obstetrics manual, *Sanka hatsumō* (1795) by Katakura Kakuryō (1750–1822), included illustrations on the use of forceps that had been invented in England. Yet, for reasons that are not completely clear, these instruments were not widely adopted by Japanese physicians.[59] Historian of medicine Ogata Masakiyo attributes this to Kagawa Mantei's invention of the aforementioned *tan-gan-ki*, which Ogata thinks was a very effective tool.[60] It is also conceivable that European obstetric instruments were not practical in terms of technological requirements and cost of manufacture.

The widespread reputation of the Kagawa School prompted many physicians to want to learn its techniques, and some Kagawa School head masters established private academies. The rise of these academies points to a change in the way medical knowledge was disseminated. Such training was provided not only to a small number of chosen disciples, but also to people from diverse backgrounds who could afford to attend the academies. Mainly by completing required courses, disciples earned a certificate that was issued at the time of graduation. A close personal master–disciple relationship was no longer the central feature of instruction at the Kagawa House's *ryū* operation.

Kagawa Mankyō (1734–1793), founder of the Kyoto branch of the Kagawa House, established an academy that admitted over a dozen disciples on a yearly basis in the late eighteenth and early nineteenth centuries.[61] This school's student register [*monseki*] suggests that a total of 950 disciples studied at this academy between 1769 and 1875.[62] Many disciples came from areas relatively close to Kyoto, including the Kyoto-Osaka region, the Chūgoku region, west of the Kyoto-Osaka area of the main island (Honshū), and Shikoku island. However, a considerable number came from domains located in the northeastern section (Tōhoku) of the main island, the Kantō region centered in Edo, and Kyūshū island, indicating that the reputation of the Kagawa School had spread nationwide (see Table 3.1).[63]

The Kyoto Kagawa academy operated like many other private academies in medicine and various branches of scholarship.[64] Disciples were divided into students-in-residence and commuters, and it generally took between two and two-and-a-half years to complete a standard curriculum. Different tuition rates were specified for students in residence and commuters. In 1858, the former paid 100 bronze coins[65] twice a year, supplemented by 1 *ryō* of silver at the beginning of each year and 3

Table 3.1 The number of disciples of the Kyoto Kagawa House

Region year	*Kinki* 近畿	*Chūgoku* 中国	*Shikoku* 四国	*Kyūshū* 九州	*Chūbu* 中部	*Kantō* 関東	*Hokkaiō* 北海道 *Tōhoku* 東北	*Total*
1769–1779	49	37	13	17	25	4	16	161
1780–1789	48	33	15	11	35	5	3	150
1790–1799	21	31	10	5	23	5	2	97
1800–1809	34	36	5	12	30	6	12	135
1810–1819	33	35	11	6	19	3	4	111
1820–1829	20	13	0	1	9	3	2	48
1830–1839	15	7	4	0	17	3	9	55
1840–1849	22	12	11	13	17	2	8	85
1850–1859	17	7	3	7	24	1	8	67
1860–1869	6	0	2	9	12	1	8	33
1870–1875	2	0	0	1	0	0	0	3
Total	267	211	74	82	211	33	72	950

Of a total of 950 disciples, 95 are clearly identified as domanial physicians who were sent for obstetrics training by local domains

Source: Sutō, ***Kagawa ryū kaiseijutsu no kenkyū***, 88, originally published in Sugitachi, "Kagawa Gen'etsu to Kagawa ryū sanka no hatten," in Yamada and Sugitachi, eds., ***Sanron, sanron yoku, dokusan ron***

monme of silver[66] as a summer and winter gift. The entrance fees for commuters were set at 400 *hiki*[67] for two courses of instruction. At the entrance ceremony, each commuting student was also required to pay the following to his or her main master: entrance fees [*Sokushū*], 100 *hiki*; fees for drinks [*Sake ryō*], 300 *hiki*; banquet fees [*Sakana ryō*], 50 *hiki* and 1 *ryō* of silver; fan fees [*Sensu ryō*], 1 *ryō* of silver; fees for teacakes [*Kashi ryō*], 2 *monme* of silver.[68] Secondary instructors received a lesser amount, but the fees were divided similarly. In addition, students gave each servant 100 bronze coins. They paid 1 *ryō* as an honorarium at each seasonal festival for the months of March, May, September, and November, and 50 *hiki* as a summer and winter gift in July and December. Upon leaving the academy, disciples were required to pay additional fees to receive their certificates.[69]

One plausible reason for the tuition fees to be phrased in euphemistic terms was to disguise the commercial nature of the transaction. That is to say, they were paid under the pretense that they were gifts and not fees. The fixed amounts specified for each payment type indicated the degree to which obstetric knowledge had become a commodity for sale, yet also a reluctance to openly admit to the economic nature of the transaction. Whether or not Kagawa masters and disciples were consciously aware of the ongoing commodification of medical knowledge, they must have undeniably had a good understanding of its commercial value, for these fees were quite large, although not prohibitive for affluent commoners and domanial physicians with generous scholarships from the local governments they served.[70]

The institutionalization of academies with fixed tuition fees most likely facilitated the entry of commoners and women into the profession, since the provision probably made the school more open to disciples from diverse backgrounds who possessed the necessary financial means.[71] It is difficult to know the ratio of disciples from each status group by reviewing the names listed in the academy's student registry because almost all the names presented in the list include a family name. This was because physicians customarily used a family name, even though the shogunate in principle specified the official use of family names as a privilege granted only to warriors. There were exceptions, however. A man named Shigeichi, the son of a nurseryman named Heijirō lacking a family name, was admitted to the academy in 1856.[72] The evidence indicates that the commercialization of medical instruction may have contributed to an opening of opportunities for commoners to obtain the prestigious formal training provided at the academy instead of being limited to the indirect and unstructured training given by practicing Kagawa School physicians.

Moreover, the Kagawa academy permitted women as disciples and officially initiated them into the Kagawa ryū. The students' registry indicates that nine women joined the training program between 1792 and 1861.[73] Biographies published after the 1868 Meiji Restoration also indicate the presence of female disciples in the Kagawa School. The author of one such biography, a midwife named Miyama, claimed that the Nakaoka family produced talented female obstetricians for nine generations, beginning in 1741 and continuing until the turn of the twentieth century. In particular, a seventh-generation Nakaoka physician named Kosen gained mastery in obstetric skills after receiving her training under a renowned Kagawa School physician. Because of her outstanding reputation, she was summoned by the deputy governor of Osaka castle to attend to the deliveries of women there. Miyama also suggested that some Nakaoka doctors adopted capable female disciples as their heirs.[74] Thus, there were female obstetricians with a high level of professional competence during the Tokugawa period, indicating an absence of institutionalized sex discrimination in the medical field.

The State, Physicians, and Reproductive Surveillance

Kagawa School physicians pursued their innovations without active state intervention. The vast majority did not hold any official position within the shogunate or domanial governments, nor did they receive any financial support. Although some daimyō houses sought the services of Kagawa physicians for the women in their families, such recruitment did not necessarily entail an appointment to an official position. Nor did Kagawa School physicians play a significant role in implementing the reproductive surveillance that the shogunate and some domains introduced in the eighteenth and early nineteenth centuries. In these pronatalist reproductive campaigns, shogunal and domanial authorities sought to regulate people's reproductive practices by indoctrinating and punishing women, their families, and community leaders for abortion or infanticide but not by policing medical practitioners who assisted with such practices. Some domanial authorities made sporadic attempts to mobilize physicians to inspect dubious cases of miscarriages and stillborn births, but the weight of physicians' medical opinion remained marginal within the overall system of pregnancy investigation.

As the 1767 *bakufu* edict to warn against infanticide demonstrates, from the mid-eighteenth century onwards, fertility control that was performed in the form of abortion and infanticide raised serious concerns on the part of state officials and intellectuals, who viewed these practices as the cause for a decline in population.[75] The shogunate and the domains realized that any reduction in the size of the farming populace would lead to an erosion of their tax base. At the same time, they viewed acts of abortion and infanticide as a grave infringement of the Confucian moral tenets that constituted the ideological foundation of Tokugawa rule. Thus, state administrators in some domains and shogunate districts implemented campaigns to inculcate peasants in the evils of infanticide and the moral and material merits of having many children. In particular, state authorities produced popular pamphlets of didactic tales, the contents of which were preached to peasants by state officials, village headmen, and Buddhist priests.[76]

Realizing the limited effectiveness of moral persuasion, some domains and shogunate districts resorted to a policy of reproductive surveillance with a provision to grant compensation for bearing a child. A system of pregnancy supervision usually required families to report pregnancies to an inspection official, who would record each case in a registry. From the time of registration until full term of pregnancy, women were subject to regular inspections by administrators, who were dispatched by a shogunal magistrate or a domanial government. It was required that miscarriages, premature births and stillbirths be reported to administrators, commonly through the auspices of a neighborhood organization called the five people group [*goningumi*] and village officials. An official investigation would then be conducted. The birth of a child also had to be reported, after which the family of a newborn would receive a certain amount of money or rice as remuneration. A monthly allowance continued to be granted, in many cases until the child was several years old. Depending upon the area, the provision for such child support varied in terms of form, amount, and duration.[77] This policy of reproductive surveillance relied on the vigilant eyes of people in the community to ascertain the accuracy of pregnancy reports. In addition to the pregnant woman and her family, the neighbors, the five people group to which the woman's family belonged, and the village headman were held responsible for failing to report pregnancies, miscarriages, and stillbirths.[78] Thus, the family and the village community, not medical professionals, were the major means through which state authorities enforced reproductive surveillance.

In the Tsuyama and Sendai domains, where reproductive control was probably more rigorous than in most other domains, physicians were mobilized for the inspection of miscarriages and stillbirths.[79] The historian

Sawayama Mikako suggests, however, that a physician's investigation of a pregnancy that did not carry to full term in Sendai domain consisted solely of an oral interview with the woman. There was no physical examination.[80] In most other domains and shogunal districts, such inspections were conducted by state officials without the assistance of physicians. This indicates that within the Tokugawa penal system, a medical professional's testimony was not required for the state to define a case as a crime, despite sporadic efforts on the part of state authorities to use physicians' expertise and authority. The role played by physicians in the pregnancy investigation system was thus peripheral.

In the absence of state supervision, the vast majority of physicians and midwives performed infanticide and abortion instead of working together with state authorities to prevent them. The shogunate and some domains made occasional efforts to punish midwives and physicians for their involvement in abortion. It seems, however, that state authorities brought suit against midwives who assisted with abortion only when the case involved the death of the mother.[81] The number of cases in which a midwife was punished for performing an abortion was also very small.[82] Thus, it may be that the state generally lacked the will and the means to circumvent the pervasive practice of abortion performed by midwives or abortion providers. The flourishing of abortion businesses in Edo from the eighteenth century is reflected by the many satirical verses called *senryū*, which referred to the practices of abortion doctors.[83]

The following *senryū* shows that many concubines of the shogun, retired shoguns, high-ranking samurai, and prostitutes used their services: "*Otsubone no onnna isha to wa sumanukoto.*" There are two ways to translate this: "Apologies for a high-ranking concubine to use the service of an abortion doctor (presumably for having affairs with men other than the shogun)," and "A high-ranking concubine cannot get away with it just by seeing the abortion doctor when she becomes pregnant from an affair with a lover." The key to reading this verse is the pun on the word, "*sumanu*" that has the double meaning of "sorry" and "cannot get away with it just by." Another verse shows that the business of abortion providers was quite lucrative: "*Chūjō wa mugottarashii kura wo tate*": "Abortion doctors build cruel storehouses," indicating they accumulate wealth by running a "cruel" business.

Another indication of the widespread nature of abortion was a complaint from a group of cotton-spinning entrepreneurs in the Yamato region, in today's Nara prefecture, about female workers who reportedly engaged in illicit sexual liaisons and regularly sought assistance from abortion specialists called "*Oroshi-ya*."[84] In this way, while some domains and *bakufu* districts succeeded in regulating reproductive practices in castle towns and farming areas where rigorous surveillance was possible, most failed to control the sexual and reproductive practices of city dwellers and migrant laborers.

Physician-assisted abortion was mentioned in the Kagawa House regulations for its disciples, where it was described as an unlawful operation frequently performed for profit by physicians outside the Kagawa School circle. A pledge not to perform abortions was one of the six cardinal oaths that physicians were required to take when they were initiated as disciples of the Kagawa School.[85] Such a prohibition was most likely intended, in part, to distance Kagawa physicians from disreputable abortion doctors.[86] Thus, it was through the efforts of the Kagawa ryū establishment and not those of state authorities that physicians were directly cautioned against assisting in abortions. In the absence of an effective state policy to prohibit them from providing the service, however, they could always perform abortions discreetly. To what extent Kagawa physicians adhered to the admonition issued by the *ryū* establishment is difficult to determine.

In this chapter and in Chap. 2, I explore the development of the field of reproductive medicine in the Tokugawa period. In a society dominated by samurai, physicians were placed in a subordinate position to the samurai elite. The medical field, however, saw a significant development and expansion that occurred in conjunction with a briskly growing economy and a flourishing of both popular and high culture. An increasing number of Japanese physicians engaged in the study of Chinese medical classics, and many began to popularize the understanding of the human body and medicine they found in these texts. This helped make the medical profession more respectable.

The publication and wide circulation of childbirth manuals, such as Katsuki Gyūzan's *Fujin Kotobukigusa*, reinforced the Chinese classical medical understanding of the body both within and outside the fields of obstetrics and gynecology. The rise of medicine as a profession and an increase in the influence of medical doctors over lay patients generated a new understanding of the physiology and anatomy of the body and of the cosmology in which the body is located. In this process of medicalization, the expectant mother's body became an important object of treatment and management by medical experts.

In the mid-eighteenth century, the rise of the Kagawa School led to a radical transformation in the field of obstetrics. Borrowing methods from the Ancient Practice School, Kagawa School physicians challenged the conventions established by the Goseihō School. Instead of relying on concepts of the body and therapeutic processes elaborated in Chinese medical classics, Kagawa School obstetricians developed their methods based on their own observations and experiments. The Goseihō School understood the body through cosmological categories that systematically linked physiological phenomena to the shifting climate of the larger universe outside the individual body. The Kagawa School, in contrast, understood the body as a self-contained, machine-like entity inscribed with new anatomical knowledge, similar to the model used in the modern European medical tradition. This shift entailed the rise of a body marked with reproductive organs and pelvic anatomy specific to each sex. In other words, this sexualization of the body was part of a larger process of medicalization that remade women's bodies as objects of medical study and treatment.

The increased medicalization of women's bodies, however, had little effect on shogunate or local domain authority governance of reproduction and women's bodies. Indeed, one feature that characterizes reproductive surveillance from the Tokugawa period is the lack of medical language to describe women's bodies and reproduction. Legal discourse on abortion and infanticide did not take into account the health of either the mother or child. Physician involvement in government surveillance of reproduction was marginal at best.

The measures for reproductive supervision that the shogunate and domain authorities formulated were predominantly prohibitive rather than productive in nature. Policy measures, for the most part, consisted of a series of bans aimed at preventing abortions and infanticides. The state made almost no effort to help ensure a healthy pregnancy, proper development of the fetus, or the safe delivery of the child, and little effort was put into preventive health measures for infant mortality and morbidity. There was no concept of public health in modern terms in the area of maternal and infant health and, thus, no organized effort to protect and enhance the health of mothers and infants. Some domains provided funds to encourage families to produce more children, but this was meant to subsidize the cost of childrearing not to enhance the health of the mother during pregnancy or, more generally, that of women who were reproductively active.

Because they were regulated by prohibitions and taboos, women and their families from the early modern period remained passive subjects rather than voluntary participants in government reproductive policies. The medicalized surveillance that would emerge in the modern period, on the other hand, was accompanied by recreating individuals as active, motivated agents of self-surveillance and improvement. In Tokugawa childbirth and women's manuals, we find passages devoted to Confucian encouragement of self-cultivation in the mother; the bulk of advice for the mother in these writings, however, takes the form of straightforward prohibitions. Mothers are assumed to be passive, submissive subjects who reluctantly obey bans and taboos. It was only after the establishment of the Meiji state that women became thinking, self-motivated citizens who regarded "proper" maternal and infant health as an important duty for the nation and society.

Modern efforts by the Japanese government to regulate fertility involved a partnership between the state and medical experts. It is true that obstetricians and gynecologists from the Tokugawa period could use their expert knowledge to exert a certain amount of influence on mothers and their families. Authorities from the shogunate and various domains, on the other hand, seem to have implemented reasonably successful measures for preventing infanticides in rural areas, if not abortion practices in large cities.[87] But government authorities did not enlist physicians or appropriate expert medical knowledge for the purpose of reproductive supervision. In other words, both the state and medical specialists attempted to increase their control over women's reproductive bodies, but these two strands of authorities did not combine or cooperate in any meaningful way.

After the 1868 Meiji Restoration, the state began to work on maternal health in close collaboration with obstetricians, gynecologists, and pediatricians. This does not necessarily mean that the state was able to tighten its reproductive surveillance or make it more successful. Historian Ishizaki Shōko suggests that the Meiji government did not allocate any significant budget to the improvement of maternal health. Fabian Drixler argues that there was an increase in the number of infanticides during the Meiji period owing to the termination of pregnancy surveillance programs and the elimination of childrearing subsidies that many domains had implemented during the Tokugawa period.[88] The importance of a new alliance between the state and medical experts, however, is not the effectiveness of the new reproductive policies but the creation of a new matrix of power governing women and infants. This new conceptual and institutional template allowed the Japanese state to become more interventionist in its control over the reproduction of its citizens from the early to mid-twentieth century. The Japanese government's reproductive policies during the twentieth century oscillated widely—from pronatalism to anti-natalism, and back to pronatalism once more. But there has been a consistent partnership between the state and organized medicine that has been central to the proliferation of modern bio-power, and this partnership also helped create government reproductive policies that often privileged the interests of the state and experts over those of women.

In the chapters that follow, I discuss the restructuring of the field of reproductive medicine as well as reproductive practices and policies that took place after the Meiji Restoration. The changes that occurred in reproductive medicine were part of the larger transformation of the medical field that occurred in tandem with the formation of a modern nation state. The introduction of the idea of public health and modern techniques of governance, such as universal household registration and the gathering and managing of statistical data, marked an entirely new way of managing citizens' bodies. In Chap. 4, I explore this transformation of the field of health and medicine in the latter half of the nineteenth century, then discuss the new types of maternal and infant health policies and practices, focusing on the rise of modern childbirth practice involving licensed midwives trained in modern medicine.

Notes

1. There are different opinions about how to define "Kohō" and "Koihō." In this study the notion of the Ancient Medicine School encompasses both factions.
2. Thomas Kuhn, *The Structure of Scientific Revolutions* (Chicago: The University of Chicago Press, 1962).
3. Michel Foucault, *The Order of Things: An Archaeology of the Human Sciences* (New York: Vintage/Random House, 1973); and *The Archeology of Knowledge* (New York: Harper Colophon, 1972).
4. Maruyama Masao, *Nihon seiji shisōshi kenkyū* (Tokyo: Tokyo Daigaku Shuppankai, 1952): 71–139. For an English translation of this work, see Masao Maruyama, *Studies in the Intellectual History of Tokugawa Japan* (Princeton, NJ: Princeton University Press, 1974).
5. The author of this text is Chōki [Ch Zhang Qi, 142–220] who is also referred to as Chō Chūkei [Ch Zhang Zhongjing]. Paul U. Unschuld,

Medicine in China: A History of Ideas (Berkeley, Los Angeles, and London: University of California Press, 1985): 168–169.

6. Fujikawa Yū, *Nihon igakushi* (Tokyo: Shinrisha, 1941): 296–299; Sakai Shizu, *Nihon no iryōshi* (Tokyo: Tokyo Shoseki, 1982): 237–238.
7. Sakai, *Nihon no iryōshi*, 92–93. Wada Masatsugu, "Kakumei no iketsu, Yoshimasu Tōdō," Tōa Igaku Kyōkai, ed., *Nihon no kanpō wo kizuita hito-bito, Kanpō no Rinshō*, Vol. 9, No. 11–12 (December 1962): 90–94.
8. For the section of Yoshimasu Tōdō, Sakai, *Nihon no iryōshi*, 240–248; Fujikawa, *Nihon igakushi*, 352–355.
9. *The Yellow Emperor's Inner Canon*, one of the greatest medical classics of Chinese medicine was reportedly written between the third and first century BCE. This work is divided into *Somon* [Ch *Su-wen*], which discusses basic theories, and *Reiku* [Ch *Ling-shu*], which elaborates the methods of acupuncture and moxibustion. For an English translation, see Ni Maoshing, *The Yellow Emperor's Classic of Medicine: A New Translation of the Neijing Suwen with Commentary* (Boston: Shambhala, 1995).
10. Kyoto-fu Ishikai, ed., *Kyoto no igakushi* (Kyoto: Shibunkaku, 1980): 958–959; Sōda Hajime, *Zusetsu nihon iryō bunkashi* (Kyoto: Shibunkaku, 1989): 127, 165, 168.
11. Tōyō was able to obtain permission to conduct a dissection because he was a friend and teacher to domanial physicians of the Obama domain daimyō, Sakai Tadamochi, who was the governor of Kyoto [*shishodai*] at the time. A request was presented to the daimyō by these physicians to the domain, and permission was granted by Tadamochi. Achiwa Gorō, *Kindai nihon no igaku: seiyō igaku juyō no kiseki* (Kyoto: Shibunkaku, 1982): 94–105; Sakai, *Nihon no iryōshi*, 241–242; Sōda, *Zusetsu nihon iryō bunkashi*, 159–169; Kyoto-fu Ishikai, ed., *Kyoto no igakushi*, 962–970; Ōtsuka Yasuo, "Yamawaki Tōyō," *Kinsei Kanpō igakusho shūsei 13: Gotō Gonzan, Yamawaki Tōyō* (Tokyo: Meicho Shuppan, 1979): 18–34.
12. Achiwa, *Kindai nihon no igaku*, 98–101. Japanese physicians' and scholars' curiosity in understanding the mechanism [Jp *karakuri*] of European-made clocks, music boxes or clockwork dolls is not unrelated to the type of intellectual interest that Tōyō and other physicians demonstrated in their desire to figure out the inner workings of the human body. This interest was not a monopoly of physicians of Dutch medicine. See Timon Screech, *The Lens within the Heart: The Western Scientific Gaze and Popular Imagery in Later Edo Japan* (Honolulu: University of Hawai'i Press, 2002). See, particularly, Chap. 3, "Mechanics and Motions," 61–93.
13. For an early work on this topic, see Uiriamu D. Jonsuton (William D. Johnston), "Jūhasseiki nihon no igaku ni okeru kagaku kakumei: Ranpō hatten no tame no shisō teki na zentei, Part 1 and 2," Nihon ishigaku zasshi, Vol. 27, No. 1 (1981): 6–20; and No. 2, 131–156.

14. On a development in Confucian and scientific thought that entailed the revival of the notion of *ri* in the late Tokugawa period, see also Albert Craig, "Science and Confucianism in Tokugawa Japan," in Marius B. Jansen, ed., *Changing Japanese Attitudes Toward Modernization* (Princeton: Princeton University Press, 1965): 133–160. Ng Wai-ming presents an interesting discussion on how Chinese cosmological categories that derived from *The Book of Changes* [Jp *Ekikyo*; Ch *Yijing*] mediated the adoption of European scientific knowledge. See Ng Wai-min, *The I Ching in Tokugawa Thought and Culture* (Honolulu: Association for Asian Studies and University of Hawaii Press, 2000).
15. After the death of its founder, Gen'etsu, the Kagawa School was divided into two major branches. One of the branches served a daimyō house on Shikoku island for generations. The other branch in Kyoto gained immense prestige by serving the imperial court and attracted many disciples.
16. For existing works on the Kagawa School, see my discussion in the introduction. Feminist studies on the Kagawa School other than the ones I have discussed include: Sutō Mikako, "Edo jidai no shussankan, taijikan," in Fujiwara Noboru, ed., *Nihon seikatsu shishi kenkyū, Vol. 2.* (Morioka: Seikatsu-shisō Kenkyū kai, 1990): 61–94; and " 'Umu' shintai no kindai," *Gendai shisō* 19-3 (March 1991): 220–229.
17. Gen'etsu's biography is documented in Fujikawa, *Nihon igakushi*, 371–373; Fujikawa Yū, Kure Shūzō, and Masuda Tomomasa, eds., *Nihon sanka sōsho* (Kyoto: Shibunkaku, first published in 1895, reprinted in 1971): 5–6; Ogata, *Nihon sanka gaku shi* (Tokyo: Maruzan, 1918, reprinted by Kagaku Shoin, 1980); 108–109.
18. Fujikawa, *Nihon igakushi*, 372; Ogata, *Nihon sanka gaku shi*, 108. The Hikone domain did not allow, at the time, sons of concubines to inherit the father's family name and status.
19. *Sanron* is written in the *kanbun* style (Japanese written in the form of classical Chinese) that was customarily used in scholarly writing. This most likely helped enhance Gen'etsu's methods' prestige. For the significance of writing scholarly work in the *kanbun* style, see Sakai Shizu, "Edo jidai no seiyō igaku no juyō," in Yoshida Tadashi, ed., *Higashi ajia no kagaku* (Tokyo: Keisō Shobō, 1982): 5–49.
20. Gen'etsu's stipend was about half that received by noted scholars such as Arai Hakuseki (1657–1725) while in the service of the Kōfu domain during his early career, and was comparable to that received by some middle ranking retainers serving Chōshū domain, according to information dated 1852. Kate Wildman Nakai, *Shogunal Politics: Arai Hakuseki and the Premises of Tokugawa Rule* (Cambridge, MA: Council on East Asian Studies, Harvard University, 1988): 29–38; Kimura Motoi, *Kakyū bushi ron* (Tokyo: Hanawa Shobō, 1967): 50–53, 58–76.

21. Sanka Bunken Dokushokai, ed., *Heiseiban sanron, sanron-yoku* (Tokyo: Iwata Shoin, 2008): 18–19.
22. Ibid.
23. By "cross bones," Gen'etsu most likely referred to a section of pubic bones. Sanka Bunken Dokushokai, ed., *Heiseiban sanron, sanron-yoku*, 72–73.
24. 1 *sun* = 3.03 centimeter = 1.193 inch; 10 *bu* = 1 *sun*.
25. Gen'etsu also disapproved of the use of the birth chair after delivery, which was another custom recommended by many childbirth manuals including *Fujin kotobukigusa*. The cover of this book portrays this custom.
26. Sanka Bunken Dokushokai, ed., *Heiseiban sanron, sanron-yoku*, 30–31.
27. Gen'etsu remarked, "I looked at the illustrations of internal anatomy that the red-haired barbarians brought ['*kōi tsutauru tokorono naikei zu*']."
28. Sanka Bunken Dokushokai, ed., *Heiseiban sanron, sanron-yoku*, 24–27.
29. As I discuss later in this chapter, there were female obstetricians who were trained in the Kagawa School tradition.
30. Sanka Bunken Dokushokai, ed., *Heiseiban sanron, sanron-yoku*, 26–27.
31. The conception meridian was the only meridian that Gen'etsu refers to in *Sanron*. His discussion of this meridian, however, does not demonstrate a systemic understanding of traditional Chinese medicine.
32. Modern anatomy does not recognize sexual difference in terms of the size of this "gap" that is connected by the pubic symphysis, but identifies sexual disparity in the form of the pubic arch shaped like an inverted V. The angle forming this pubic arch is more acute in the male pelvis, while the female pelvis generally pertains to a broader and rounder pubic angle. Elain N. Marieb and Jon Mallatt, *Human Anatomy*, Second Edition (Menlo Park, CA: Benjamin/Cummings, 1997): 173–178.
33. Sanka Bunken Dokushokai, ed., *Heiseiban sanron, sanron-yoku*, 24–25.
34. This work was a Japanese translation of a Dutch version of the anatomy text, *Anatomische Tabellen* (*Third Edition*, 1732) by the German natural scientist Johann Adam Kulmus (1689–1744). It was translated into Dutch by the physician Gerardus Dicten and was published as *Ontleedkundige Tafelen* in Amsterdam in 1734. *Kaitai shinsho* was the first Dutch medical text whose full translation was published in Japan. Nihon gakushi-in, *Nihon igaku shi, Vol. 1* (1955): 152.
35. Shigehisa Kuriyama, "Between Mind and Eye: Japanese Anatomy in the Eighteenth Century," in Charles Leslie and Allan Young, eds., *Paths to Asian Medical Knowledge* (Berkeley, Los Angeles, and Oxford: University of California Press, 1992): 21–43.
36. Ogata, *Nihon sanka gaku shi*, 304.
37. Medical historians speculate that Gen'etsu attended Yamawaki Tōyō's lectures at his residence in Kyoto, and saw Dutch texts in Tōyō's possession. The historian Ogata Masakiyo suggests the possibility that Gen'etsu was inspired to invent his iron hook by viewing illustrations in a Dutch obstet-

rics text written by Hendrik van Deventer (1651–1724). Ogata points out, however, that the illustration of the hook was presented by Deventer to warn against its use. Ogata assumes that because Gen'etsu was unable to read the text, he thought the hook was recommended for obstetrical purposes. Ogata, *Nihon sanka gaku shi*, 133–135.

38. Judith Walzer Leavitt, "Birthing and Anesthesia: The Debate over Twilight Sleep," in Leavitt, ed., *Women and Health in America* (Madison: University of Wisconsin Press, 1984): 175–184; Richard W. Wertz and Dorothy C. Wertz, *Lying-In: A History of Childbirth in America* (New Heaven: Yale University Press, 1977): 116–119, 137, 178–197, 289. In Japan, the use of anesthesia was pioneered by "external medicine" (*geka*, surgery) physicians who had learned certain methods from Dutch physicians in Nagasaki in the late sixteenth century and early seventeenth centuries. They provided local anesthesia by means of a prescription of *mandarage* (thorn apple). The physician Hanaoka Seishū (1760–1835) was the first to perform general anesthesia with a formula that consisted of *mandarage* and aconite, which he used for mastectomy operations for treating breast cancer, beginning in 1804. Sōda, *Zusetsu nihon iryō bunka shi*, 227–232; Nihon Gakushi-in, Nihon Kagakushi Kankōkai, ed., *Meiji-zen igaku shi, Vol. 4* (Tokyo: Nihon Gakujutu Shinkōkai, 1964): 807–808.
39. Sanka Bunken Dokushokai, ed., *Heiseiban sanron, sanron-yoku*, 84–85.
40. Ibid., 84–85.
41. Modern medical statistics suggest that abnormal fetal positions are found in 3.5% of births; breech presentations account for 3% and transverse presentations for 0.5%. *Nanzandō igaku daijiten* (Tokyo: Nanzandō, 1966): 514, 934.
42. Sanka Bunken Dokushokai, ed., *Heiseiban sanron, sanron-yoku*, 84–87.
43. Sanka Bunken Dokushokai, ed., *Heiseiban sanron, sanron-yoku*, 86–87.
44. Ibid., 86–89.
45. One of these texts was a presentation of Gen'etsu's oral lectures, *Fujin sanzen sango fukushin shujutsu hō* [*Methods of Diagnosis and Operations for Women before and after Childbirth*, the transcriber and hand-written year unknown], reprinted in Fujikawa et al., eds., *Nihon sanka sōsho*, 133–143; Oku Ressai, *Kaisei kōhō hiketsu* [*Keys to the Use of the Hook for Restoration*, publication date unknown], ibid., 764.
46. Ochiai, "Edo jidai no shussan kakumei," "Kinsei matsu ni okeru mabiki to shussan," and "The Reproductive Revolution at the End of the Tokugawa Period"; Sutō "Edo jidai no shussankan, taijikan," in Fujiwara Noboru, ed., *Nihon seikatsu shisō kenkyū, Vol. 2* (Morioka: Seikatu-shiso Kenkyūkai, 1990): 61–94; "'Umu' shintai no kindai," *Gendai shisō*, Vol. 19-3 (March, 1991): 220–229.
47. Ogata, *Nihon sanka gaku shi*, 224.

48. Sutō Mikako's discussion on the major obstetric texts of the Kagawa School and others includes their organization and subject headings. See Sutō Mikako, *Kagawa-ryū kaiseijutsu no kenkū: Edo jidai no shussan kan, taiji kan ni kansuru ichi kōsatsu* (M. A. Thesis, Ochanomizu University, 1990): 37–126. The two texts of the school founded by Hiruta Katsumei also demonstrate that Hiruta generally followed the classification of childbirth established by Kagawa Gen'etsu. See *Sanka shinron* (1815) and *Hiruta sensei sanjutsu kunkai* (publication year unknown) in Fujikawa et al., eds., *Nihon sanka sōsho*, 790–843.
49. In his obstetrics manual, *Sanka shinron* (1819, a text different from the Hiruta School), physician Tateno Ryūtei (date of birth and death unknown) remarked that eight or nine out of ten obstetricians belonged to the Kagawa School tradition. Kyoto-fu Ishikai, *Kyoto no igakushi*, 1103.
50. *Sanka kibun* (the Bunka era, 1804–1814) reprinted in Fujikawa et al., eds., *Nihon sanka sōsho*, 213–217.
51. Some research in Folklore Studies suggests that the Japanese did not consider a child under seven years of age as a human being. Some scholars refer to this concept to explain why the practice of infanticide, commonly called *mabiki* (the culling), was pervasive. The act of infanticide was also often referred to as "returning" [*kaesu*] from the world of humans to the world of gods [*kami*]. For the notion of the human life cycle in the Japanese folk custom, see Herman Ooms, "A Structural Analysis of Ancestral Rites and Beliefs," in William Newell, ed., *Ancestors* (Chicago: Aldine, 1976): 61–90. Chiba Tokuji and Ōtsu Tadao extensively discuss the Japanese folk perception of the fetus and the child in Chiba Tokuji and Ōtsu Tadao, *Mabiki to mizuko* (Tokyo: Nōson gyoson bunka kyōkai, 1983). William LaFleur analyzes the perceptions of the fetus in certain Buddhist thought in both the early modern and modern periods. Some recent scholarship in history and anthropology has questioned the theory that age seven was the definitive point at which the child came to be perceived as a full human, and that infanticide was acceptable; however, most scholars still maintain that the fetus and the baby were not considered as fully human in early modern Japan. William LaFleur, *Liquid Life: Abortion and Buddhism in Japan* (Princeton, NJ: Princeton University Press, 1992). For a good summary of various views on this issue, see Sawayama Mikako, *Shussan to shintai no kinsei* (Tokyo: Keisō Shobō, 1998): 20–24. For a recent English-language work on Tokugawa practices of infanticide, see Fabian Drixler, *Mabiki: Infanticide and Population Growth in Eastern Japan, 1660–1950* (Berkeley: University of California Press, 2013).
52. Ogata, *Nihon sanka gaku shi*, 355–362, 372–374.
53. Ibid., 353–359. Ogata, *Nihon fujinka-gaku shi* (Tokyo: Maruzen, 1914, reprinted, Tokyo: Kagaku Shoin, 1980): 275–279.

54. Fujikawa et al., eds., *Nihon sanka sōsho*, 616–620.
55. Ogata, *Nihon sanka gaku shi*, 279–280. The Kyoto branch of the Kagawa House, which was established by Gen'etsu's natal son Mankyō (1734–1793) did not serve any daimyō house; however, some of its head masters received the official rank of physician for serving in the imperial court. During the Tokugawa period, in addition to the officers at the Health Ministry [Tenyaku-ryō], there were 30 hereditary medical families that provided court physicians and medical administrators, and another ten medical families that were recruited on a temporary basis due to their superior skills. The head masters from the Kyoto Kagawa House were among the ten medical houses whose services were sought after by the imperial family. Gen'etsu's grandson, Kagawa Mantei (1772–1833) first served the court as a masseur at the age of 45 in 1815, and later gained the title of medical officer specializing in illnesses specific to women and in charge of the education of midwives and nurses serving within the court [*Nyoi hakushi*] with the sixth rank. Mantei's son Mansō (1796–1864) began his career as a court physician with a very low rank in his late twenties but eventually gained the title of *Nyoi hakushi* and the fifth rank. Mansō's heir Mansai (1830–1891) rose to the sixth rank by his late twenties. The ranks received by these Kagawa head masters were comparable to those granted to other medical houses that had newly obtained status as a court physician, but were, on the whole, slightly lower than the ranks that were given to the hereditary medical houses. Kyoto-fu Ishikai, ed., *Kyoto no igakushi*, 1245–1342; Sōda, *Zusetsu nihon iryō bunka shi*, 25–29; Aoyagi Seīchi, *Shinryō hōshū no rekishi* (Kyoto: Shibunkaku Shuppan, 1996): 116–117.
56. There is virtually no record of the fees that Kagawa School physicians generally charged for assisting a delivery. It is plausible that fees varied greatly depending upon the reputation of individual physicians and the wealth of clients. Until the mid-nineteenth century, it was also common that patients remunerated physicians in the form of summer and winter gifts instead of paying for each visit.

    As of 1842, a type of female physician called *onna isha*, who was most likely a professional midwife and abortion specialist, reportedly charged between one and two *bu* (a quarter *ryō* and half *ryō*) for a visit to a pregnant woman. The *onna isha* also very often provided in-patient care at her residence. She charged one-and-a-half *ryō* for a patient to stay at her house for seven days, including fees for medical services and food. These fees, however, were lowered for poor patients. Aoyagi, *Shinryō hōshū no rekishi*, 136–185. Kagawa School obstetricians were most likely more socially established than these female midwives, but it is not certain if their higher social position always led them to collect higher fees.

57. Bunpaku's book was based on readings of Japanese, Chinese and Dutch texts. The publication year of this text is still unclear. Although some sources indicate that it was published during the Meiwa era (1764–1771), medical historians Kaji Kanji and Fujī Naohisa think that this was unlikely because the text demonstrates an unmistakable influence of Kagawa Gen'etsu's *Sanron*. They speculate that it was published during or after the Bunka era (1804–1817). Nihon Gakushi-in, Nihon Kagakushi Kankōkai, ed., *Meiji-zen Nihon igaku shi, Vol. 4*, 143–144.
58. Fujikawa et al., eds., *Nihon sanka sōsho*, 136–141.
59. The German physician Philipp Franz Barthazar von Siebold (1796–1866), who resided in Japan during 1823–1829 and 1859–1862, brought European style forceps, which a few of his disciples put to use. Ogata Masakiyo, *Nihon sanka gaku shi* (Tokyo: Maruzan, 1918, reprinted by Kagaku Shoin, 1980): 478–481. For an English-language study on Siebold and his Japanese disciples, see Ellen Gardner Nakamura, *Practical Pursuits: Takano Choei, Takahashi Keisaku, and Western Medicine in Nineteenth-Century Japan* (Cambridge, MA: Harvard University Asia Center, 2006).
60. Ibid., 282–283. Ogata also argues that the introduction of *tangan-ki* led to the renewed popularity of the Kagawa School in the field of obstetrics.
61. Ogata, *Nihon sanka gaku shi*, 156–185.
62. Kyōto-fu Ishikai, ed., *Kyōto no igakushi* (Kyoto: Shibunkaku, 1980): 1119.
63. Ogata, *Nihon sanka gaku shi*, 156–185.
64. For the operation of various private academies of Tokugawa Japan, see Richard Rubinger, *Private Academies of Tokugawa Japan* (Princeton, NJ: Princeton University Press, 1982).
65. 100 *mon*. 1 gold *ryō* = 4000 *mon* = 4 *kan*.
66. 60 *monme* of silver was equivalent of approximately1 *ryō* although the exchange rate was subject to the fluctuation of the market price of gold and silver.
67. The currency unit *hiki* was commonly used for the measurement of gifts and donations, including tuition fees for private academies. 1 *hiki* = 10 *mon*. 400 *hiki* = 1 *ryō*.
68. In the earlier period, it was common at the time of entrance into the academy for students to offer their teachers gifts in the form of food and drink instead of cash payment—a practice that continued throughout the Tokugawa period in small academies or those located in rural areas. Rubinger, *Private Academies of Tokugawa Japan*, 70–71. This particular Kagawa academy required cash payments instead of gifts; however, cash payments were made under the pretense that they were to be used for buying food and drink. Fees for a fan generally indicate instruction fees. Rubinger notes that in some cases students from warrior families continued the practice of offering a fan to his teacher. Ibid, 70.

69. The payment to their master at the time of graduation included: fees for issuance of the certificate, 100 *hiki*; banquet fees, 100 *hiki*; fan fees, 1 *ryō*; fees for teacakes, 1 *ryō*. Another instructor was to receive: banquet fees, 500 *hiki*; fan fees, 1 *ryō*; fees for teacakes, 2 *monme* of silver. They were also required to give 100 bronze coins to each servant.
70. The total fees paid by the disciples of the Kyoto Kagawa academy seem to be more than those for most private academies of medicine and other branches of scholarship. For example, while the Dutch School physician Itō Genboku's academy in Edo was known for its relatively expensive fees, its entrance fees included 200 *hiki* for *sokushū*, and 450 *hiki* for other miscellaneous fees. Confucian scholar Hirose Tansō's academy in today's Ōita prefecture, which attracted a large number of students, most likely set its *sokushū* at 100 *hiki* complemented by seasonal gifts, each of which would cost approximately 100 *hiki*. Rubinger, *Private Academies of Tokugawa Japan*, 70–71, 121–122, 232.Historians Mitamura Engyo and Shinji Yoshimoto estimate that carpenters made a daily wage of 5 *monme* 4 *bu* (442 *mon*); however, they had to pay for all the living expenses for themselves and their families on their own. (In this currency system measured in silver, 1 *monme* was divided into 10 *bu*.) Male servants for warrior houses earned a yearly allowance of between 2 and 3 *ryō* 2 *bu* in the early nineteenth century. Female servants, on the other hand, made between 1 and 3 *ryō*. See Inagaki Fumio, ed., *Edo seikatsu jiten* (Tokyo: Seiabō, 1959): 110–113; Kodama Kōta, ed., *Kinseishi handobukku* (Tokyo: Kondō Shuppansha, 1972): 318–319.
71. New students had to produce a reference from a guarantor to be initiated as a Kagawa School disciple. Nonetheless, this requirement did not seem to deter commoners whose families did not practice medicine from being admitted to the academy.
72. Ogata, *Nihon sanka gaku shi*, 182.
73. Ibid., 161–185. There were several renowned female physicians in the Tokugawa period. Most had been born into a physician's family and obtained medical knowledge from their fathers and other family members, and by reading books on medicine and other branches of Chinese scholarship. Ogata Masakiyo refers to the following female obstetricians from the Tokugawa period in his *Nihon sanka gaku shi*: Morikawa Hoyū, who in 1823 published an obstetrics text entitled *Sanka shinan* written by her deceased teacher Ōmaki Shūsei; a woman physician named Ine, the daughter of Philipp Franz Barthazar von Siebold and his Japanese wife, Taki, who obtained medical training from Siebold's Japanese disciples; Yamada Hisao, who published an obstetrics manual entitled *Uka hatsumō zukai* in 1852. Ogata, *Nihon sanka gaku shi*, 427–428, 481, 831–835.

74. Miyama, "Kyūdai sanba gyō no eizoku shitaru nakaoka ke no ryakureki," *Josan no shiori* 80 (January 1903): 26–29.
75. See, for example, Sakurai's brief introduction to this issue in Sakurai Yuki, "Kinsei no ninshin shussan gensetsu," *Rekishi hyōron*, No. 600 (April, 2000): 33–35; and Drixler, *Mabiki*. Many historians have suggested that peasants in Tokugawa Japan practiced infanticide, not simply to reduce the number of children in the family, but to have a specific number of male and female children with an adequate age difference between them. Such a practice emerged from rational calculations designed to maximize benefits to their families. See, for example, T. C. Smith, *Nakahara: Family Farming and Population in a Japanese Village, 1717–1830* (Stanford, CA: Stanford University Press, 1977); Sawayama, *Shussan to shintai no rekishi*, 20–27; Ōta Motoko, "Joron kyōdō kenkyū no kadai to hōhō oyobi tōtatsuten ni tsuite," in *Kinsei nihon mabiki kankōshi shiryō shūsei* (Tokyo: Tōsui Shobō, 1997): 19–23.

    For demographic studies of the Tokugawa period, see Hayami Tōru, *Rekishi jinkō gaku no sekai* (Tokyo: Iwanami Shoten, 1997); Laurel L. Cornell, "Infanticide in Early Modern Japan? Demography, Culture, and Population Growth," *Journal of Asian Studies* 55: 1 (February 1996): 22–50; and "Retirement, Inheritance, and Inter-generational Conflict in Preindustrial Japan," *Journal of Family History* 8 (Spring 1983): 55–69; L. L. Cornell and Akira Hayami, "The *Shūmon Aratame Chō*: Japan's Population Registers," *Journal of Family History* 11: 4 (1986): 311–328. Drixler draws on this earlier work.
76. Drixler discusses this topic in *Mabiki*. See also Kanazu Hidemi, "'Datai mabiki' kō," in Kinugasa Yasunobu, ed., *Kinsei shisō shi no genzai* (Kyoto: Shibunkaku, 1995): 401–421; Sawayama, *Shussan to shintai no rekishi*, 188–218; Yatabe Mariko, "Akago yōiku shiyōhō ni tsuite," in Watanabe Nobuo, ed., *Miyagi no kenkyū, Kinsei hen II* (Osaka: Seibundō, 1983): 250–267; LaFleur, *Liquid Life*, 102–118.
77. Sakurai, "Mabiki to datai," 102–119; Sawayama Mikako, " 'Umu' shintai to kaitai shussan torishimari," Edo no Shisō Henshū Īnkai, ed., *Edo no shisō, Vol. 6* (Tokyo: Perikansha, 1997): 7–18; and also Sawayama, *Shussan to shintai no rekishi*, 95–187; Takahashi, *Datai mabiki no kenkyū*, 153–170, 183–185, 209–223, 231–240, 245–261; Yatabe, "Akago yōiku shiyōhō ni tsuite," 238–249.
78. For example, the Tsuyama domain specified that the punishment for the mother for abortion or infanticide was a fine of 2 *hyō* of rice, and seven days' confinement. (1 *hyō*= approximately 0.5 bushel.) On the other hand, the next-door neighbor and the five-people group were liable to five days' confinement. In the case of failing to report a pregnancy, the woman was fined 3 *hyō* of rice and received seven days of confinement in manacles; the

next door neighbor and the five-people group were also to pay 3 *hyō* of rice and confined for five days; the head of the group was fined 2 *hyō* of rice; the village headman was to give 1 *hyō* of rice as penalty. Sakurai, "Mabiki to datai," 114–115.

79. Sawayama, "'Umu' shintai to kaitai shussan torishimari," 18–23; and *Shussan to shintai no rekishi*, 158–187.
80. Ibid., 176.
81. Sakurai, "Mabiki to datai," 119–120.
82. Ibid.
83. See, for example, Okitsu Kaname, *Edo senryū onna hyakkei* (Tokyo: Jiji Tsūshin sha, 1994): 79–83.
84. Sakurai, "Mabiki to datai," 123–126.
85. The six oaths included the following clauses: (1) You should not forget gratitude toward the Kagawa House which gratefully shared with you the foremost obstetrics skills. (2) At the client's house you should strictly adhere to the Kagawa School methods and should not combine other methods with them. You should also take utmost care to provide services equally for the rich and poor. (3) You should not share with others, including your relatives and friends, the teachings and operation methods that were transmitted orally (from your Kagawa School masters). The treatment techniques should also not be demonstrated freely without deliberation. (4) The prohibition of abortion. (For the exact wording, see the main text.) (5) There are differences in competence even among those of you who have received a certificate. When you find a junior disciple with remarkable skills, you should not act condescendingly toward him or her just because you are senior, but should admire his or her expertise. (6) You are not allowed to practice until you have obtained a certificate. Those who violate this promise will be ousted from the Kagawa School. Those who practice in the name of the Kagawa School without being initiated into it should also discontinue their practice at once. Ogata, *Nihon sanka gaku shi*, 156–157.
86. Although the practice of abortion doctors was denounced as a mercenary act, the Kagawa School's operation itself was structured by market forces. The Kagawa head masters sought to represent their practice as enacting the Confucian tenet of *jin* [benevolence] by saving the life of the mother and that of the child, and not as a money-making business. It might be beneficial to think about this issue in terms of an overall historical trend toward medical practices becoming highly profitable businesses by the mid-Tokugawa period. The emphasis many physicians placed on their moral mission might have stemmed from the fact that they were so profitable.
87. See Drixler, *Mabiki*.
88. Ibid.

CHAPTER 4

The State, Midwives, Expectant Mothers, and Childbirth Reforms from the Meiji Through to the Early Shōwa Period (1868–1930s)

In earlier chapters, I show how obstetricians, domanial governments, and the shogunate exerted influence over biological reproduction during the Tokugawa period. After the 1868 Meiji Revolution, Japan's new leaders began to build a modern nation state, and overarching reforms of the medical system were a part of this larger development. This led to the emergence of new networks of reproductive surveillance, replacing the existing Tokugawa system. Central to this shift was the establishment of a health bureaucracy that regulated medical practices and administered public health as well as a modern medical institution in line with the Western model. The broader historical and global phenomena that accompanied the establishment of a modern medical and health administration in Japan were the ever-expanding movements of goods and people in the age of empires. With an influx of foreign visitors and their increased contact with Japanese locals after the 1850s, Japan experienced serious outbreaks of epidemic diseases—cholera in particular. The signing of unequal treaties with Western nations, beginning in the 1850s, deprived Japan of the authority to regulate the entry of foreign visitors and prevented it from implementing a rigorous naval quarantine system. This led the Meiji government to develop active domestic public health campaigns. New government efforts to improve maternal and infant health were a part of much larger medical and public health reforms. The establishment of modern institutions and practices, including those in the fields of medicine and public health, gave rise to what Foucault calls "bio-power" that works in

Y. Terazawa, *Knowledge, Power, and Women's Reproductive Health in Japan, 1690–1945*, Genders and Sexualities in History,
https://doi.org/10.1007/978-3-319-73084-4_4

conjunction with the governance of medicalized bodies by the state and organized medicine.

The Establishment of New Medical and Public Health Systems

The institutionalized mechanisms by which bio-power emerged first appeared in the late Tokugawa period to prevent the spread of epidemics—smallpox, for example. While the smallpox vaccination project in general was voluntary and thus, unsystematic, the Tokugawa shogunate carried out compulsory vaccination trials on the Ainu—a discriminated minority—based in today's Hokkaidō, on the northern island of Japan. The trials foreshadowed the government initiated vaccination drive during the Meiji period.[1] Combating the cholera epidemics that first arrived in Japan in 1822 and caused repeated, devastating damage thereafter, however, was a higher priority for Japan's spirited public health effort than tackling smallpox.[2]

To control cholera epidemics, Japanese public health bureaucrats initially attempted to implement a rigorous quarantine system during the 1870s (the early Meiji period) but eventually shifted their focus from quarantine to public health measures. Although the Meiji government had engaged in ongoing negotiations with foreign nations to implement effective quarantine, the unequal treaties prevented the government from enforcing the quarantine it preferred. This is shown in the 1879 "Hesperia Incident," which provoked outrage from both Japanese officials and the public. In the incident, a German envoy ignoring Japanese quarantine regulations instructed the ship's captain to sail into Yokohama harbor where foreign ships were forbidden to enter without epidemic inspections. The envoy forced entry into Yokohama harbor under the protection of a warship.[3]

Concurrently, Japan failed to obtain support from Western powers at international health conferences, the first of these having been held in 1881 in Washington, DC. Discussions at such conferences tended to focus on measures to protect Western Europe from the influx of cholera, which was assumed to have spread from Asia, the Middle East, and the Balkans. Issues related to China and Japan, did not attract much attention from diplomats who were present.[4] In other words, Japan along with other Asian nations failed to acquire the support from Western powers necessary to enforce adequate quarantine. Given this setback, Meiji health officials had no choice but to launch vigorous public health campaigns targeting the domestic population to prevent and contain cholera epidemics. Cholera epidemics played a defining role in the development of Japanese public health, and the government used laws and systems that were originally created to control cholera to address typhoid fever, dysentery, diphtheria, influenza and other contagious diseases.

An alternative strategy was used to handle another major contagious disease, syphilis. Lacking resources and the will to implement an anti-venereal disease policy that involved the testing of sexually active men and women (which would have been more efficacious but risked inciting fierce protests from politically powerful citizens, who were typically men) the government singled out female sex workers as the target of public health control. Authorities assumed that prostitutes were more prone to venereal disease than others and were the source of its spread. This venereal disease (VD) check-up system for sex workers originated first in Nagasaki under the rule of the Tokugawa shogunate as a response to concerns of Western diplomats and ship captains that Japanese prostitutes were infecting their countrymen with syphilis.[5] The demand for exams was reinforced by a general distrust of Japanese standards of hygiene by Western powers and the stigma of the semi-colonial status that Japan was forced to accept. This VD examination system was eventually expanded for all licensed prostitutes.

An important part of this system was that it allowed the government to obtain information systematically about the identity and residence of prostitutes along with their health status. Authorities meticulously collected knowledge of and had firm control over licensed prostitutes whom they classified differently and segregated area-wise from "normal" women. As I discuss in this and the next two chapters, while government health officials from the Meiji period onward were concerned about the health of "normal" Japanese women and focused overwhelmingly on their reproductive capabilities, they put prostitutes in a separate category where reproductive potential was of no relevance. Their bodies were used for a different purpose than the biological reproduction of the next generation of citizens, and the spread of eugenics ideas had an adverse effect on safeguarding sex workers' reproductive rights. The pre-World War II Japanese state generally endorsed pronatalism, but not in the case of sex workers. That the state ignored the "reproductive" health of sex workers and their right to have children, frequently caused egregious predicaments for the women. In the Epilogue, I further address this issue, tying the issue of the reproductive lives of sex workers to modern bio-power.

The modern medical and public health systems established in the early Meiji period tied together these various attempts to control epidemic diseases. The important health guidelines, the *Isei* [the Medical Ordinance], drafted in 1874, show how the government sought a centralized system for overseeing medicine and public health.[6] Although the *Isei* guidelines were not law with binding powers, the government in due course created laws using this document as a guideline. Central to the new medical system was the introduction of the licensing system for physicians, midwives, and pharmacists. To train qualified medical practitioners, the government also set up standards for medical education in line with the Western model.

The significance of the *Isei* guidelines lay in the establishment of the health administration solidly within the government bureaucracy. What was initially called the Medical Bureau and later renamed the Hygiene Bureau was first located in the Education Ministry. By 1875, it was moved under the supervision of the Home Ministry, with only the division of medical education being kept in the Education Ministry.[7] The leading medical bureaucrat, Nagayo Sensai, viewed the state of health and hygiene in Japan at the time to be far below what he desired, inferior to the standards of European nations and the USA, whose public health systems he observed as a member of the Iwakura Mission (1871–1873). He considered the measures presented in the *Isei*, including the development of extensive local public health offices, inspirational, forward-looking goals that the nation should strive to achieve.[8] Although the development of local health administrations lagged somewhat, the central Hygiene Bureau became the pivotal locus for gathering health data and making medical and public health policies.

To formulate and implement policies described in the *Isei* document as well as new laws on epidemic control, the health administration of the Meiji government required data about the population. The Hygiene Bureau had to rely on other key mechanisms of a modern government, namely, the system of citizen and resident registration and the compilation of national and local statistics about the general population. In Meiji Japan, the government replaced the more decentralized registration systems of the Tokugawa period with a revised version of the ancient system of household registration. This new family registration, the *koseki*, became highly controversial in the twentieth century and beyond, due to the reinforcement of the patriarchal household (*ie*) system and the continuous discrimination against outcast groups (e.g., the burakumin) through records in their family ancestry.[9] The family registration records were the basis for carrying out compulsory home sterilizations, detentions, and patient segregation for public health purposes. They were useful for investigating and reporting disease outbreaks and educating the public about epidemics.[10]

Equally vital for pursuing new public health policies and collecting data on individual citizens was the processing of the data into statistics that would provide a macro perspective on the nation's population. The Meiji government created the statistics division first within the Grand Council of the State [Dajōkan] in 1881, consolidating units for data gathering scattered within the government's bureaucracy. The Cabinet took over the supervision of this statistics office from 1885 onward.[11] Local- and national-level statistical data were essential for formulating and enforcing measures for epidemic control and for distributing available public health resources.

The newly established health bureaucracy carried out surveillance and reform of its citizens' health with the support of other modern state institutions, including the military and public schools. The institutions that functioned in conjunction with the family registration system were not only useful for spreading nationalism among ordinary citizens, but they also helped monitor and enhance the nation's health.[12] The military and schools screened the physique and health of soldiers and students and attempted to "improve" their health and physical capabilities. The health bureaucracy was a site where the government accumulated health statistics, classified bodies into categories for different uses and treatments, and pursued health campaigns. While these institutions inspected and sought to improve the health of individual citizens, they offered important venues where the government collected health data on given population groups to produce what Michel Foucault calls the micro and macro knowledge of the bodies of citizens—that is, knowledge of each individual and different group.

To work on public health issues the Japanese government classified bodies into different categories. For public health policy analysis and development, the population was divided primarily into different age groups and into healthy and diseased categories. The body's sex, which slotted each body as either female or male, was a fundamental marker for dividing the population. Although it did not pay notable attention to the issue of sexual difference for controlling many epidemic diseases, the government handled women's health issues very differently than men's. Among women's health issues, the government paid close attention to their reproductive health.

Changing Power Relations Involving the State, Medical Experts, and Expectant Mothers in Meiji Japan

The government's strong interest in controlling its citizens' biological reproduction and fertility was certainly not entirely new. The Meiji government created new reproductive policies; more important, however, it pursued such policies in the entirely different institutional set-up that I outline earlier. The reorganization of the medical field according to the Western model was not the only change that altered the nature of government reproductive policy. Other modern government institutions such as the health bureaucracy, statistics office, and the new system of household registration helped formulate and implement new reproductive policies. Further, schools emerged as a new institution that the government and organized medicine could use to control the bodies and health of its citizens. The establishment of these modern institutions and practices changed the relationship between the government and individuals, giving rise to "bio-power."

Bio-power works through modern institutions and practices that both confine and guide individual behaviors. Peculiar to this new type of power is its reliance on methods developed in various scientific and medical fields that researchers use to study individual bodies and minds, and define their physical and mental traits. Scientific discourses along with legal regulations form a web that categorizes and situates individuals in social positions. Experts in relevant fields investigate and train individuals at the micro level to grasp the characteristics of various groups in the population of the nation at the macro level. The Meiji era restructuring of the medical and public health systems was a major example of this modern system of governance.

The Meiji reorganization of the medical field changed how access to authorized medical knowledge took place. No longer was it regulated through the system of *ryū* (schools or factions) as it had been during the Tokugawa period. Professional physician groups and major medical universities, both developed with strong government backing, managed the reorganization. While the Tokugawa regime neglected licensing the medical profession and showed little interest in developing public health programs, Meiji officials relied heavily on newly imported European and American concepts of public health. They actively controlled the medical field and made efforts to improve individual and national health. The process of increasing state intervention in medicine reshaped the political relationship between the state and individuals.

These developments, an example of bio-power, changed the nature of state reproductive and fertility control. They occurred because the modern Japanese state developed its reproductive policy parallel to an expansion of the state and organized medicine's ability to procure and use knowledge about reproductive health. During the Tokugawa period, shogunate and domanial authorities attempted to regulate fertility, abortions, and infanticide through legal bans, moral prohibitions, and rewards in the form of child support funds. Efforts at controlling fertility by improving maternal and child health care were non-existent. Medical professionals (e.g., physicians, midwives, and nurses) were, for the most part, absent from the state's efforts at regulating biological reproduction at the national level. During the post-Meiji Restoration era, the government shifted fertility control from a reliance on prohibitive regulations and moral suasion to medicalized maternal health care. While research techniques and statistics advanced, government officials, public health researchers, and reformers gauged the health of mothers and infants as aggregates. On an individual level, physicians and trained midwives began to monitor the health of individual mothers and children more extensively. Women's reproductive decisions came to be affected more and more by the suggestions of physicians, midwives, and public health nurses, many of whom espoused government health care directives and nationalist sentiments. At the same time, government reproductive policies, which the state pursued increasingly through maternal health care programs rather than legal regulations (e.g., anti-abortion and anti-birth-control laws), exerted a growing influence on individual reproductive practices.

Central to the intrusion and spread of bio-power in the field of reproductive medicine in Meiji Japan were childbirth reforms. Trained midwives—sometimes called "new midwives" [*shin sanba*] or "modern midwives" [*kindai sanba*]—were educated by licensed obstetrician-gynecologists and qualified by government-regulated licensing examinations. And they were at the core of the reforms. In the absence of affordable medical care and physical exams for the vast majority of the population, childbirth constituted a rare occasion in which women from less-privileged classes could receive professional medical care. In this sense, it was through the practices of trained midwives that organized medicine made an increasing number of women's bodies its target for medical observation and treatment. School-educated midwives spread new ideas and practices not only about pregnancy and childbirth but also about medicine, the body, and public health. Additionally, they helped mothers and their families report the birth and death of the newborn to local municipal offices and the police. Trained midwives were instrumental in co-opting mothers and their families into modern, state- and organized medicine-promoted health practices, which extended the web of modern bio-power.

Childbirth reforms, however, did not occur only because government officials desired to serve the national government and elite physicians, but they also involved multiple players and various interests. While the national government occupied a vital position in implementing education policies and supervising midwives, new childbirth practices would not have been adopted unless other actors—local physicians, influential local politicians and educators, trained midwives, women clients, and their families—promoted the new practices. Modern childbirth practices developed and expanded through complex ways as these diverse powers and interests interacted and converged.

While the state and organized medicine's surveillance of pregnancies and childbirth increased between 1868 and the early twentieth century, those policies were far from homogeneous or coherent. The reproductive and fertility controls that emerged by the 1930s were a culmination of policy measures, health reforms, and changes in women's everyday life. Despite a formal commitment to improving women's health, early implementation of policy was not always inspiring. Meiji government resources to increase trained midwives were far from generous, which held back the pool of formally trained midwives until the late 1910s. Yet, new midwife-assisted childbirth practices spread as a set of diverse interests and desires converged with government efforts to modernize childbirth customs.

Players engaged in their own agenda without aiming precisely to strengthen reproductive surveillance. A range of developments in reproductive medicine and maternal health led to the expansion of modern bio-power. The medical profession and modern medical knowledge steadily intervened in women's reproductive lives. With the assistance of trained midwives, organized medicine and divisions of the government began gathering data on women's reproductive health, and finally the spread of new childbirth customs helped "tame" expectant mothers into subjects willing to follow the health instructions of midwives, physicians, and government officials. This development was not a simple top-down process of the government telling mothers what to do; rather, the state occupied a strategic position for the spread of new, midwife-assisted childbirth as well as intensifying reproductive surveillance.

Early Meiji Regulations

From its earliest years the Meiji state sought to control the nation's reproductive capabilities. The government issued a series of edicts to regulate childbirth, abortion, and infanticide, but it would be erroneous to presume that key officials and the physicians who headed major medical associations had a cohesive plan to establish a system of reproductive control. Those who formulated and implemented policy had a range of motivations and goals for policing and controlling reproductive life. A model that views the state as a monolithic entity does not bear the weight of evidence. It is to a decentralized view that we turn.

The criminalization of abortion under the Meiji legal code demonstrates that Meiji reproductive policy originated from diverse cultural and ideological sources. Existing studies illustrate that the prohibition of abortion was rooted in the Tokugawa tradition of reinforcing Confucian-based ideology. Historian Ishizaki Shōko, points out that the Tokugawa state opposed abortion, infanticide, and abandoning children because they constituted a serious infringement of the Confucian moral code, specifically, the moral relationship between the father and child.[13] In addition, the Tokugawa state prohibited abortion because it assumed that some occurred as a result of illegitimate sexual unions. During the early Meiji years, the moral stricture against illicit sex weighed heavily on reproductive policies. For example, Yasuda Muneo suggests that when turn-of-the-century newspaper reporters in the Kumamoto area wrote articles on abortion, they stressed the adulterous nature of relationships that resulted in unwanted pregnancies and abortions. Later they would focus more on the act of abortion as a crime rather than the illegitimacy of the sexual union in question.[14] Whether aversion to abortions was a reaction to parents who denied their obligations or a moral condemnation of adultery, officials and the educated public based their thinking on Confucian ethics.

The elevation of abortion to a serious illegal act was codified in the 1880 criminal code, which replaced earlier Meiji laws. Since the 1880 criminal code was modeled after the French and German codes, penalties for obtaining or performing illegal abortions were heavily influenced by European values and practices, although Japanese government officials incorporated ideas rooted in Japanese customs that were widely practiced at the time. Ishizaki, for example, points to the absence in the Japanese legal code of the idea of the fetus as a human being (unlike in the European tradition). Despite the illegality of abortion, the Meiji codes had relatively light penalties for obtaining and assisting abortions compared to their European counterparts. According to articles 330 and 331 in the criminal law enacted in 1883, women who had induced an abortion or those who helped obtain one could be imprisoned from one to six months. If a woman died during an abortion, the person assisting could be imprisoned from one to three years. When physicians, midwives, or pharmacists assisted in abortions, the punishment was one grade more severe.[15] These punishments were inconsequential compared to, for example, those mandated in Germany, which stipulated that women who aborted a fetus on their own were to be imprisoned for more than five years and that those assisting would receive a penalty of up to ten years' imprisonment.[16] Nonetheless, it is undeniable that the Japanese criminal code incorporated European and Christian legal and cultural traditions against abortion.

While criminalizing abortion gave the state one path to regulating citizens' reproductive lives, another new method it employed for supervising pregnancy, childbirth, abortion, and infanticide was placing physicians and midwives—potential abortion providers and infanticide practitioners—under state control. In addition to monitoring the reproductive practices of local communities in collaboration with state authorities, midwives also became major players in the dissemination of new health and hygiene ideas and practices mandated by Japan's nationalist ideology.[17] The impetus to license physicians and midwives originated in the strong nationalistic concern of the Meiji state and organized medicine. Although they continued to use the language of Confucian morality to give guidance to medical doctors and midwives and to lecture on the evils of abortion and infanticide, state officials and leading physicians drew on the new language of health and hygiene along with appeals to nationalism to discuss the need for supervising citizens' reproductive life.

State initiatives to regulate midwife practices nationwide began with an 1869 edict prohibiting them from prescribing drugs for and assisting in abortions, describing these activities as "something that should never happen." Ishizaki suggests that endeavors by the government to prevent abortions were primarily rooted in state-endorsed Confucian ideology rather than in a desire for population increase. Preceding the nationwide Meiji laws was an 1868 Tokyo prefectural edict prohibiting midwife and pharmacist collaboration in abortions, which presented the procedure as a breach of the tenets of the Meiji government's "Edicts by the Five Bulletin Boards" [Gobō no keiji]. Meiji authority declared that nurturing the populace and promoting the five human ways of Confucian tradition were an urgent task and that those who supplied drugs to induce abortion would be punished.[18]

After issuing this early edict to direct midwives' practices according to Confucian ethical norms, the Meiji state, before long, attempted to control midwives by turning them into educated "medical" professionals. The Hygiene Bureau's 1874 Medical Ordinance [*Isei*] laid out a licensing and examination system that required midwives to obtain a local government license. As in other areas of medicine, the government pursued its policy of adopting European medical practice standards for midwifery. To practice midwifery, candidates had to pass physician-administered qualifying examinations in anatomy, physiology, and pathology. In addition to passing the exams, applicants had to demonstrate their practical skills by performing ten normal and two abnormal deliveries in the presence of an obstetrician-gynecologist. The *Isei* directives also specified the age of those who could practice midwifery as over 40 years; they did not specify, however, that midwives had to be women.[19]

The 1874 *Isei* guidelines consisted of broad policy statements directing prefectural authorities rather than legislation to be enforced. Thus, each prefectural government took on the task of implementing the new policies on midwifery on its own terms.[20] Local administrators knew at the time—the mid-1870s— that a majority of physicians lacked a solid background in European medicine, and even fewer practicing midwives possessed any such knowledge. Many midwives were illiterate and had learned their professional skills from older midwives by assisting them in their work. Under these circumstances administrators waived examinations and issued temporary licenses after reviewing each midwife's qualifications. They issued licenses on the condition that licenses would be invalid if midwives moved to other towns or villages; older practicing midwives rarely relocated.[21] Even when midwives took examinations, administrators were relatively lenient, often conducting oral tests with simple questions on anatomy and physiology. Consequently, health officials could maintain midwives in each locality.

Some prefectural governments established midwife-training programs to educate practicing midwives about European medical knowledge. In many cases, however, these efforts produced few results. For example, Niigata prefecture offered a seminar series for midwives at the publicly funded Niigata Hospital. Midwives were required to attend seminars six times a month, and to encourage them, the prefectural government issued an edict prohibiting midwives from practicing if they failed to attend the seminars. Since most midwives had to rely on a limited transportation system that lacked even a railroad, it is most likely that only a small number of all practicing midwives in Niigata prefecture attended the seminars. Furthermore, since many midwives were illiterate, instructors were unable to have them learn from textbooks. Thus, teachers would read out loud from textbooks on anatomy and physiology and have midwives memorize the content.[22]

This state of affairs indicates that examinations and seminars were largely a formality. The real significance of these measures was that they allowed the state to identify midwives and record names and where they lived. That is, registering midwives laid a foundation for supervision by prefectural and municipal offices as well as local police. At the same time, it was a means through which midwives were compelled to recognize the authority of newly established state institutions that could intervene and regulate their practice, including the prohibition of abortion and infanticide.

The *Isei* guidelines stipulated the subordination of midwives to physicians by defining procedures midwives could perform. Midwives were prohibited from providing "medical treatment," including using "obstetric instruments"—forceps and scalpels—and prescribing medication except in emergency situations.[23] In the case of complicated deliveries requiring surgical operations, midwives were to contact licensed physicians, who had a monopoly on the necessary surgical tools and drugs.[24] Thus, midwives were integrated into the medical establishment as low-ranking medical practitioners who were expected to work in conjunction with licensed physicians.[25]

The Making of Modern, Nationalist Midwives

In addition to government officials, leading obstetrician-gynecologists also emerged as enthusiastic advocates for educating midwives in European medicine. In the late 1870s these physicians began to publicize vigorously the need for establishing midwife-training centers. By the 1910s numerous midwife schools had been created nationwide with both public and private funding. During the early years of their implementation, the quality of the programs varied because there was no standard curriculum or state-specified training period. The duration of training ranged anywhere from six months to three years. Despite the disparity among programs, midwife training during this period shared one common feature: it was fervently nationalistic.[26]

One of the earliest midwife-training programs was established in Tokyo Prefectural Hospital in Shiba, Tokyo, in 1876, by obstetrician-gynecologist and hospital vice president Sakurai Ikujirō (1852–1916). When the hospital closed for budgetary reasons in 1882, Sakurai established a private midwife school called Kōkyō-sha adjacent to his maternity hospital. Hamada Gentatsu (1854–1915), professor of medicine at Tokyo Imperial University's School of Medicine, presented a proposal to the Education Ministry in 1889 to create a midwife-training program. His proposal was accepted and a midwife school was established at his university in the same year. A few years later, in 1893, Ogata Masakiyo (1864–1919) founded a midwife-training institute at Ogata's maternity hospital in Osaka. He also presented his proposals on the education and supervision of midwives to the director of the Home Ministry's Hygiene Bureau.[27] Physicians in both public and private institutions, thus, promoted midwife education.

In an environment where training in European-style midwifery was deemed urgent for the nation, many prefectural governments established midwife schools and provided scholarships to midwives in training. Obstetrician-gynecologists in private practice also founded midwife-training institutes. The large number of early twentieth-century privately funded midwife-training centers suggests that such educational projects were economically viable and even profitable. By 1915, 135 midwife-training schools existed nationwide. The total number of graduates of each school varied from a few students to more than 1000, depending on the year and the scale of the training program. Of 135 programs, 115 were privately owned; the rest were funded and operated by either local administrative units or the national government.[28]

Home Ministry bureaucrats, for their part, recognized the need to more rigorously regulate midwives. In 1899, the Ministry issued a 20-article Midwifery Ordinance [*Sanba kisoku*], accompanied by two additional sets of rules: one for supervising qualifying examinations [*Sanba shiken kisoku*] and another establishing a midwife registry [*Sanba meibo tōroku kisoku*]. After issuing these edicts, the Home Ministry took over the preparation of qualifying examinations from prefectural governments, although prefectural governments continued to be responsible for administering the exams. Under this arrangement, the examinations became much stricter than those formulated by local examiners.[29] Many of the precepts delineated by the Midwifery Ordinance, such as the division of labor between midwives and obstetricians, had already been implemented to varying degrees at the prefectural level following the 1874 *Isei* guidelines; the newly introduced regulations, however, brought national uniformity and increased rigor to the overseeing of midwives.

New regulations also stipulated that practicing midwives should be *women* at least 20 years of age. The midwife profession was thus officially feminized in conformity with the gendered hierarchy in medicine. Although the government allowed women to become physicians in 1885—after repeated petitions by women wanting to practice medicine—the number of female physicians remained small. Moreover, because women were not admitted to Tokyo Imperial University's School of Medicine, the nation's most prestigious medical school, they could never become first-rate practicing physicians or medical researchers. In general, women were expected to perform supporting roles as nurses or midwives within the medical field.

Obstetrician-gynecologists who supported official and unofficial projects to train midwives viewed themselves as bringing "progress" and "enlightenment" to the Japanese people. They found the presence of the large number of so-called "old midwives" [*kyū-sanba*] or "amateur midwives" [*shirōto sanba*] troubling.[30] For example the obstetrician-gynecologist Nikawa Toshio deplored the underdeveloped state of midwife training when compared with the great feats that Japan had accomplished in other areas since the Meiji Restoration.

> Our country reversed its closed-door policy and began to interact with European nations and the United States three decades ago, which seems to be an ancient time, looking back now. Since then we have engaged in active exchange with other countries. The wave of enlightenment coming to the East renewed the appearance of many things. Progress has been made day after day, and month after month. It is almost as though there is not a thing that has not received the benefit of such progress. Medicine, in particular, has achieved unprecedented progress. However, only the area of midwifery does not match the level of enlightenment that [Japanese] society has attained. The practice of midwifery looks the same as it was in the old days. The government has not issued rigorous regulations for supervising the practice [of midwives].[31]

Arguing in a similar vein, Ogata Masakiyo scornfully described the so-called "old midwives" as illiterate, old, frail, and stubborn, and unsuitable for taking on the serious responsibilities entailed in childbirth.

> There are still midwives who take on such important responsibilities though they are illiterate old women with bent backs and walking canes.

> They stick to old customs without much reasoning and lack the will to adopt new knowledge. On the contrary, there are also midwives who have mastered all the techniques, including proper methods for sterilization based on the most advanced scholarly knowledge.[32]

Ogata continued by arguing that midwives who learned the most advanced scientific methods would be able to save the lives of mothers and children during emergencies, unlike uneducated midwives, who would certainly endanger both.

Nikawa and Ogata considered a midwife's work to be important for achieving national goals by protecting and enhancing the health of mothers and children, particularly male children. They believed that proper prenatal care and treatment during childbirth would safeguard healthy male bodies that were indispensable for pursuing the state policy of "enriching the nation and strengthening the military" [*fukoku kyōhei*].

> The basis for enriching the nation and strengthening its military is to take care of our people's health. The strength and weakness of the people's health depends on the health of babies. The health of infants originates in their condition during the time when they are inside their mothers' wombs. Thus the health of mothers and children not only affects the fate of their families, but that of the nation. Therefore midwives have a tremendous responsibility.[33]

Physicians often defined the ultimate purpose of promoting mothers' health as the production of healthy and intelligent men to serve in the military and engage in other professions. Women's health was important only as it related to the improvement of their reproductive functions.

> Could we possibly nurture new young men [*shin-seinen*] for a new Japan which must renew itself to demonstrate its power to the world comprised of powerful countries in Europe and the Americas which have been watching us ever since our nation triumphed in the Sino-Japanese War? Wouldn't anyone who is concerned about our nation shed tears [seeing the neglect of women's and children's health]?[34]

Nikawa continued:

> When young men are strong and healthy, the state and family will be kept safe. [On the other hand,] when young men are weak and sickly, they will be endangered. [Therefore,] the responsibility of those who rear children is enormous.[35]

Nikawa was writing this around 1898, a few years after the Sino–Japanese War (1894–1895) and during a time when European powers and the USA were increasingly aware of Japan's emergence as an imperialist power in Asia. He thus stressed the need to build solid foundations, including measures to promote women and children's health, to demonstrate the country's power in the international arena. What mattered most, however, was to nurture the health of young men who were to build a new and strong nation. Within this ideological framework, women had secondary roles—primarily as nurturing mothers for young male children but also as professionals working in fields such as midwifery, nursing, and education. The feminization of the midwife's profession held a particular significance for reinforcing this gender ideology.

The idea that midwives had a nationalist mission was widely promulgated through training programs, textbooks, and midwifery journals.[36] Sasakawa Misu's ten-page booklet advising midwives of their basic moral obligations emphasized orderliness, purity, gentleness, accountability, endurance, kindness, concentration, and care. And it reminded them of their duty to the state.[37] The goal was for midwives to "sacrifice themselves" for the prosperity of clients' families and the nation.

> The quality of a midwife's work could influence the rise and fall of a person and a family. Speaking in broad terms, [for midwives to provide good services] is also one of the fundamental elements of the policy of enriching the nation and strengthening the army…[A midwife] should sacrifice her body [*shintai*] and life [*seimei*] to apply herself to fulfill her professional duty following these thirteen moral tenets that I have described.[38]

The type of devotion demanded of midwives was comparable to what was demanded of young men volunteering for the military. The moral qualities emphasized in midwife education, however, were attributes usually associated with femininity. Sasakawa, who was a female midwife teacher, made no reference to typically masculine characteristics, such as bravery, a quality midwives particularly needed.

The nationalistic fervor of midwife education found much support among midwives. In the first two decades of the twentieth century, when the memory of the Russo–Japanese War (1904–1905) remained fresh in the national consciousness, many young women were already zealous patriots. Some chose midwifery to serve the nation, much like those who decided to become nurses and school teachers. Many midwives, however, were not merely passive followers of state officials and physicians, but passionate and resourceful activists who energetically promoted health reforms for the nationalist cause. At the personal level, this helped them gain respect, professional satisfaction, and self-esteem.

Despite the enthusiastic campaign led by the government and organized medicine to promote "modern" midwives, it was not easy for them to gain entry into the market for midwives. Many pregnant women and their families preferred to hire "old" midwives. The lower fees charged by untrained midwives was one reason. More important, though, these clients felt that old midwives had more experience than the younger trained midwives, some of whom were barely 20 years old and had never given birth themselves.[39] Further, with the promulgation of the Home Ministry's new guidelines for midwife education, many young school-educated midwives lacked the actual experience of assisting childbirths. Some had not witnessed any actual deliveries during their training and had only been allowed to practice with paper models.[40]

Additionally, few trained midwives practiced in rural areas before the 1920s. For example, the first midwife-training institute established in Nagano prefecture, located in the mountainous inland area of Japan's main island, started operating in 1905 and trained only two or three prospective midwives per year.[41] Despite that the examination system in the prefecture had been established in 1884, the percentage of trained midwives rose from 1 to only 9% between 1883 and 1901. The great majority were old midwives practicing with temporary licenses. Even in 1894, ten years after examinations had begun, there were only 12 trained midwives in the entire prefecture.[42] In Miyazaki prefecture on Kyūshū island, midwife-training programs targeting practicing old midwives had been carried out by local physicians' groups, which provided bi-monthly seminars in each county until a publicly funded midwife institute was finally established in 1909. Thereafter, the school began to produce about 20 graduates each year.[43] In spite of the underdeveloped state of midwife education, the prefecture managed to produce a small number of successful candidates who passed the qualifying examinations. In 1901, ten candidates out of 13 trainees obtained a license, although in the following year only two women applied for the exam. In 1903, two of the three women who took the examinations became new midwives.[44]

If we look at the statistics of new and old midwives nationwide, the number of new midwives increased dramatically from 1900 onward (see Table 4.1).

Table 4.1 The numbers of old and new midwives in Japan (1900–1950)

Year	*Old midwife*	*New midwife*
1900	24,862	228
1901	24,632	823
1902	24,170	1539
1903	23,712	2247
1904	22,997	3223
1905	21,967	4031
1906	21,259	5128
1907	20,545	6132
1908	19,827	7130
1909	18,968	8252
1910	18,173	9501
1911	17,488	10,874
1912	16,934	12,441
1913	15,725	14,309
1914	15,272	15,776
1915	14,088	17,766
1916	13,304	19,536
1917	12,754	21,541
1918	11,531	22,817
1919	10,769	24,466
1920	10,231	25,824
1921	9623	27,034
1922	9100	28,614
1923	8351	31,159
1924	7922	33,785
1925	7550	35,327
1926	6757	38,019
1927	6074	39,826
1928	5467	40,832
1929	4967	43,432
1930	4782	45,530
1931	4416	48,121
1932	4163	50,492
1933	3851	52,739
1934	3688	54,582
1935	3448	56,112
1936	3091	57,876
1937	2900	58,832
1938	2657	59,550
1939	2444	59,863
1940	2165	59,203
1941	2038	60,703
1942	1383	51,608
1943	941	33,763
1944	291	14,592
1945	1864	16,051
1946	1878	60,082

(*continued*)

Table 4.1 (continued)

Year	*Old midwife*	*New midwife*
1947	780	66,458
1948	649	69,705
1949	947	73,087
1950	1450	73,382

Author's note: The steep decline in the number of midwives between 1943 and 1945 is most likely the result of difficulties collecting data under wartime conditions

Source: Kobayashi Takashi and Katsushima Kimi, *Boshi hoken nōto, Vol. 1* (Tokyo: Nihon Kango Kyōkai Shuppanbu, 1972): 67–68

In that year only 228 midwives passed the state-supervised examination, but that number increased to 823 in the following year. By 1909 there were more than 8000 school-educated midwives compared to about 20,000 old midwives. The number of new midwives surpassed that of old midwives for the first time in 1914, a trend that accelerated in the 1920s and 1930s partly due to the passing away of many old midwives.[45] These nationwide trends show that, while wealthy and progressive families in large cities began to use the services of new midwives in the 1880s and 1890s, it was only in the 1920s and 1930s that their assistance became widely accepted in the countryside.

The spread of trained midwives made the profession one of the few occupations in which women could earn a stable income, though success depended on establishing a good reputation in a local community.[46] While many prospective midwives came from lower-middle-class families, they were diligent and ambitious women who had aspirations for upward mobility.[47] Despite the efforts of reform-minded doctors who endeavored to spread the idea that midwifery was a noble [*kōshō*] profession, a midwife's work was still often viewed, especially in rural areas, as lowly and identified with pollution.[48]

Childbirth Reforms

The extent to which the services of trained midwives became available or accepted varied greatly by geographic area. In large cities, such as Tokyo, where school-educated midwives were heavily concentrated, professional midwives had already been present before the rise of midwives trained in modern medicine. During the 1910s and early 1920s new childbirth practices became the norm. In contrast, in rural areas, childbirth reforms did not occur until much later. Trained midwives first arrived in many such localities in the 1920s and 1930s and in some places years after the end of World War II. Even though the number of trained midwives practicing in the countryside gradually increased, many who introduced new childbirth methods had to struggle for acceptance from rural mothers and their families. Their efforts were often supported by local state or quasi-state agencies through officials working for the local government or police, as well as educators teaching at primary and middle schools. Nonetheless, trained midwives constantly had to strategize about, negotiate with, and make significant concessions to their clients to change childbirth customs that conflicted with new standards of health and hygiene.

Underlying many existing childbirth customs that school-educated midwives attempted to change was the idea of birth pollution. Not all traditional childbirth practices were detrimental to the health and comfort of birthing women, but pollution taboos associated with childbirth played a large part in making the birthing environment inhospitable by obliging women to follow rigid and painful regimens. Moreover, in many rural areas across Japan, older, untrained midwives took up the task of handling and disposing of pollution so that their clients would not have to. Because of this tradition, it was not uncommon that low status women (e.g., the burakumin outcasts or the *eta* and *hinin* in the Tokugawa period) predominated among midwives in many rural areas. Eradicating ideas of birth pollution would improve the status of midwives and make midwifery a respectable profession. It would also make childbirth practices conform to the norms of hygiene set by the medical establishment.[49]

Blood and afterbirth were the materials most severely shunned as polluted and polluting. To avoid contamination, the families of birthing women had to reject or minimize contact with birthing discharges. Families prepared birthing beds in locations secluded from the rest of the family. The extent to which the birthing woman and her newborn were segregated varied according to time and location. During the Meiji period, in many farming villages, families commonly prepared a birthing bed in a dark room at the back of the house. It would often be the young couple's bedroom or a storage room (*nando*, see Fig. 4.1 for the floor plan of a farm house).

To avoid soiling the *tatami* (straw) mat floor, family members often temporarily removed one or two *tatami* mats, and the expectant mother located herself on straw, ashes, or rags that covered section (see Fig. 4.2 for an example of a birth room).

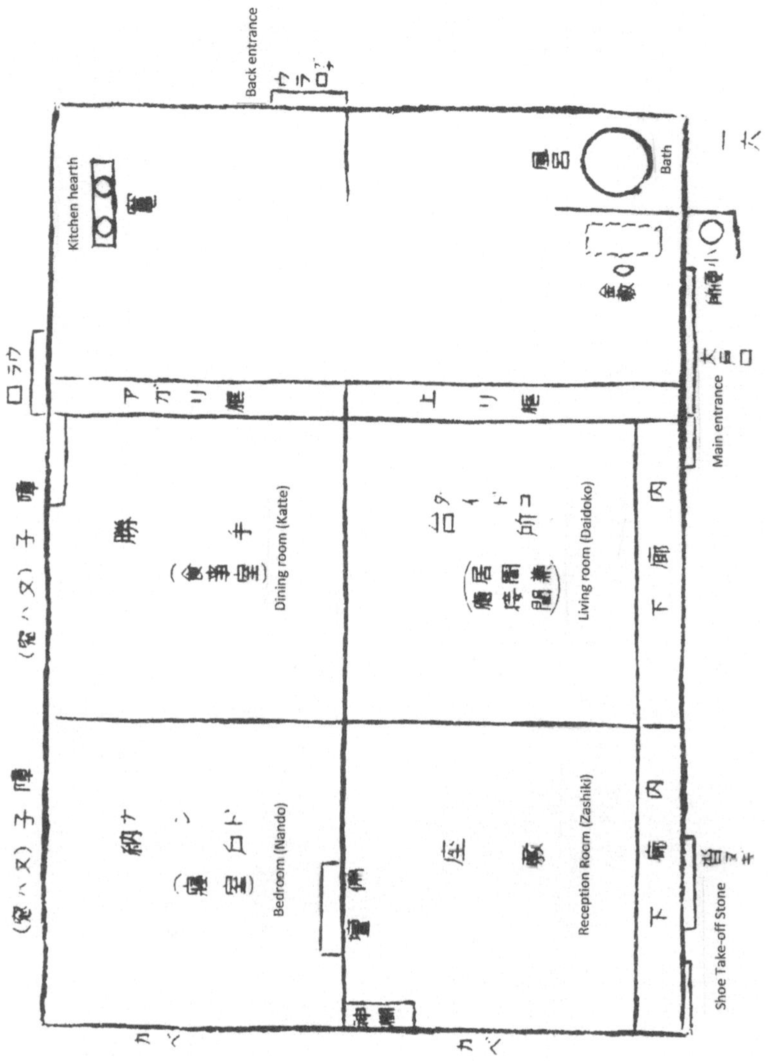

Fig. 4.1 The layout of a typical farmhouse (Aichi prefecture) (Source: Matsushita, *Sanshū okugōri san'iku fūzoku zue*, 16)

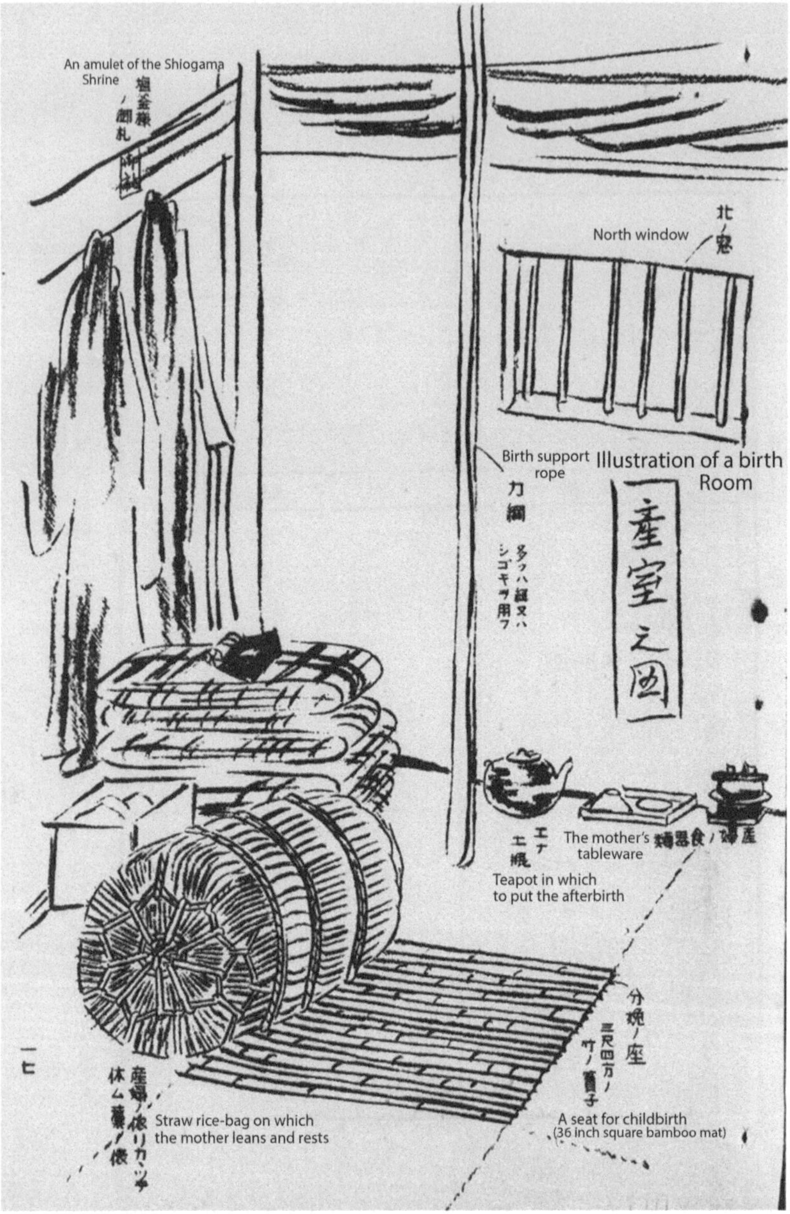

Fig. 4.2 Illustration of a birth room (Source: Matsushita, *Sanshū okugōri san'iku fūzoku zue*, 17)

The dirt-floor entrance section (*doma*) of the home where the kitchen was located was an alternative place to set up a birthing bed. In the northeast of Japan, anthropologists report, women sometimes delivered babies in a horse or cow stable away from the main house.[50] Historians have recently criticized folklorists for overstating the use of birth houses (*ubuya* or *san'ya*) specifically built for "traditional" childbirths. Giving birth in a birth house certainly seems to be a rare rather than a common occurrence, but oral interviews conducted by folklorists and anthropologists substantiate that women customarily gave birth in a birth house in certain rural areas as late as the 1920s.[51] Indeed in a recent study Fushimi Yūko discusses the birth-house custom on Ibuki island in the Inland Sea where the traditional birth house, which existed until 1970, was turned into a maternity hospital attended by a trained midwife in the 1930s. In this case, although the maternity hospital helped introduce a modern, hygienic childbirth practice, the continued segregation of the expectant mother and newborn led to the endurance of the notion of birth pollution.[52]

The degree to which mothers and their families observed the mourning period, waiting for the removal of birth pollution also varied greatly (see Fig. 4.3 for the mother's bathing in the sea as a ritual cleansing of pollution).

By the early twentieth century, many families, including those residing in rural areas, abandoned the post-partum segregation and confinement of the mother and child that was supposed to prevent contamination by birth pollution. In oral history interviews by anthropologists and historians, many women who gave birth during the pre-World War II period attest that they were obliged to engage in arduous farm labor or housework until they went into labor, and returned to the regular work routine shortly after childbirth. One of the women anthropologist Yoshimura Noriko interviewed relates that she felt she had no choice but to perform the regular daily work assigned to her so as not to displease her mother-in-law, even though she was suffering from a severe headache.[53]

For those who participated in the mourning ritual, birth pollution would dwindle day by day. The mother and child were confined to a birthing room and the mother was required to keep sitting on the birthing bed for one to three weeks following childbirth. Yoshimura along with folklorists suggest that the mourning period lasted 40–70 days.[54] The extent to which such a long mourning period was observed is difficult to determine, owing to the paucity of sources about childbirth practices in rural Japan prior to the

Fig. 4.3 Mother going down to the beach [*Hamaori*] (Source: Matsushita, *Sanshū okugōri san'iku fūzoku zue*, 41–42). Author's note: According to Matsushita, it was customary to bathe the mother in the sea or a river to rid her of ritual pollution 15 days after childbirth. The new mother was accompanied by either her mother or a midwife before she could leave her sequestered life for a normal one. The timing of bathing differed from region to region. By the time Matsushita's book was published in the 1930s, the prohibition of bathing by midwives and physicians made this custom obsolete; the belief in the idea of pollution was tenacious, however, and this practice was often converted into taking a bath in a tub

twentieth century. Traditional mourning entailed the segregation of women in an often inhospitable birthing area, but it offered the advantage of creating a valuable rest time during which birthing women—in many cases, younger women oppressed by their families—were free from a demanding work routine. If women were segregated in a birth hut, they were also free from their in-laws' constant vigilance, and thus, they could give birth and rest in a more relaxed environment. Regardless of where a birthing bed was prepared, the mourning period after childbirth was often the only period when rural women could take an extensive rest with public approval. As such, the women whom the folklorist Kamata Hisako and her colleagues interviewed declared that they "enjoyed childbirth" [*tanoshi katta*] or "were happy" [*ureshi katta*] to give birth and rest after childbirth.[55]

In the context where childbirth was understood as a polluting event, midwives' services included managing and removing polluted material. Beyond assisting in the delivery and cutting and tying the umbilical cord, the midwife had to wipe away the defiling childbirth blood and the discharge from the mother and baby. It was also often the midwife who washed the blood-stained rags, burned the soiled straw or straw mats, and buried the afterbirth in the vicinity of the mother's family house. Folklorists suggest that in regions where pollution taboos were more rigidly observed, food and drink for the new mother and the baby had to be prepared by a fire that was lit separately from the one used to cook the family's meal. In this case, it was not uncommon that midwives took up the responsibility to cook food for the mother and the newborn. A school-educated midwife lamented that midwives were often viewed only as "those who handle and dispose of dirty materials" [*obutsu shori sha*].[56] This age-old perception of midwives was still widely held around the turn of the twentieth century in rural areas.

Outcast groups provided midwives in certain rural areas. Anthropologist Nishikawa Mugiko reports that in rural Ishikawa prefecture midwives practicing before the mid-1920s were from a local outcast group called "*tōnai*" or "*tōnai kawata.*" In this area, the *tōnai* people not only served as midwives, but also as physicians and funeral workers. They took up the role of handling pollution-producing events involving disease and death. Serving their clients, *tōnai* midwives, doctors, and funeral workers were obliged to avoid physical contact with family members other than the birthing woman or ill patient. They were prohibited from entering the rooms laid with *tatami* mats; instead, they carried out their work on the dirt-floor section of the house. When a family offered food, they could not use the family's dishes or utensils; they either brought their own rice bowls and chopsticks or ate rice balls with their hands going so far as to avoid touching serving trays that might then touch a family's tableware.[57] Pollution taboos operated as an important system for organizing people's practices involving birth, death, and illness and for sustaining discriminatory customs. This respect of taboos resulted in people's viewing and treating school-educated midwives as defiling agents during the early years of their practice.[58]

To understand the meaning of birth pollution and the role of the midwife who attends to it, we must situate the idea and role in a larger cosmology of ancestor worship where an individual soul undergoes a cycle of life and afterlife. Within this system, birth and death mark the most important rites of passage during which the soul floats from the body and dwells in a liminal realm. The transition between life and afterlife following birth or death is a period of uncertainty. Because power is believed to be released during the transition, it is a time of danger. Pollution taboos develop to manage the danger that could harm a family's well-being. Within this cosmology, birth pollution is vested with a powerful force hazardous to those who come into contact with it. The midwife who is designated to process and dispose of pollution possesses a special power by which she mediates between human and supernatural realms. Describing how a midwife picks up a baby at the time of delivery, the Japanese widely used the words, "[she] pulls [*hiku*] the infant." The midwife pulls the child into "this world" from the province of liminal existence. With her special expertise, the midwife helps the mother and infant pass safely through the period of mourning and eventually resume the normal routines of human life. *O-shichiya*, the celebration that takes place seven days after the birth of the child, is an important rite of passage marking the reception of the child into the human realm. It may be precisely because of their unique power dealing with liminality that midwives were outcasts, shadowy figures in a segregated borderland.[59]

Alongside a midwife's power to intervene in life and death transitions she carried out infanticide. The language of killing a newborn, "returning" [*kaesu*] or "pushing [the child] back" [*oshikaesu*] into the world from which it has come, reflects the cosmology of ancestor worship. The child, once dead, clearly becomes polluted matter and family members must avoid contact with it; thus, the midwife is relegated to the job of dealing with the child's corpse. It is difficult to assess whether the deceased newborn was simply discarded with the childbirth waste of blood-soaked rags and straw or was buried in a special location or grave. In so far as local cultural norms did not recognize the newborn as fully human, it is unlikely that the family or the midwife perceived infanticide as homicide. Whether a family asks its midwife to kill the child or let it live, she is obliged to carry out the delivery, the disposal of childbirth waste (which may include the infant's dead body), and last but as important, the ritual cleansing of the house and bodies that are affected by birth pollution.

Carrying out childbirth reforms between the late nineteenth and early twentieth centuries entailed changing the age-old customs, perceptions, and roles of the midwife. Trained midwives attempted to persuade families to use a bright, warm, well-ventilated *tatami*-laid room as the birthing room. Although families in rural areas frequently resisted their suggestions, trained midwives eventually succeeded in convincing mothers to avoid dirt floors and give birth on futon mattresses covered with oil paper in a *tatami* room.[60] The "new" midwives also replaced rags with absorbent cotton, gauze cloth, or bleached cotton cloth for treating the mother and infant's bodies. Some midwives used rags along with sanitary medical products, but they instructed the mother or her family to sterilize them prior to childbirth either by boiling in water or drying them in direct sunlight.[61] Finally, trained midwives introduced the antiseptic sterilization of patients' bodies, the hands of the midwife, medical equipment, and materials to be used for childbirth. It took time and effort to accustom mothers to the use of antiseptic solutions.[62]

Giving birth while lying on your back was a sweeping change in childbirth practice that accompanied modern midwifery. Previously, women gave birth in a variety of positions: kneeling, sitting, squatting, crouching on hands and knees, and in some cases, standing. Even when a midwife or an obstetrician assisted a birthing woman, it was rare that a normal delivery occurred in a supine position.[63] Women were supposed to go through labor and deliver in a kneeling position while holding onto an attendant or straw bundles, rolled up bedding, or a rope hung from the ceiling for support.[64] When more than one attendant was available, it was not uncommon that one pushed the woman's back from behind to facilitate the delivery.[65] It was also customary for an attendant to put her hand on the perineum area to prevent its laceration. Mothers without any assistance received instructions to protect the perineum with a heel while in the kneeling position.[66] (See Figs. 4.4, 4.5, and 4.6 for a pictorial portrayal of traditional childbirth presented in a 1937 folklorist text.)

Giving birth in a sitting or squatting position has advantages and disadvantages. The force of gravity facilitates the delivery of the child, and a mother is able to pick up the newborn safely without assistance. Moreover, women were allowed a significant amount of freedom and flexibility in choosing the position in which they could give birth. Based on her discovery of diverse birthing positions, anthropologist Yoshimura suggests that, instead of following a rigid rule, women gave advice to one another regarding the most comfortable position for delivering a child. On the

Fig. 4.4 Childbirth assisted by a traditional midwife (Source: Matsushita, *Sanshū okugōri san'iku fūzoku zue*, 21). Author's note: Matsushita remarks that in cases in which the midwife is old or childbirth is difficult to handle, the birthing mother's mother or somebody who is close to the mother in labor assists

Fig. 4.5 A midwife bathing a newborn (Source: Matsushita, *Sanshū okugōri san'iku fūzoku zue*, 25)

Fig. 4.6 The aftermath of childbirth. From the text: The mother should tie the maternity band tightly and sit in a correct manner on a bamboo mat covered with rugs (Source: Matsushita, *Sanshū okugōri san'iku fūzoku zue*, 50)

other hand, there is a problem with the kneeling position. The tendency of the child to move downward rapidly and with a greater force due to gravity, the very factor that allows an easy delivery, often results in severe laceration of the perineum.[67] Most birth attendants likely lacked expertise to sew this wound and had no choice but to let it heal naturally. Injuries of this type could cause bleeding and infection, and lead to fatal medical problems such as puerperal fever.

What birthing position to use became a serious point of contention between mothers and school-trained midwives. Protests led midwives to relent to mothers who wanted to give birth in a kneeling or sitting position, but school-educated midwives were loath to abandon their training. For example, in an oral history interview, retired midwife Takejima Mii remarks that when she first began her practice in 1926, she let mothers who refused to lie down give birth in a conventional position, but she insisted mothers lie down once they had delivered the baby to break the practice of sitting upright for days after childbirth.[68] In cases where trained midwives commanded reasonable respect and authority in a given local community, women do not seem to have objected to lying down during labor. This does not mean women fully approved of the new childbirth method. Women from the Inland Sea region identified the new method of childbirth as more laborious (*shindoi*) than what they were used to, although they did not seem to have shared this feeling with their midwives out of courtesy.[69]

Although some mothers, especially those who had had children with the assistance of a traditional midwife, did not readily agree to give birth lying down, rural mothers seem to have generally been appreciative of the services of trained midwives. If a family agreed to use a warm and pleasant *tatami* room as the birthing room, mothers could avoid drafty, cold rooms that were a feature of winter deliveries. Trained midwives also endeavored to break age-old local food taboos during pregnancy and after childbirth, and their interventions led families to prepare more nutritious food for mothers. It is difficult to determine to what extent rural mothers with limited education understood the value of antiseptics and sterilization. Yoshimura's interviewees, however, expressed their appreciation that trained midwives used "white, sanitary" cotton instead of rags made of old clothes and linens.[70] In retrospect, the reaction of rural mothers toward their treatment by trained midwives shows how members of the marital family generally mistreated young mothers even when they were pregnant or giving birth. Modest rural families commonly did not give much care or attention to the health and comfort of mothers and their newborn babies. Trained midwives, on the other hand, brought comfort and attention to these often ill-treated and overworked young wives.[71]

Rural families prioritized maximizing a wife's labor time, tending to view childbirth as a burden and nuisance that took young wives away from farm and domestic work. Pressured to engage in laborious daily routines even when pregnant, expectant mothers regarded frequent pregnancies burdensome. Some of Yoshimura's interviewees attest that they loathed pregnancies to the extent that they attempted to induce abortions by engaging in particularly strenuous labor or by jumping from a high place.[72] Hata Meitsu (b. 1891), the hard-working wife of a settler in Hokkaidō, experienced 11 pregnancies. She recounts how, at one point, she tried to induce an artificial miscarriage by eating a large quantity of *konnyaku* (yam cake), following a neighbor's suggestion, but this method turned out to be ineffectual.[73]

If farming families were unwilling to allow expectant mothers to reduce their daily work, they certainly did not appreciate the time away from work and modest costs of visiting a midwife for prenatal counseling. Thus, when rural women wanted to consult with a midwife, they had to do so discreetly, minimizing the displeasure of their in-laws. For example, women had to visit a midwife on rainy days, when they were prevented from engaging fully in farm work.[74] Considering in-laws' and husbands' inattention, trained midwives provided valuable support to young wives and they became allies.

Even if they preferred to use trained midwives' services, in most cases it was not the choice of expectant mothers to employ them. Mothers and families followed other neighborhood families in choosing midwives. Less affluent families may have avoided hiring trained midwives because of the comparatively higher fees, but most likely a lack of money did not make it impossible to use their services. Many trained midwives offered poor families their services practically free of charge or for token gifts to be given at mid-summer or at the end of the year.[75] The most important reason that rural families refrained from hiring trained midwives was simply that there were few modern midwives available in many rural areas until the 1930s or 1940s.

Considering rural families' widely shared inattentiveness to the health and protection of mothers and their newborn babies and the little or almost no power young mothers could exert in selecting a midwife, the spread of trained midwives could not have happened without state intervention, national-level policymaking, and childbirth-reform campaigns run by local governments. From the early Meiji period, the government established a hygiene division within the police department in each municipality to address local health and hygiene issues such as epidemic control and health education. Between the turn of the century and the 1920s, bringing trained midwives into rural localities was an important goal to which local hygiene officers aspired. The sociologist Ochiai Emiko suggests that administrators of local governments felt pressure to have at least one school-educated midwife in their district. Local governments encouraged childbirth reforms by recruiting bright female students and offering them government-sponsored scholarships to attend midwife schools. Officials assumed that they would return to their natal villages to open practices.[76] Following the earnest solicitation of a local official, Saitō Ei, for instance, decided to attend midwife training with a scholarship provided by the county government.[77] Takejima Mii also learned about a newly established midwife-training center in the nearby city of Kanazawa from a police officer in her village who was recruiting female students as prospective midwives.[78] In both cases, in addition to the urging from a local administrator or police officer, a push from school principals crucially persuaded candidates to attend midwife schools and set up midwifery practices.

Many local governments also supported new midwives by pressuring old midwives to discontinue their practices. For example, when Takejima Mii gave notice to the police station in Monzen that she would begin her midwifery practice, the police officer in charge suggested that he would shut down the town's *tōnai* midwife, who did not have a license.[79] Local state authorities were not only supportive of the reforms that new midwives endeavored to implement, they also supervised licensed midwives so that they would follow official regulations. Until the national government instituted the public health nurse system in the late 1930s and early 1940s, it also often mobilized trained midwives for state-sponsored public health projects such as physical check-ups and vaccinations.[80]

The state demonstrated its strong support for new midwives by having local governments set up midwife service projects with public funding so that less affluent villagers could have access to a trained midwife. In one such plan implemented in the town of Kawada in Nagano prefecture, a trained midwife was employed for a monthly salary of 45 yen. Townspeople paid 1 yen per delivery to the town's government for a midwife's services, instead of the normal 5 yen charge for a licensed midwife in private practice. Formally trained midwife Kawamura Kiyo (b. 1906) opened a private practice in this poor farming town in 1927 but only attended to ten clients in her first year because few villagers could afford her services. After the town employed her the following year, the number of clients rose to 60.[81] Developing the project to employ Kawamura involved a local administration and police effort to reinforce control over unlicensed midwives.[82]

Implementing childbirth reforms in rural areas was difficult due to little incentive or insufficient money for families to hire a trained midwife. Moreover, unless they had a solid familial tie to a given locality, school-educated midwives were generally unwilling to open practices in unfamiliar farming villages. Rural recruitment was also difficult, as very few young women in rural villages had the chance to attend midwife school. When women from isolated farming areas received midwife training in urban areas, some never returned to their natal villages. This is why government support was vital in bringing trained midwives to the countryside. But even with such support the spread of childbirth reforms in rural Japan was a slow and gradual process.[83]

Reproductive Surveillance Through Midwives

The spread of school-educated midwife services was instrumental in reinforcing state surveillance of the reproduction of its citizens. The training inculcated in midwives a pronatalism that celebrated the birth of a child, making them think that abortion was repulsive and infanticide unthinkable. The state reinforced their training through local midwives' associations, which it monitored and used to prevent abortions and infanticide. Childbirth reforms implicated midwives in gathering data about and the micro-management of women's bodies. The close working relationship between local government and trained midwives suggests that the state had access to client data gathered by midwives. State reproductive surveillance was rather haphazard before Japan entered the prolonged wars years in the late 1930s and the establishment of the Ministry of Health and Welfare in 1938 in particular, but the system of maternal and infant health care and record-keeping that had been established by the early twentieth century became the basis for the government's more intrusive system of control.

Midwives were involved in their own governance. Between the late 1880s and early 1910s they formed associations at the prefectural or county level with the support of local police and government. At regular meetings the local associations reminded licensed midwives that assisting in abortions and infanticide was a crime.[84] Licensed midwives were often automatically included in local associations, which were frequently presided over by a notable physician or police officers responsible for local health and hygiene. In Fukushima prefecture, though, local organizing predated the establishment of midwife associations. For example, public health officers at major police stations organized meetings with local midwives and instructed them on issues of hygiene before the midwives' association was formed in the region in 1912, when it received a mandate from the local public health administration to carry out continued education and supervision of midwives. The Kitamurayama County Midwives' Association held annual meetings at which a physician or prefectural public health officer gave a lecture. The meetings also became a forum for midwives to exchange information about their work experiences and to establish regular fees and other rules.[85]

The 1899 rulebook for the Osaka midwives' association reveals how midwives' associations carried out policing functions. In addition to advancing scholarship on midwifery and protecting their interests, the organization stated its interest in reforming midwives' morals.[86] The section stipulating pledges specifically stated that members should avoid engaging in "immoral acts" [*haitoku no kōi*].[87] Although the book never clarified what constituted immoral acts, the term most likely referred to abortion and infanticide. The association's language resembled that found in some state edicts against abortion. The word *haitoku* connoted a violation of the Confucian tenet of following humane ways, a moral code central to the ideology of the family state, which the government endeavored to promote. Within this ideological construct, which conceived of the emperor as a father figure and his subjects as his children, killing an infant was equated with murdering the emperor's own child, whose survival and well-being were perceived to be crucial to the prosperity of the nation.[88]

State officials and leading physicians extended their control over midwives' illegal activities as well as unlicensed midwives' practices by establishing a system of mutual policing.[89] The Osaka association's 1899 rulebook stipulated that midwives should inform the association if they discovered members had violated state regulations or engaged in immoral acts. Members were obligated to inform the association's headquarters about the illegal practices of unlicensed midwives. If they failed to fulfill these duties, they had to pay a fine to the association. Moreover, the names of those who seriously violated the group's regulations would be publicized in newspapers and journals.[90]

Several midwives' journals that began circulating around 1899 were also instrumental in educating midwives about the prohibition on abortion and infanticide. For example, the March 1900 issue of *Josan no shiori* publicized a case of infanticide by an unlicensed midwife in Kisarazu in Chiba prefecture. The article notes that following the mother's request the midwife killed the baby by plugging its nose with rags and then buried its body. Based upon the investigations of a physician as well as the police, the midwife was sentenced to nine years of penal servitude. The article concluded that such a horrific act was something only demons would do.[91]

While childbirth reforms helped curtail abortion and infanticide, the new forms of surveillance went beyond preventing them. Following standardized procedures, trained midwives monitored their clients for an extended period and accumulated medical records. While midwives were micro-managing individual bodies they were simultaneously cooperating with the government's macroscopic monitoring of the populace through the collection of vital statistics.

The new midwives introduced a regimen of medicalized management, which entailed monitoring mothers' bodies for an extended period with regular prenatal and postpartum care.[92] The old generation of midwives generally intervened only at the time of delivery, except for a ritual of binding with a pregnancy sash during the fifth or seventh month of pregnancy (Fig. 4.7).

While traditional midwives would attend celebrations after the child's birth, such as the seventh day ritual following childbirth, they rarely examined a mother unless she was suffering from a severe postpartum illness. School-educated midwives, on the other hand, conducted prenatal diagnoses beginning a few months before delivery as well as at least one postpartum follow-up examination. The prenatal diagnosis was aimed at discovering any symptoms or anatomical anomalies that were likely to cause problems during delivery. When they found a serious problem, they referred the client to an obstetrician.[93] The postpartum visit was to ensure the satisfactory recovery of the mother and the healthy growth of the child. Trained midwives usually included daily visits after childbirth for two weeks to bathe the newborn. Examining the mother and child over an extended

Fig. 4.7 The maternity sash ritual (Source: Matsushita, *Sanshū okugōri san'iku fūzoku zue*, 4). Author's note: According to Matsushita, on the day of the zodiac sign of the dog in the fifth month of pregnancy, a celebration in which the maternity sash is put on takes place. A midwife puts on the sash, and although the mother's natal family sends gifts, this is not an occasion where the family invites guests

period allowed midwives to gather detailed information about the health of the mother, the fetus, and the newborn, with the aim of securing the intervention of higher-level medical professionals when necessary.

These surveillance techniques were enhanced by another standard practice that midwives followed: maintaining medical records on each patient. Textbooks published at the turn of the twentieth century often included a sample clinical chart for each patient (see Chart 4.1 for example).[94]

On the chart the midwife recorded the name, address, and age of the patient; the number of previous births; time and duration of the last menstruation; medical problems the patient had experienced; expected date of delivery; dates of the midwife's visits and the patient's condition at the time of each visit; date and time of delivery and the labor process; condition of the newborn child and the afterbirth; major procedures used during delivery; and dates of the midwife's visits during puerperium.[95] In Ishikawa prefecture licensed midwives were not only required to fill out a chart for each client but make the charts available to the police for at least ten years.[96]

Clinical charts facilitated midwife practices and state surveillance. They allowed medical professionals and state authorities to monopolize the knowledge gathered about bodies. That same knowledge could not be readily accessed by the clients themselves. In this sense, the clinical charts contributed to reducing women's autonomy vis-à-vis medical professionals and the state. Although the state did not thoroughly control the production and use of knowledge about women's bodies, it was vested with authority to access such information and to make use of it. Maintaining standardized clinical records allowed knowledge about each body to be documented and classified and facilitated the systematic accumulation of knowledge about the entire population.

The further spread of childbirth reform occurred partly with the establishment in the 1920s and 1930s of a publicly funded midwife and district nurse system and a system of government maternity books issued in the early 1940s (see Chap. 6). This maternity passbook project, built upon the history of close collaboration between trained midwives and the state, strongly encouraged mothers to request a midwife or physician to take down their childbirth records and share them with the government examination, classification, and recording system already established for new midwives' practices.

Licensed midwives also played a significant role in helping the state to collect accurate information about childbirth and failed pregnancies. The

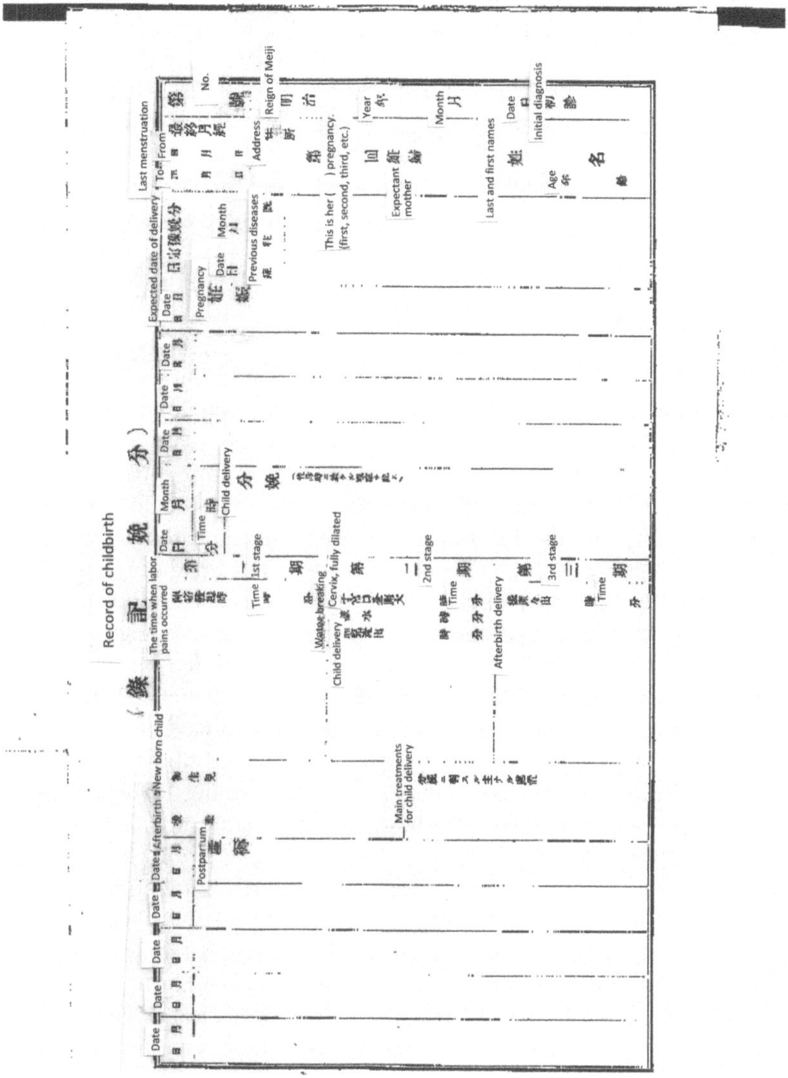

Chart 4.1 A midwife's chart (Source: Tōjō Ryōtarō and Doi Mamoru, *Shinsen sanba gaku* (Tokyo: Maruzen, 1906), Appendix)

registration law required the father of a newborn child to notify the local municipal office about each birth. Midwives were legally obliged to assist registration when a household member failed to do so.[97] In some localities, state officials expected physicians and midwives to instruct their clients' family to report the birth of a child immediately.[98]

Beginning in 1880, midwives and physicians were obligated to give their clients a death certificate so that they could present it to the local municipal office in the case of miscarriage or stillbirth that occurred after the third month of pregnancy. The certificate required them to record the name and occupation of the father, duration of pregnancy, date, time, and place of delivery, sex of the fetus, and whether the dead child's mother was a legal wife, a concubine, or an unmarried woman.[99] In Ishikawa prefecture, midwives were not to provide such a certificate if there were any doubts about the cause of death, in which case they were required to inform the police.[100] In addition to reporting cases of stillbirth, the state required medical professionals to report the deaths of infants under the age of five, including a description of the cause of death.

Because neither physicians nor licensed midwives were involved in every stillbirth or infant death, the state knew of only a fraction of the actual number. Yet, records of stillborn fetuses and infant deaths facilitated state attempts to police women's reproductive activities because the measure guaranteed a systematic disclosure of failed pregnancies and infant mortality. Ishizaki argues that the Hygiene Bureau did not necessarily intend to prevent abortion and infanticide but simply wanted more accurate statistics on infant mortality.[101] Regardless of the Hygiene Bureau's intentions, it is probable that such a requirement would have preempted midwives and their clients from performing or seeking abortions and infanticide, inasmuch as they were aware that either failing to report an unfruitful pregnancy or an infant death, or to present a suspicious report, could lead to a police investigation.

The state required a report about a fetus's death partly to identify which child would be the heir of a household. Article 968 of the Civil Code specified that even before its birth a fetus could be designated as an heir. If the child was stillborn, its inheritance rights were transferred to another child or family member. Procuring a certificate from a medical professional as a witness to the death of a fetus who had been designated a household heir was crucial for establishing the legality of a new heir.

That the legal obligations of midwives were specified in the registration and civil codes suggests that they were instrumental in gathering accurate information about each household [*ie*]. They performed an important legal and ideological function in support of the Meiji state. The household head's patriarchal control over other family members was an extension of the ideology of the family state in which the emperor was regarded as a paternal figure who protected and presided over the people of the Japanese empire. This ideology made it crucial for state authorities to support the system at a household level and to supervise the proper distribution of rights and responsibilities among family members. Thus, the state's endeavors to monitor each household and its members, including failed pregnancies, constituted an important part of social control. Licensed midwives provided essential support for this type of state surveillance over families.[102]

The Hygiene Bureau gathered information about childbirth through municipal offices and compiled it into national statistics, which were used to define the nation's health problems and shape its public health policy. From the 1880s onward, infant mortality was perceived as a significant indicator of the state of the nation's health. The cooperation of licensed midwives was essential both for information gathering and for promoting the state maternal and infant health policies.

To situate this study of pre-1930s childbirth reforms within existing feminist scholarship on the modern Japanese state's exercise of reproductive control, I would like to reconsider the widely accepted argument that the criminalization of abortion was the foundation of state control over women's reproductive activities—what historian Fujime Yuki calls "the system of abortion prohibition law" [*Dataizai taisei*].[103] Fujime interprets the mobilization and control of trained midwives as part of the state's attempt to increase population by preventing abortions. Despite her significant contribution to the history of reproductive politics in modern Japan, I would like to reevaluate her emphasis on the importance of abortion prohibition law and her tendency to view the state as a monolithic agent shaping all aspects of the system of reproductive surveillance according to that law.

My study focuses on other actors as well as the state, all with different motivations for popularizing the services of trained midwives. In addition to officials of central government, those who held a vested interest included obstetrician-gynecologists, municipal administrators, notable figures in local communities, young women who aspired to become midwives, the families of expectant mothers, and pregnant women. For obstetricians, control over women's childbirth practices—

mediated by the midwives who they had trained and supervised—expanded their professional influence and authority. For local municipal officers, politicians, and educators, the presence of school-educated midwives in their communities was visible proof of their modernizing efforts. Midwives had ambitions of their own, ranging from serving the state to obtaining independence, wealth, status, and a sense of satisfaction that a professional career could provide. The case of Shibahara Urako, who revolted against the state through her involvement in the birth-control movement in later life, suggests that midwives were not always subservient to the state or to physicians. As for the families of pregnant women, they desired not only what they considered to be a high standard of care but also the prestige attached to the services of professional midwives; being able to afford the expensive fees and other necessities associated with new childbirth practices was a testament to the family's financial affluence. Finally, even though some women felt that giving birth in the supine position was more "fatiguing" compared to giving birth in the traditional kneeling position, they appreciated new midwife-assisted childbirth methods on the whole since their bodies were treated with what they thought were "hygienic" methods and implements. Thus, the new childbirth culture prevailed as many of these different interests converged and reinforced one another.

These arguments do not imply that I support a pluralist model that ascribes to the state only a minor role. In the extensive project of training and regulating midwives, the central government, through the auspices of the Hygiene Bureau, occupied a commanding position that enabled it to direct and encourage its plans. Its authority was reinforced by its open access to the professional medical knowledge made available by medical experts, some of whom served as its officials. The state also possessed financial resources to fund midwife training. And its legislative and judicial powers, in conjunction with its local mobilization of the police, created a web of regulatory devices that circumscribed the practice of old midwives and created ample space for new midwives to make their practices viable.

What I would like to stress is the diffuse nature of the new reproductive surveillance, a position which contests Fujime's emphasis on the centrality of the abortion prohibition law in reshaping reproductive activities and surveillance. I have shown not only that there were diverse actors at play but that a variety of specific procedures were implemented to control midwives and women's reproductive bodies. My study demonstrates how the state supervision of midwives and expectant mothers was carried out through the application of disparate and often mundane techniques initiated by various agents at different time periods. These included:

1. Registering midwives.
2. Organizing them into midwives' associations.
3. Educating them at training centers and through midwifery journals.
4. Subordinating midwives to physicians in terms of their professional responsibility.
5. Requiring midwives to issue a certificate for the dead fetus and infant.
6. Creating prefectural and national statistics for infant mortality
7. Linking midwives with the local administration, police, and physicians.
8. Introducing the new birthing method and thereby depriving women of their capacity to deliver a child on their own or assisted by untrained birth attendants.
9. Providing prenatal and postnatal care.
10. Using clinical charts.

The introduction of these practices was in no way planned or systematic. Some were initiated by state administrators and Japanese physicians while others were embedded in the modern midwifery method adopted from Europe. State reproductive surveillance was tightened through a range of organizational and medical practices constituting new childbirth culture.[104]

Monitoring and managing women's bodies and their health in the late nineteenth and early twentieth centuries not only occurred within the field of reproductive medicine, but also through women's education and health-reform activism. As well as being circulated by trained midwives, new scientific knowledge of bodies and reproduction started to be disseminated by way of schools and print media. On one hand, educators and health reformers sought to recreate women's bodies and improve their reproductive capacity by using new discourses and techniques, as we see in the implementation of physical education programs and the reforming of young women's health habits, clothes and hairstyles. On the other hand, women who were educated in girls' schools that taught health science and adopted reformed clothes began to engage in resistances and collusions in ways that not infrequently betrayed the expectations of their educators.

The next chapter covers this topic, looking at the expansion of bio-power concerning women's reproductive health in areas broader than maternal health care. It discusses how women themselves became important players in modernizing their understanding of their bodies and health, either consciously or inadvertently, contributing to the diffusion of modern bio-power.

Notes

1. Kōzai Toyoko, "Ainu wa naze 'Yamani ni nigeta' ka: Bakumatsu ezochi ni okeru 'waga kuni saisho no kyōsei shutō' no okuyuki," *Shisō*, No. 1017 (January, 2009): 78–101. For English-language publications on this topic, see Brett Walker, "Epidemic Disease, Medicine, and the Shifting Ecology of Ezo," in Michael Weiner, ed., *Race, Ethnicity and Migration in Modern Japan, Vol. 1: Race, Ethnicity and Culture in Modern Japan* (London: RoutledgeCurzon, 2004): 397–425; Ann Jannetta, *The Vaccinators: Smallpox, Medical Knowledge, and the "Opening" of Japan* (Stanford, CA: Stanford University Press, 2007): 161–162. A classic study on this subject includes, Nakano Misao, "Wagakuni saisho no kyōsei shutō," *Itan*, No. 6 (December 1954): 1209–1210. For more recent work, see, for example, Kikuchi Isao, "Hōsō ryūkō to ainu shakai: Jūkyūseiki zenki no jinmei sōshitsu to ezochi kaihatsu," *Rekishi kagaku*, No. 171 (December 2002): 1–11; Kaiho Yōko, "Kinsei ishikari basho hōsō ryūkōshi nōto: Bunka 14-nen wo chūshin ni," *Sapporo no reikishi*, No. 53 (August 2007): 22–33. For a succinct account of the development of government-led compulsory smallpox vaccinations in early Meiji, see Kōzai Toyoko, "Yobō sesshu to iu 'eisei': shutō no rekishi no hanshō kara," *Gendai shiso*, 38:3 (March 2010): 226.
2. On Japan's shifting policies on cholera control during the nineteenth century, see William Johnston, "The Shifting Epistemological Foundations of Cholera Control in Japan, 1822–1900," *Extrême-Orient, Extrême-Occident*, Vol. 37 (2014): 171–196.
3. Aoyagi Seīchi, *Kindai iryō no akebono* (Kyoto: Shibunkaku Shuppan, 2011): 307–311; Imai Shōji, "Hesuperia gō jiken ni tsuite," *Rekishi kyōiku*, 12:1 (January 1964): 35–40; Inoue Kiyoshi, *Jōyaku kaisei* (Tokyo: Iwanami Shoten, 1955): 42–43. For a recent English language study on this, see Harald Fuess, "Informal Imperialism and the 1879 'Hesperia' Incident: Containing Cholera and Challenging Extraterritoriality in Japan," *Japan Review*, No. 27 (2014): 103–140.
4. Oazaki Kōji, "Bankoku eisei kaigi to kindai nihon," *Nihonshi Kenkyū*, No. 439 (March, 1999): 135.

5. For English-language discussions of prostitution and VD examinations in nineteenth-century Japan, see Susan Burns, "Bodies and Borders: Syphilis, Prostitution, and the Nation in Japan 1860–1890," *U.S.–Japan Women's Journal, English Supplement, No. 15* (1998): 3–30; Ann-Marie Lynne Davis, *Bodies, Numbers, and Empires: Representing "the Prostitute" in Modern Japan, 1850–1912* (Ph.D. Dissertation, University of California, Los Angeles, 2009). Frühstück discusses the military's attempt to control syphilis among its soldiers through the licensed prostitution system and VD examinations for prostitutes. Sabine Frühstück, *Colonizing Sex: Sexology and Social Control in Modern Japan* (Berkeley: University of California Press, 2003): 35–49.

   Fukuda Mahito, "Kenbai no hajimari to baidoku no gensetsu," in Fukuda Mahito and Suzuki Noriko, eds., *Nihon baidoku shi no kenkyū* (Kyoto: Shibunkaku, 2005): 140–145; Miyazaki Chiho, "Nihon saisho no baidoku kensa to roshia kantai," in Fukuda and Suzuki, eds., *Nihon baidoku shi no kenkyū*, 175–221. For the history of licensed prostitution and the government-mandated VD exams of sex workers, see Hayashi Yoko, *Sei wo kanri suru teikoku: Kōshōseido ka no "eisei" mondai to haishō undō* (Osaka: Osaka Daigaku Shuppankai, 2017).
6. For the significance of the *Isei* document and the public health system that Nagayo and other health officials created, see Susan Burns, "Constructing the National Body: Public Health and the Nation in Nineteenth-Century Japan," in Timothy Brook and Andre Schmid, eds., *Nation Work: Asian Elites and National Identities* (Ann Arbor: University of Michigan Press, 2000): 17–28. Ruth Rogaski also presents a useful discussion of ideas about the new health and medical system developed by the major architects of the Meiji medical system, such as Nagayo Sensai and Gotō Shōjirō. Ruth Rogaski, *Hygienic Modernity: Meanings of Health and Disease in Treaty-Port China* (Berkeley: University of California Press, 2004): 136–164.
7. Kōseishō Imukyoku, ed., *Isei hachijiū-nen shi* (Tokyo: Insatsukyoku Chōyōkai, 1955): 484.
8. Kasahara Hidehiko and Kojima Kazutaka, *Meiji ki iryō eisei gyōsei no kenkyū* (Kyoto: Mineruva Shobō, 2011): 27–45, 55–58, 133–139, 55–58.
9. For the establishment of the koseki system in early Meiji and its outcomes, see, for example, Fukushima Masao, "Meiji yonen kosei hō no shiteki zentei to sono kōzō," in Fukushima Masao, ed., *Kosei seido to "ie" seido* (Tokyo: Tokyo Daigaku Shuppankai, 1959): 94–169; Ninomiya Shūhei, "Kindai koseki seido no kakuritsu to kazoku no tōsei," in Toshitani Nobuyoshi, Kamata Hiroshi, and Hiramatsu Hiroshi, eds., *Koseki to mibun tōroku* (Tokyo: Waseda Daigaku Shuppanbu, 2005): 146–164.

10. See, for example, Kobayashi Takehiro, *Kindai nihon to koshū eisei* (Tokyo: Yūzankaku, 2001): 1–12, 39–57; Ozaki Kōji, "1879-nen korera to chihō eisei seisaku no tankan: Aichi-ken wo jirei to shite," Nihonshi kenkyū, No. 418 (June 1997): 41–48.
11. For a general history of statistics in Japan, see Shimamura Shirō, *Nihon tōkei hattatsushi* (Tokyo: Nihon Tōkei Kyōkai, 2008).
12. For a discussion on the physical examination for screening conscripts, see Ījima Shigeru, *Nihon Senpei Shi* (Tokyo: Kaihatsu Sha, 1938); Frühstück, *Colonizing Sex*, 26–35. An example of a health manual intended to be read by soldiers is: Uyama Dōseki and Tamura Shunji, eds., *Heishi no eisei* (Tokyo: Eiseigaku Kenkyū Kai, 1909). Hōgetsu Rie explores health surveillance of children with a focus on school dental hygiene during the early twentieth century in *Kindai nihon ni okeru eisei no tenkai to juyō* (Tokyo: Tōshindō, 2010). Sawayama Shinichi presents a detailed history of trachoma prevention and treatment through the school health system. Sawayama Shinichi, *Gakkō hoken no kindai* (Tokyo: Fuji Shppan, 2004). See also Frühstück, *Colonizing Sex*, 49–54.
13. Ishizaki Shōko, "Nihon no dataizai no seiritsu," Rekishi hyōron, No. 571 (November 1997): 58–60.
14. Yasuda Muneo, "Meiji-ki no sanba yōsei ni tsuite," *Shikyō*, No. 35 (September 1997): 15–18.
15. Koizumi Eīchi, *Dataizai kenkyū* (Tokyo: Ganshodō Shoten, 1934): 55–57.
16. Ishizaki, "Nihon no dataizai no seiritsu," 66.
17. Until the turn of the twentieth century, many expectant mothers in Japan gave birth without the assistance of a professional midwife or a physician. When midwives were used, it was often specifically for the purpose of assisting in abortion and infanticide. This is one of the reasons that the Meiji government repeatedly criticized traditional, untrained midwives and pressed them to discontinue their practice. A physician's note on the practice of midwifery published in the journal *Josan no shiori* [*Guides for Midwifery*] indicates that it was not at all unusual for patients to ask a physician or a midwife to assist in abortions. The writer declared that this "pernicious custom" was finally in the process of being eradicated. *Josan no shiori* 5 (October 1900): 22.
18. Ishizaki, "Nihon no dataizai no seiritsu," 54–55.
19. Kōseishō Imukyoku, *Isei hachijiū-nen shi* (Tokyo: Insatsukyoku Chōyōkai, 1955): 205–206; Ogata, *Nihon sanka gaku shi*, 1054–1055; Usami Hideki, "Meijiki no sanba kisoku: Shiga-ken no jirei," *Shakai kagaku*, Vol. 45 (March, 1990): 8–9.
20. Kōseishō Imukyoku, *Isei hachijiū-nen shi*, 206–208.

21. Ogata, *Nihon sanka gaku shi*, 1054–1055; Murakami Nobuhiko, *Meiji joseishi: On'na no shokugyō, chūkan-kōhen* (Tokyo: Riron-sha, 1971): 55–57; Usami, "Meijiki no sanba kisoku," 11–15.
22. Murakami, *Meiji joseishi: On'na no shokugyō, chūkan-kōhen*, 55–57.
23. Ibid., 9; Kōseishō Imukyoku, *Isei hachijiū-nen shi*, 205–206; Ogata, *Nihon sanka gaku shi*, 1055. Medical practitioners were able to purchase imported forceps and drugs as foreign trade expanded in the early years of Meiji, but the government had not yet rigorously regulated the purchase and use of these products. These regulations seem to have been a response to this situation.
24. Homei Aya discusses the medicalization of Meiji midwifery, focusing on the rise of the category of "normal" and "abnormal" births, and the division of labor between obstetricians and midwives. Aya Homei, "Birth Attendants in Meiji Japan: The Rise of the Biomedical Birth Model and a New Division of Labour," *Social History of Medicine* 19, No. 3(2006): 407–424; and Homei Aya, " 'Seijō san' to 'Kindai eisei': Kindai sanba no senmon bunya wo meguru poritikkusu," *Seibutsugaku shi kenkyū*, No. 70 (December, 2002): 1–16.
25. For the attempts by midwife associations nationwide to obtain the rights to administer injections and prescribe medication in the 1920s and 1930s, see Kimura Naoko, *Shussan to seishoku wo meguru kōbō: sanba josanpu dantai to sankai no hyakunen* (Tokyo: Ōtsuki Shoten, 2013): 69–108; Ōide Harue, "Mikan no sanshi hō to sanba no kindai," in Shirai Chiaki, ed., *Umi sodate to josan no rekishi: Kindaika no nihyaku-nen wo furikaeru* (Tokyo: Igaku shoin, 2016): 52–74.
26. For a useful recent study on midwife education during the Meiji period, see Ogawa Keiko, "Seiyō kindai igaku no dōnyū to sanba no yōsei," in Shirai, ed., *Umi sodate to josan no rekishi*: 26–46; Kimura, *Shussan to seishoku wo meguru kōbō*: 19–67.
27. Nihon Sanka Fujinka Gakkaishi Hensan Īnkai, ed., *Nihon sanka fujinka gkkaishi* (Tokyo: Shindan to Chiryō-sha, 1971): 10; Ogata, *Nihon sanka gaku shi*, 1177, 1205, 1711.
28. Ogata, *Nihon sanka gaku shi*, 1714–1721.
29. Ibid., pp. 1216–1223; Usami, "Meijiki no sanba kisoku," 31–39. For a detailed discussion of the Midwife Ordinance, see Julie Rousseau, *Enduring Labors: The "New Midwife" and the Modern Culture of Childbearing in Early Twentieth Century Japan* (Ph.D. dissertation, Columbia University, 1998), 69–113. See also Ogawa, "Seiyō kindai igaku no dōnyū to sanba no yōsei," 38–39.
30. In many regions, "old midwives" were commonly called *toriage bāsan* (the old woman who induces the child into this world). Many other

names were used depending upon the region. Onshi Zaidan Boshi Aīkukai ed., *Nihon san'iku shūzoku shiryō shūsei* (Tokyo: Daīchi Hōki Shuppan, 1975): 219–221.
31. G. Leopold and P. Zweifel, Nikawa Toshio, trans., *Josan-gaku* (Tokyo: Maruzen, 1898): 1–2.
32. Ogata Masakiyo, *Josanpu-gaku kōgi* (Tokyo: Maruzen, 1906): 2. Ogata strongly endorsed the use of the term *josanpu*, which he coined to indicate midwives, instead of the word *sanba*, which literally means "old women for childbirth" and was associated with old-fashioned midwives. The term *josanpu* was not accepted into common usage until the post-World War II period. Toward the end of the twentieth century, the Japanese medical community adopted the new term *josanshi* to avoid gender discrimination.
33. Ibid., 3.
34. Fukui Shigeko, *Ninpu no barazono* (Osaka: Ogata Byōin Josanpu Gakkai, 1898): 2.
35. Ibid.
36. Murakami notes that publicly funded midwife schools generally made it mandatory for students to take courses on moral training [*shūshin*], during which they were required to read the Confucian text *On'na shōgaku* and all four volumes of the government authorized ethics textbook *Meiji kōsetsu-roku*. Murakami, *Meiji joseishi: On'na no shokugyō, chūkan-kōhen*, 58.
37. Sasakawa Misu, *Sanba jūsan kai* (Niigata: Sasakawa Misu, 1892).
38. Ibid., 11.
39. Kondō Shōichi, *Joshi shokugyō an'nai* (Tokyo: Hakubun-kan, 1906): 314–315.
40. Murakami, *Meiji joseishi: On'na no shokugyō, chūkan-kōhen*, 57–58.
41. Yanagisawa Fumiaki, *Naganoken Meiji iji-shi* (Ueda: Ueda-shi Ishikai Fūzoku Shiryōkan, 1977): 204.
42. Yanagisawa, *Naganoken Meiji iji-shi*, 202–203. It should be noted that there were many other unlicensed local midwives who were not included in these statistics. As an example of the number of new midwives in cities, in 1885, within the 15 wards of Tokyo prefecture there were 117 trained midwives and 404 untrained midwives. Ishizaki Shōko, "Meiji-ki no seishoku wo meguru kokka seisaku," *Rekishi hyōron*, No. 600 (April 2000): 41.
43. Tashiro Itsurō, ed., *Miyazaki ken ishi*, Vol.1 (Miyazaki: Miyazaki ken Ishikai, 1978): 1815.
44. *Josan no shiori* 102 (November 1904): 22–29. By the end of 1902 there were a total of 397 registered midwives in this prefecture, including untrained ones. *Josan no shiori* 103 (December 1904): 3.

45. Nishikawa Mugiko, *Aru kindai sanba no monogatari: Noto, Takejima Mī no katariyori* (Toyama: Katsura Shobō, 1997): 328–329. Again, we should keep in mind that there would have been a number of unlicensed midwives who were not included in these statistics.
46. Murakami, *Meiji joseishi: On'na no shokugyō, chūkan-kōhen*, 65. Kondō, *Joshi shokugyō an'nai*, 314–315.
47. See, for example, Suzuki Yūko, ed., *Adachi on'na no rekishi: Ashibue no uta* (Tokyo: Domesu Shuppan, 1989): 156–162. In contrast, nurses tended to come from families of higher status than those of midwives during the early years after the nursing profession became available to women. Indeed, many daughters of middle-class and upper-middle-class families wanted to serve as nurses in the late nineteenth century and in the 1910s. Murakami, *Meiji joseishi: On'na no shokugyō, chūkan-kōhen*, 220–243.
48. For example, see Takahashi Tatsugorō, "Joshi no kōtō no shokugyō to shite josanpu gyō wo suisensu," *Josan no shiori* 96 (May1904): 22–27.
49. See, for example, *Josan no shiori* 8 (January 1897): 13; Nishikawa, *Aru kindai sanba no monogatari.*
50. Takada Jūrō, *Ninshin, shussan, ikuji ni kansuru kyōdo yamato ni okeru minzoku* (Nara: Nara-ken shakai jigyo kyokai, 1938, Reprinted in *Nihon kodomo no rekishi sōsho 12*, Tokyo: Kyūzansha, 1997: 27–28; Matsushita Sekijin, *Sanshū okugōri san'iku fūzoku zue* (Nagoya: Seibunkan Shoten, 1937): 16–18, 20–21; Kamata Hisako, Miyasato Kazuko, Suganuma Hiroko, Fujikawa Yūko, and Sakakura Yoshio, *Nihonjin no koumi kosodate* (Tokyo: Keisō Shobō, 1990): 122–125; Nishikawa, *Aru kindai sanba no monogatari*, 125–127; Yoshimura Noriko, *Osan to deau* (Tokyo: Keisō Shobō, 1985): 195–196; Murakami, *Meiji joseishi: On'na no shokugyō, chūkan-kōhen*, 53; Fujita Makoto, *Osan kakumei* (Tokyo: Asahi Shinbunsha, 1988): 154; Emiko Ochiai, "Aru sanba no nihon kindai" in Ogino Miho et al., eds., *Seido to shite no "on'na": Sei, san, kazoku no hikaku shakai shi* (Tokyo: Heibon sha, 1990): 53; Kikuchi Sakae, "Jitaku de undeita hitobito: nōsangyoson no taikensha no katari kara," Shirai, ed., *Umi sodate to josan no rekishi*: 140.
51. Yoshimura, *Osan to deau*, 112–113; Kamata Hisako, Miyasato Kazuko, Suganuma Hiroko, Fujikawa Yūko, and Sakakura Yoshio, *Nihonjin no koumi kosodate* (Tokyo: Keisō Shobō, 1990). Challenging the scholarship of the mainstream tradition of folklore studies led by Yanagita Kunio on the birth house, which tends to see the primary purpose of the birth house as a way to contain birth pollution, Akamatsu Keisuke argues that one of the reasons that the custom of the birth house survived was because it provided a place where infanticide and abortion could be performed

without the knowledge of state authorities. Akamatsu Keisuke, *Hijōmin no minzoku bunka* (Tokyo: Akashi Shoten, 1986): 87.

52. Fushimi also presents cases in Fukui and Yamagata prefectures where trained midwives assisted births at local birth houses from the Taishō period through to the mid-1960s. Fushimi Yūko, "Sanya, kegare, shussan no shisetsuka," in Shirai, ed., *Umi sodate to josan no rekishi*: 149–154. For an extensive study on this issue, see Fushimi Yūko, *Kindai nihon ni okeru shussan to ubuya* (Tokyo: Keisō Shobō, 2016).
53. Yoshimura, *Osan to deau*, 165.
54. Yoshimura, *Osan to deau*, 112–113; Takada Jūrō, *Ninshin, shussan, ikuji ni kansuru kyōdo yamato ni okeru minzoku*, 48–50; Fujita, *Osan kakumei*, 114–117.
55. Kamata Hisako et al., eds., *Nihonjin no koumi kosodate*, 127.
56. *Josan no shiori* 8 (January 1897): 13.
57. Nishikawa, *Aru kindai sanba no monogatari*, 102–103, 105–119.
58. Ibid., 137.
59. Kamata Hisako, "Sanba: Sono miko teki seikaku ni tsuite," *Seijō bungei*, 42 (1966): 47–60; Honda Masuko, "Arau onna kō," in *Shōjo fuyū* (Tokyo: Seidosha, 1986): 181–202.
60. Nishikawa, *Aru kindai sanba no monogatari*, 141, 144; Yoshimura, *Osan to deau*, 154, 163, 171; *Fujin eisei zasshi* 138 (May1901): 23–24. *Josan no shiori* 4 (September 1897); ibid. 52 (September 1900): 1–2, 6; ibid. 83 (April 1903): 2–6; ibid. 107 (April 1905): 10
61. Nishikawa, *Aru kindai sanba no monogatari*, 126–128, 141, 144; Yoshimura, *Osan to deau*, 154,163, 171; Kamata et al., *Nihon-jin no koumi kosodate*, 125; *Josan no shiori* 52 (September 1899): 2; and 107 (April 1905): 10.
62. Nishikawa, *Aru kindai sanba no monogatari*, 143; *Josan no shiori* 22 (March 1898): 1–4; ibid. 2 (September 1900): 3–4; ibid. 84 (May 1903): 1–8.
63. Kamata et al., eds., *Nihon-jin no koumi kosodate*, 122–123, 128–129; Yoshimura, *Kodomo wo umu*, 16–19; and Os*an to deau*, 111–114, 146–148, 196, 210; Ochiai, "Aru sanba no nihon kindai," 280–281; Fujita, *Osan kakumei*, 44, 130.
64. Takada, *Ninshin, shussan, ikuji ni kansuru kyōdo yamato ni okeru minzoku*, 27; Matsushita, *Sanshū okugōri san'iku fūzoku zue*, 20–21; Kamata et al., eds., *Nihon-jin no koumi kosodate*, 122–125, 128–129; Nishikawa, *Aru kindai sanba no monogatari*, 126–127; Yoshimura, *Osan to deau*, 111, 116–117; Murakami, *Meiji joseishi: On'na no shokugyō, chūkan-kōhen*, 54.
65. Yoshimura, *Osan to deau*, 111, 146–147, 196; Ochiai, "Aru sanba no nihon kindai," 280–281. In many regions it was common for the husband

to play the role of birth attendant. Matsushita, *Sanshū okugōri san'iku fūzoku* zue, 22; Kamata et al., eds., *Nihon-jin no koumi kosodate*, 125; Yoshimura Noriko, *Kodomo wo umu* (Tokyo: Iwanami Shoten, 1992): 24–27. In some communities, however, men's entry into the birthing room or birth house was entirely prohibited. Matsushita, *Sanshū okugōri san'iku fūzoku* zue, 22; Yoshimura, *Kodomo wo umu*, 15.

66. Yoshimura also notes that it was often the case that a family member of the expectant mother called in a neighbor skilled in childbirth *after* the mother had delivered the infant on her own. The birth attendant's role was to cut the umbilical cord and remove the afterbirth. Yoshimura, *Osan to deau*, 146–148. For various positions that the mother took for delivering the baby, see also Kikuchi Sakae, "Jitaku de undeita hitobito: nōsangyoson no taikensha no katari kara," Shirai, ed., *Umi sodate to josan no rekishi*: 143–144.
67. Yoshimura, *Osan to deau*, 226–227.
68. Nishikawa, *Aru kindai sanba no monogatari*, 141.
69. Yoshimura, *Osan to deau*, 163, 210, 230.
70. Ibid., 155–156, 163, 223. Rags washed with boiling water or dried in the sun were indeed hygienically acceptable, though.
71. Ibid., 165; Ochiai, "Aru sanba no nihon kindai," 288.
72. Yoshimura, *Osan to deau*, 148, 165–166.
73. Takahashi Mieko, *Daichi ni kizanda seishun: hokkaidō wo hiraita on'na tachi* (Tokyo: Nihon Keizai Hyōron-sha, 1985): 241.
74. Ochiai, "Aru sanba no nihon kindai," 288.
75. Ibid., 288–289, 296–297.
76. Ibid., 265–267; Ishizaki, "Meiji-ki no seishoku wo meguru kokka seisaku," 48.
77. Ochiai, "Aru sanba no nihon kindai," 266–267.
78. Ibid., 227–279, 310.
79. Nishikawa, 97. In fact, Mii wanted the experienced *tōnai* midwife to continue her practice and relayed that wish to the police officer. Mii did not believe that the police actually prohibited the *tōnai* midwife from practicing; nevertheless, within a year after Mii began her practice, the only remaining *tōnai* midwife ceased to work. This greatly helped Mii expand her influence over her clients, who were left with no alternative.
80. It was also common for midwives to provide simple medical care in rural areas where physicians were scarce. Ochiai, "Aru sanba no nihon kindai," 294–295.
81. Fujita, *Osan kakumei*, 128–133.
82. Ibid., 132–133. The dilemma the government faced in dealing with unlicensed midwives was demonstrated, for example, in an incident in

1929, when seven untrained midwives were arrested for lacking a license, only to be released within several days due to an insufficient number of midwives in Shizuoka prefecture. Ochiai, "Aru sanba no nihon kindai," 316, originally cited in Nihon Josanpu-kai Shizuoka Shibu, ed., *Shizuoka-ken josanpu kai shi* (1966).

83. For more on the popularization of new childbirth methods, see Aya Homei, "Midwives and Medical Marketplace in Modern Japan," *Japanese Studies*, Vol. 32, No. 2 (2012): 275–293.
84. One of the earliest of these prefectural midwives' associations was organized in Kyoto by the prefectural governor in 1875, followed by those in Tokyo and Osaka in 1888. Murakami, *Meiji joseishi: On'na no shokugyō, chūkan-kōhen*, 64; *Josan no shiori*, 14 (June 1897): 23–27; Ochiai, "Aru sanba no nihon kindai," 296–297; Nishikawa, *Aru kindai sanba no monogatari*, 289–290. It was often the prefectural administration that promoted the establishment of midwives' associations. For example, Niigata prefecture issued an edict in 1900 to outline the rules for local midwives' associations. *Josan no shiori*, 48 (May 1900): 25–26. According to that edict, Niigata City Midwives' Association, headed by physician Hasegawa Kanji, was established in 1901. It was assisted by several prefectural police officials and a number of local physicians. *Josan no shiori*, 11 (April 1897): 20–24. These local midwives' associations were organized into a nationwide umbrella organization, called the "Nihon sanba kai" in 1927. Nishikawa, *Aru kindai sanba no monogatari*, 290.
85. During the Taishō period, midwives led a nationwide movement to take over the management of the midwives' associations rather than leave them in the hands of state administrators and physicians, a move which reflected the rise of democratizing trends during that period. As midwives took charge of their associations' activities, the nature of many associations changed from being a state agency into a union of skilled workers. Ochiai, "Aru sanba no nihon kindai," 296–299.
86. Fujimura Takahide, ed., *Osaka sanbakai kaisoku* (Osaka: Osaka Sanbakai Honbu, 1899): 4.
87. Ibid., 17.
88. This idea was manifested, for example, in midwife Shibahara Urako's view of infants in relation to the emperor. Fujime Yuki, "Aru sanba no kiseki: Shibahara Urako to sanji chōsetsu," *Nihonshi kenkyū*, 366 (February 1993): 109.
89. Despite the vigilance of police and trained midwives, many abortion specialists, who were called *oroshi bāsan* or *sashi bāsan*, practiced underground. Fujime points out that it was common for waitresses who worked at hotels (and perhaps at cafes as well) to use the services of such abortion

providers when they became pregnant as a result of providing sexual services to customers. One abortionist who practiced in Hiroshima prefecture in the mid-1910s told a waitress who sought her service that she was willing not only to perform abortion for 6 yen but to teach the waitress the abortion method for 40 yen. Fujime Yuki, *Sei no rekishi gaku* (Tokyo: Fuji Shuppan, 1997): 133.

Such abortionists seem to have commonly inserted roots of ground cherry into the woman's vagina. The reason why abortionists were often called "*sashi*" [insert] *bāsan* was due to the act of "inserting." Folklorist Kanzaki Nobutake describes an interview with an unlicensed physician working in a pleasure quarter in Nagoya. The physician remarked that this procedure was one of the safest methods for abortion: An alkaloid that is comprised of ground cherry roots petrifies the fetus within a few days. Various abortifacients seem to have been sold under the guise of medication for inducing menstruation. Kanzaki Nobutake, *Kikigaki yūkaku Narikomaya: Fushigina basho no fōkuroa* (Tokyo: Kōdansha, 1989): 139–141; Fujime, *Sei no rekishi gaku*, 133.

90. Fujimura, *Osaka sanbakai kaisoku*, 19–21. The fine was specified as up to 5 yen.
91. *Josan no* shiori 46 (March 1900): 21–22.
92. Ochiai's interview with Saito Ei shows how the midwife's diagnosis during pregnancy was instrumental for the expectant mother's peace of mind. Ochiai, "Aru sanba no nihon kindai," 288.
93. Ochiai, "Aru sanba no nihon kindai," 288.
94. This was included as the Appendix of Tōjō Ryōtarō and Doi Mamoru, *Shinsen sanba gaku* (Tokyo: Maruzen, 1906).
95. Not all midwives recorded their patients' conditions precisely in this format. Nonetheless, most attempted to record some basic information about their clients and newborn children.
96. Nishikawa, *Aru kindai sanba no monogatari*, 296.
97. Kinoshita Seichū, *Sanba-gaku kōgi* (Tokyo: Nankodo, 1902), Appendix, 16–17.
98. *Josan no shiori*, 11 (April 1897): 20–21.
99. Kinoshita, *Sanba-gaku kōgi*, Appendix, 16–18. Ogata Masakiyo discussed the midwife's responsibility for accurately identifying and recording whether the child was born by a rightful wife or by a concubine to clarify the child's inheritance rights. *Josan no shiori*, 27 (August 1898): 25–26.
100. Nishikawa, *Aru kindai sanba no monogatari*, 296.
101. Ishizaki, "Meiji-ki no seishoku wo meguru kokka seisaku," 43–44.
102. For the ideological function of the *koseki* registration system, see Kano Masanao, *Senzen "ie" no shisō* (Tokyo: Sōbunsha, 1983): 38–59.
103. Fujime, *Sei no rekishi gaku*, 117–140.

104. It would be interesting to look at the standardization of childrearing, the area adjacent to childbirth. Since the early Meiji period, many manuals on "scientifically validated" childrearing methods were published, and "specialists" on this subject, including educators, physicians, and psychologists, provided their views through their writing and public lectures. On how the discourses and practices of childrearing developed in the early twentieth century, see, for example, Kathleen Uno, *Passages to Modernity: Motherhood, Childhood, and Social Reform in Early Twentieth Century Japan* (Honolulu: University of Hawai'i Press, 1999); and Mark Jones, *Children as Treasures: Childhood and the Middle Class in Early Twentieth Century Japan* (Cambridge, MA: Harvard University Asia Center, 2010).

CHAPTER 5

Women's Health Reforms in Japan at the Turn of the Twentieth Century

As I mention at the beginning of Chap. 4, the transition from the early modern to modern health care system, and the spread of bio-power that accompanied it, involved a wider transformation than reconstructing the medical field. The improvement of the health of citizens was also sought in areas other than the field of medicine—including education and military training—by employing modern scientific and medical knowledge. With the creation of the modern nation state in the age of modern imperialism, advancing the health of the nation emerged as an important goal, and by the late nineteenth century many health reformers and educators began to engage in spirited debates on the broader issues of Japanese women's health, beyond the childbirth reforms I discuss in Chap. 4. The spread of health-reform activism at the turn of the twentieth century led to further expansion of modern bio-power through education in schools and in the print media outside the medical field. Health activists, such as medical professionals, government officials, and educators, introduced new gender and racial stereotypes that portrayed Japanese women's bodies as remarkably frail and defective. At the same time, they believed they could improve these bodies by adopting new, scientifically approved health practices from Europe and North America.[1]

The health-reform movement helped expand modern forms of social control over bodies, an idea discussed by Michel Foucault and developed by feminist scholars. They claim that power works directly on bodies,

Y. Terazawa, *Knowledge, Power, and Women's Reproductive Health in Japan, 1690–1945*, Genders and Sexualities in History,
https://doi.org/10.1007/978-3-319-73084-4_5

incorporating them into disciplinary practices formulated by modern scientific and medical knowledge. Their scholarship pays particular attention to the newly introduced minute discursive mechanisms that redefine and manage bodies to make them more productive and useful.[2] The health-reform movement in Japan during the late nineteenth and early twentieth centuries deployed a series of these discursive mechanisms to analyze, evaluate, and discipline women's bodies. Women internalized these disciplinary practices, which were often expressed in various forms of regulation of bodily habits. This chapter focuses on physical education and clothing reforms as instances of the operation of such power, which ultimately relied on the authority of scientific knowledge of human bodies that was considered valid at the time.

In addition to transforming women's daily regimen and providing physical education to schoolgirls, the health-reform movement strove to change women's understanding of their own bodies. More and more women were exposed to medicalized knowledge about their bodies by attending schools, reading magazines, or seeing physicians trained in modern Western medicine. Further, new standards for physique, fitness, and beauty that were introduced by health science courses and popular magazines made women self-conscious subjects, scrutinizing their bodies and striving to modify themselves to attain those ideals. Both European and Japanese health reformers, steeped in racial bias and lacking sensitivity about diversity, often presented an idealized body of a European woman as the model. This made Japanese women feel a constant sense of inferiority and, thus, created a new type of oppression caused by a desire to possess a body like that of a woman of European descent.

At times, women engaged in imitating the dress and behavior of flappers of the Roaring Twenties. Those behaviors undermined the intentions of medical professionals and educators, but these forms of resistance took shape within bodies and perspectives that had already been reshaped by modernizing discursive practices. Since this resistance was located inside the web of the discursive power it was meant to subvert, the results were frequently ambiguous. In other words, women were located within a web of knowledge and power. Their desire, passion, and behavior were formed within this discursive space, effected by circulating knowledge and power. Within these power networks, power and resistance do not work simply to affect each other. Attempts on the part of the state and medical experts to control women and women's resistance work in multiple ways, often producing unintended collusions or subversions.

With these broad theoretical perspectives in mind, this chapter examines the new discourses about women's bodies circulating through diverse organizations, institutions, and publications. The women's health-reform movement involved various agents but emerged only with the Meiji state's tacit approval and encouragement. The Meiji state, for instance, only allowed philanthropic and activist organizations when it deemed they would serve the nation's interests. Moreover, the direct involvement of influential state officials and members of the imperial family was often indispensable for forming such groups and schools.[3] In this way, although health activists worked both inside and outside state institutions, women's health reforms, on the whole, corresponded with state policies and ideology. In other words, as I argue in Chap. 4, within the networks of expert knowledge and bio-power, the state occupied a strategically advantageous position for circulating scientific knowledge about women's bodies that promoted its interests.

Several associations, journals, and schools were key to women's health reforms in the late nineteenth and early twentieth centuries. The Japan Association for Women's Hygiene [Dainihon Fujin Eiseikai], the primary group promoting the women's health-reform movement, was established in 1888 by leading female physicians, aristocrats, and educators. It also drew the support of numerous prominent male physicians. Even though she was not involved in the actual running of the association, Princess Yoriko—a member of the imperial family and wife of Prince Komatsunomiya Akihito (1846–1903)—was its president, a fact that lent immense prestige and legitimacy to the association's efforts. Its members included female physicians, nurses, midwives, teachers, and socialites. This group was among other privately formed health organizations in the late nineteenth century, including the Japan Private Association for Hygiene [Dainihon Shiritsu Eiseikai], and the Japan Association for Physical Education [Nihon Taiikukai].[4] Health reformers' views were also disseminated among educated women through *Fujo shinbun* [*Women's Journal*], *Jogaku zasshi* [*Women's Magazine*], and *Fujin eisei zasshi* [*Journal for Women's Hygiene*],[5] and books written by individual activists.

Among educational institutions, Tokyo Higher Women's Normal School[6] and Japan Women's College, established in 1874 and 1901 respectively, took leadership in promoting women's health reforms. It should be noted that before these schools developed physical training curricula for women, missionary schools had already established physical education programs that were not directly linked to nationalist ideology. For example, in 1888 the Ferris Seminary in Yokohama introduced physical education classes taught by American instructor Mary Deyo.[7] However, the strong nationalist current in the mid-1890s led to a decrease in the number of students at the seminary. A series of edicts on educational institutions in 1899, such as the Higher Girls School Ordinance [*Kotō jogakkō rei*] and the Ordinance for Private Schools [*Shiritsu gakkō rei*], also worked to reinstate the peripheral status of missionary schools.[8] Consequently, physical education methods developed at the Ferris Seminary did not enjoy widespread influence. Indeed, most physical education was imbued with nationalist ideology because of the influence of methods developed by the Tokyo Higher Women's Normal School and Japan Women's College and implemented in public primary schools and women's middle schools nationwide.

Discourse on "Japanese Bodies" and Adopting the Eugenics Thought

In the health-reform movement of the late Meiji period, physicians scrutinized men and women's bodies by applying racial typologies and eugenics theories. In particular, they developed the idea of "Japanese bodies," which needed to be improved to reach the standard of "Western bodies," which were superior in size, health and strength.[9] The bodies of Japanese men and women were described as small, thin, weak, sickly, even deformed, when compared to those of Europeans and North Americans, but the physique and constitution of Japanese women were viewed as particularly inferior. Health reformers often referred to the concept of "heredity," which they had recently adopted from European biological sciences, to explain the smaller physique and inferior constitution of Japanese men and women. They did not conclude from this, however, that the defective characteristics of "Japanese bodies" were permanent and irrevocable. Rather, they stressed that environmental factors such as clothing, diet, living conditions, and daily habits were responsible for producing the undesirable physique of the Japanese.[10]

German physician Erwin von Baelz (1849–1913) was one of those who initiated public debate about problems with the bodies of Japanese women.[11] In 1898, Baelz gave a seminal lecture on the clothes and bodily habits of Japanese women at a meeting of the Japan Women's Association for Hygiene, where he discussed the peculiar and harmful characteristics of traditional kimonos, hairstyle, and footwear.[12] He paid particular attention to the problem of young girls wearing kimonos with long, hanging sleeves, which in his view obstructed the free movement of their hands. The long sleeves, he asserted, weighed down their shoulders and upper arms, which made their shoulders protrude forward and point downward, eventually causing poor chest development. Baelz also thought that wearing a wide *obi* sash made of stiff fabric would cause serious health problems similar to those caused by wearing tightly laced corsets. Both the sash and the corset, he claimed, shifted the positions of internal organs and deformed the shape of the body.

Baelz also pointed out that while the upper part of the kimono was loose and open, the lower part covering the legs and knees was very tight, thus impeding women's movements. This problem was exacerbated by the practice of wearing the belt below the hipbones instead of where the body narrows.[13] Women could not walk in the "correct" manner when wearing a kimono because it clung to their legs, forcing them to point their toes inward and twist their legs as they walked. As a result, Japanese women had protruding knees and anklebones, while their counterparts in the West possessed straight legs. The peculiar walking style caused by wearing a kimono also made women bend their upper body forward. Baelz's discussion conveyed a sense that the shape of Japanese women's bodies and their walking style were deviant and inferior when compared to their Western counterparts.

The debate about the peculiar characteristics of Japanese bodies by Baelz and other European and US physicians reinforced concerns about the body, health, and physical wellness of the nation's populace—concerns shared by Japanese state officials, physicians, and educational experts. Terada Yūkichi (1853–1921), a board member of the Association for Physical Education of Japan, viewed the constitution and health of Japanese bodies as generally undesirable.[14] Terada contended, however, that during the ancient period when Japan conquered neighboring countries in East Asia, Japanese people's health and "spirits" (*kiryoku*) were superior to those of their contemporary counterparts. According to Terada, the Tokugawa shogunate was largely to blame for the weakening of Japanese people's bodies. He argued that the Japanese people had lost their progressive spirit during the Tokugawa period because the Tokugawa government's closed-door policy did not allow them to compete with other nations. Added to this, many domains refrained from providing rigorous military training for their warrior retainers in an effort to demonstrate that they had no intention of rebelling against Tokugawa authority. Adopting the new scientific discourse of eugenics, Terada contended that endogamous marriages within the same village, which had been widely practiced during the Tokugawa period, were responsible for producing children with weak constitutions, hearing disabilities, and mental illnesses. It was this combination of historical developments, Terada concluded, that had resulted in the current debilitation of Japanese bodies.

Terada gave evidence from his experiences in Europe that Japanese bodies were inferior to those of Europeans. When he visited a shipbuilding factory, for instance, he discovered that European factory workers possessed greater muscular power than the few Japanese workers, who were there to learn engineering technology. According to Terada, these Japanese workers were not able to handle the same hammers that Europeans used. At the various schools he visited, Terada noticed that European students seemed to yawn less during class than Japanese students. Because of their low level of physical stamina, he thought, the minds of Japanese students tended to become exhausted sooner than those of European students.

Terada's concern about the physical wellness of Japanese people was based on a version of social Darwinism, influential among Japanese intellectuals at that time, that viewed nations as competing against one another. Superior races or nations defeated weaker ones, which were destined to perish. Terada argued that the Japanese had to compete with foreigners, Westerners, or European nations, and it was therefore imperative that they improve their physique and constitution. Terada's appropriation of social Darwinist theory was selective. Although he argued that Japanese people possessed inferior bodies and physical strength, he did not conclude that the Japanese were among those racial groups destined to decline. Terada shared the optimistic view that the serious health problems afflicting Japan could be solved by actively implementing measures developed by modern science and medicine.

Through health reformers' efforts, educated Japanese women were exposed to the idea that the European body—along with its posture and movements—was the standard that they should emulate. Japanese women who visited Europe and the USA felt keenly that their bodies and physical movements were inadequate. This awareness led them to make conscious efforts to wear Western clothes, engage in physical exercise, improve their diet, and adopt new ways of moving their bodies. One such woman who put these into practice is Inokuchi Aguri (1870–1932), who graduated from Tokyo Women's Higher Normal School and later taught at its elementary school. In 1899 the Ministry of Education sent her to the USA to study physical education for women. After a short stay in Seattle, she went to Massachusetts to attend Smith College and Boston Gymnastics School.[15] On her return to Japan in 1902, she became a physical education instructor and took a leadership role in the women's health-reform movement.

At a monthly meeting of the Japan Association for Women's Hygiene, Inokuchi described how impressed she was by the admirable physique and health of women in the USA.[16] She stated that many American women were so tall that she had to look up to see them. Even tall Japanese men were shorter than American women. While she was of average height in Japan, she was the shortest woman in her class in the USA, and to her chagrin she was called "little girl" wherever she went. Inokuchi believed that "Westerners" possessed greater muscular power than races with smaller physiques. Japanese women were simply unable to match their Western counterparts in feats of physical strength. Students at Smith College, Inokuchi recalled, had generally been in good health, unlike Japanese women, who "appeared weak with their pale faces" and frequently became sick.

Inokuchi praised American women's "straight" bodies and their style of walking: "Women over there walk regularly, pleasantly, and briskly, their steps aligned with the men." In contrast, she described her own walking style as "slow and sluggish" [*guzuguzu*] and "clumsy" [*goshagosha*]. She believed that if she continued to walk in this manner she would look ridiculous. Inokuchi sought to learn a new way of walking and new posture habits, making these efforts her primary task during her stay in the USA.

Inokuchi's comments reveal how she viewed her body as representing the nation of Japan to the US public. When she was called a "little girl," she felt ashamed and mortified not only because of her personal shame, but, in her mind, this amounted to mockery and condescension toward Japan. Inokuchi's lecture situated the female Japanese body at the intersection of scientific discourses on eugenics and genetics, making it an object of discussion from the perspectives of both national interest and prestige. The physical education knowledge Inokuchi obtained in the USA that considered the Western body as the standard also led her to develop a colonized mindset that viewed Japanese women's bodies as anomalous and substandard.

Inculcated by such an idea, Inokuchi did not defend Japanese women's physique and bodily habits, despite her nationalism, or rather because of it. She censured Japanese habits, accepting and reinforcing Western habits as the norm, and this attitude was not unique to Inokuchi. In line with ideas and suggestions articulated by male health reformers, she projected the gestures and movements that she herself engaged in as defective, but proved by example that such flawed habits could be remedied by skills developed within the discipline of physical education. One of Inokuchi's primary goals was to make Japanese women's bodies larger, better, and stronger, and thus, similar to women's bodies in the West. By making such efforts she deemed that Japan could catch up and compete with Western nations. Interestingly, she never discussed the possibility that these endeavors would be futile because the small physique of "Japanese bodies" was a hereditary characteristic, and therefore could not change drastically in a short period of time. Not unlike male Japanese health reformers, her appropriation of social Darwinist theory was selective, and her view on this issue was highly optimistic.[17]

Eventually Inokuchi became an important, pioneering female educator who developed the field of physical education for women in Japan. Holding a position as a professor at a prestigious public teacher-training school and working closely with predominantly male government officials and leading educators, her efforts in the area of physical education no doubt served foremost the state and business interests, which not infrequently conflicted with those of women. Inokuchi laid the ground for what women themselves claimed was their own education. She joined a prestigious teaching faculty at one of the leading institutions of higher education, and through research and teaching activities based on her independent thinking and her own experiences as a woman, she overcame male instructors' monopolization of education.

By the early twentieth century, armed with an education in modern health science, women with feminist beliefs began to develop their own ideas of and policies for women's health and hygiene. Unlike Inokuchi, who followed and collaborated with male authorities, they subverted or revised ideas and policies presented by male experts and government officials, as I discuss later in this book.

Molding Young Women's Bodies Through Physical Education

To address the problem of Japanese women's physical inferiority, some state officials and education experts endorsed the introduction of physical education in girls' schools. This proposal was opposed by conservative educators and state officials, who feared that Japanese women would lose their modesty and decorum by learning European gymnastics and games. Progressives prevailed in the end, however. By the turn of the twentieth century, physical education had become a requirement at almost all primary schools, girls' middle schools, and college-level institutions for women.

Inasmuch as the introduction of physical education for girls during the Meiji period was strongly linked to nationalist goals, it failed as an emancipating force for Japanese women. This stands in marked contrast to Europe and the USA, where women's participation in sport and physical exercise in the late nineteenth and early twentieth centuries was often part of the rise of feminist consciousness.[18] For the most part, Japanese educators who promoted physical training for women did not support sexual equality, women's political rights, or women's entitlement to pursue a professional career. Instead, they believed that physical education was instrumental in making women into "good wives and wise mothers" [*ryōsai kenbo*]. Contemporary scientific theories of sexual difference appropriated by Japanese educators in the areas of anatomy, physiology, biology, psychology, sexology, and anthropology were used to reinforce the assumption that women's bodies and minds were suited to specific types of work, such as maternal care, and not for professions traditionally pursued by men.

For instance, Shimoda Jirō, a prominent male figure in women's education, developed ideas by appropriating new scientific theories of sexual difference from Europe.[19] Among the many European works Shimoda referred to was English educator Henry Herbert Donaldson's (1857–1938) *The Growth of the Brain*.[20] Shimoda focused on the physical differences between male and female bodies described by Donaldson. Male bodies, in his view, are large, the skin coarse, and the bone structure robust. The bony constitution is overtly manifested and male muscles are tight. On the other hand, female bodies are small and frail, the bones are hidden, and the muscles are soft. Moreover, men's bodies are straight, in contrast to women, who have protruding chests, wide hips, and a curvy torso. Furthermore, men possess wide shoulders, large lungs, and small stomachs—qualities that make them suitable for physical activity—while women have small lungs and large hips more suitable for maintaining static postures.[21]

Shimoda cited work by English sexologist Havelock Ellis (1859–1939) to make a similar point, but noted that Ellis discussed sexual difference in terms of both the mind and body.[22] Shimoda summarized Ellis's argument in the following way: sexual difference does not manifest itself solely in the way the bodies and minds of men and women develop or how the parts of their bodies and minds relate to one another. Rather, sexual difference exists throughout their bodies, making men and women different "to the tips of their fingers." Men are progressive and susceptible to changes, while women are conservative and slow to change. Women mature and age faster than men, but they maintain more childlike characteristics than men in their adult lives.[23]

Shimoda also used work by Hermann Lotze (1817–1881) to develop his views on sexual difference.[24] Lotze argued that men's bodies are suited for physically active work, while women's bodies are built to endure hardships. Lotze refused to accept the argument that there were differences in men and women's intellectual capacity, focusing instead on differences in how men and women's minds work. Women, he argued, are good at holistic and intuitive understanding but are less capable of analyzing, while men appreciate abstract ideas and absolute truths. Because of this, Lotze concluded that women were not suited for careers as scientists or judges.[25]

Shimoda also introduced two German texts that discussed female physiology and psychology: *Die Bestimmung der Frau* [*Definition of Woman*] written in 1892 by H. Fehling, a Swiss gynecologist-obstetrician; and *Das Weib in seiner geschlechtlichen Eigenart* [*Woman in Her Sexual Peculiarity*] written in the early 1900s by Max Runge (b. 1849), a German gynecologist-obstetrician who taught at the University of Göttingen.[26] Fehling cited female-specific bodily characteristics to refute John Stuart Mill's (1846–1903) contention that sexual difference is created through socialization. The differences between men and women, Fehling argued, are fundamental and embedded in nature. For his part, Runge emphasized the importance of marriage for women to maintain a healthy body and mind. He argued that if women want to live "complete" and normal lives, they need men. The reason women try to attract men by attending to their looks is because they unconsciously believe themselves to be imperfect. Runge also contended that women are unable to fulfill professional responsibilities as competently as men because they become debilitated during and after pregnancy and childbirth. Women are endowed, however, with maternal instinct, the strongest of all womanly instincts, and are more sympathetic and loving than men. Marriage is instrumental for the full development and utilization of these invaluable qualities in women.

European medical science validated Shimoda's view that women should become wives and mothers and that women's education should be designed to nurture qualities that would prepare women to perform such roles. Providing female students with physical training was conducive to achieving this goal because women needed to develop healthy bodies to fulfill their functions as wives and mothers adequately. Shimoda insisted that physical education was more important than lessons in cultural skills such as the tea ceremony, flower arranging, Japanese poetry, or formal etiquette. He suggested that a reasonable man would not marry an unhealthy woman, even if she were accomplished and beautiful. Shimoda cited English social Darwinist Herbert Spencer (1820–1903), who argued that women should be first educated in practical subjects such as physical education, physiology, psychology, and childcare before they learned other skills. He recounted the following anecdote from Spencer's work to support his view. A woman had expended excessive energy by studying foreign languages, which critically affected the health of her unborn child. A frank doctor informed her that she could have avoided the loss of her child by not having studied so hard. Spencer posed a rhetorical question: "Could she be comforted by the fact that she could read Dante in Italian?"[27]

Another health reformer, Nagae Masanao, also recommended physical training for female students based on his belief that women should develop robust physiques to fulfill their maternal responsibilities.[28] For Nagae, however, having women engage in physical exercise was also a prescription for dealing with emotional disturbances and depression, to which he presumed women were more susceptible than men. Nagae believed that the women of his time particularly needed physical training because they had to cope with the mental stresses created by modern society.[29]

To support this view, Nagae cited Huxley's comparative examination of "civilized" and "savage" societies.[30] He claimed that there were fewer cases of mental illness among "savage" peoples than among "civilized" peoples because social customs in savage societies were simpler, having only a limited number of material things available. Nagae contended that in these societies, poor people were content to remain poor. Their intellects were limited and they could endure hardships because they lacked deep feelings and wisdom. They often appeared unbalanced to observers from more advanced cultures, but this was simply because they were slow or retarded, not because they suffered from serious mental illnesses such as madness, which were rare in primitive cultures. On the other hand, civilized people engaged in lofty and complex thoughts, which, along with other aspects of life in modern cultures, agitated people's mental functions and caused psychiatric diseases.[31] In Nagae's mind, Japan was one of the civilized nations. He did not believe there were fundamental differences between the minds and bodies of Japanese and European people. He had no hesitation in applying analyses and recommendations that had originally been developed for the Europeans to the Japanese.

Nagae blamed modern civilization for bringing an "effeminizing trend" to Japan. This trend had developed from people's access to modern means of transportation, which deprived them of vigorous physical exercise—a development he viewed as particularly detrimental to Japanese women. Having been freed from the daily physical activities of pre-modern society, Nagae contended that many contemporary women were inactive because they believed they should be docile, subservient, withdrawn, and passive. For Nagae, this was a serious problem. He believed that women needed to live an active lifestyle to maintain their health and carry out their maternal duties.[32]

While placing a high value on physical training for women, his views on women's psychology led Nagae to caution against women's engagement in intense intellectual activities. Because women were more prone to depression and mental illness than men, he believed that sustained and constant use of the brain would result in psychological breakdowns. Nagae was not necessarily arguing against women engaging in academic work. But he was careful to emphasize that they should take appropriate breaks when they studied.

Physical education also occupied an important place within the policies of women's education conceptualized by Naruse Jinzō (1858–1919), founder and president of Japan Women's College.[33] Along with Shimoda, Naruse strove to educate young women according to the "good wife and wise mother" model. Although he viewed domestic duties as the primary responsibility of adult women, he did not advocate that they live a passive and withdrawn life. On the contrary, Naruse's goal was to nurture the type of women who understood their nationalist mission and would actively take on multiple responsibilities. This included caring for aging parents-in-law, supporting their husbands within the domestic sphere, educating her children, and engaging in charity and social reform activities—duties that required women to possess strong constitutions and good health. Naruse argued that housewives in poor health were unable to fulfill their obligations, which would cause domestic problems and disrupt the social order. He reasoned that this problem could be alleviated by developing fully fledged physical education programs in women's schools.[34]

Naruse also shared other health reformers' opinion that the bodies of Japanese women were "weaker and smaller" than those of European or even Chinese women. He maintained that the bodies of Japanese women were even more inferior to their European counterparts than were the bodies of Japanese men. Naruse believed that mothers passed on more physical traits to the children than fathers because women were the ones who gave birth. For these reasons, Naruse insisted that providing physical education for women to improve their physique and health was more urgent than it was for men. Naruse went as far as to remark that the rise and fall of a nation depended on the state of women's physical education.[35]

His nationalism also led him to reflect on women's physical beauty in terms of international rivalry. While affirming that Japanese women were no less attractive than women in Europe and the USA, he contended that Japanese people should still strive for improvement. Specifically, he asserted that women should develop healthy and beautiful bodies that possessed a balanced and perfect shape, as well as skin of excellent color and quality. Women should also be trained to move elegantly. An important reason for physical education was to attain these goals.[36]

Naruse established an extensive physical education program at Japan Women's College that required students at all levels to take three hours of physical education classes per week. There were four categories within the college's physical education curriculum: regular gymnastics; games and *yūgi* (a series of dancing and marching routines); educational exercises; and gymnastics for deportment. Regular gymnastics included normal and Swedish gymnastics.[37] Games included tennis, baseball for women, cricket, field hockey, basketball, tetherball, the "scarf" game, jumping-rope, Japanese-style battledore [*hanetsuki*], and the *chidori* race.[38] The third group included cycling and practicing with the Japanese-style halberd. Finally, to discipline their posture, students were trained in Delsarte-style gymnastics, developed in France and the USA.[39]

Based on their beliefs about sexual differences in physical ability and characteristics, educational experts emphasized that physical education programs for female students should be different from those designed for male students. From the early years of the Meiji period, physical education was thought to be one of the most important subjects for male students. The 1886 Middle School Ordinance stipulated that boys' middle schools (which were equivalent to high schools in post-World War II Japan) should require students to take normal gymnastics classes three hours per week for the first three years. In comparison, physical education for girls received sporadic attention during the early Meiji period. It was not until 1895, when it promulgated the Edict on Girls' Higher Schools, that the Ministry of Education established a policy on physical education curricula for women's middle schools. According to the edict, women's middle schools were to provide a three-hour class on physical education for the first four years, and a two-hour class for the last two years. In 1899, however, the Higher Girls School Ordinance decreased the required hours for physical education to two-and-a-half hours per week.[40]

Physical education specialists also believed that women were incapable of engaging in the strenuous gymnastics methods that were meant to train men's bodies. For example, Shirai Kikurō, a gymnastics instructor at Japan Women's College, declared that women would be unable to follow the orderly movements of military-style gymnastics.[41] Some educators believed that even regular gymnastics were too rigorous and monotonous for women to engage in for an extended period.

These perceptions led some educators to endorse the use of "calisthenics" or "light gymnastics" for women's physical training.[42] Japanese physical education instructors also developed a type of light exercise called *yūgi*, which consisted of dancing or marching movements often accompanied by music. *Yūgi* were classified into *kōshin yūgi* (marching routines) and *dōsa yūgi* (movement routines), which students practiced in groups ranging from several participants to more than 25. Square dancing, which was often included in physical education programs at middle schools, was also appropriated as a type of *yūgi*. The report prepared by the Education Ministry's Committee for the Assessment of Gymnastics and Yūgi remarked that marching routines were suitable for helping women and children nurture their spirit of discipline and cooperation, develop their aesthetic sensibilities, and cultivate their sense of harmony and elegance.[43] If different types of physical training were devised for girls and boys, these differences were intended to develop feminine or masculine qualities in the bodies of each sex. As Shirai remarks: "proper exercises will make men more masculine and women more feminine."[44]

The postures and movements Japanese girls learned during these physical education courses were quite novel. Throughout the Tokugawa period, and even during the Meiji period, upper-class and certain middle-class women were taught proper posture according to formal Japanese etiquette and stipulated manners for the tea ceremony. These subjects were taught along with European-style gymnastics at girls' middle schools and college-level institutions, but some educators expressed concerns about the consequences of training women in two different body-discipline traditions. A report by the Education Ministry's Committee for the Investigation of Gymnastics and Yūgi addressed this problem, focusing especially on the novelty of the postures and movements that European physical education methods introduced.

The committee enumerated a series of contrasts between the two traditions. First, they pointed out that the correct posture for standing upright in gymnastics differed from traditional etiquette practices. In gymnastics people stood with their chests protruding, shoulders and hips held back, and their toes spread apart. Japanese etiquette on the other hand, dictated that one should stand straight with the chest protruding and hips drawn back, but that the shoulders should not be pulled back. The toes should be aligned next to one another. Second, gymnastics taught people to put their toes on the ground before their heels while walking. Japanese etiquette instructed that they should either put their feet down flat or touch the heel first. Third, gymnastics trained people to let their arms and hands move naturally when walking. Traditional Japanese etiquette taught people to place their hands lightly against their thighs, and not to move them. Fourth, gymnastic methods instructed people to move their left foot first when they start walking. According to Japanese etiquette, however, people moved their left or right foot first depending on the situation. Finally, there was a discrepancy between the two traditions as to where people should keep their eyes when they were standing or walking. The committee concluded that these differences should not present major problems and recommended that students adopt the gymnastics method for everyday practice but follow Japanese-style etiquette for formal occasions. When girls confronted situations where they had to decide between the two, they should adopt the gymnastics method because it was superior in nurturing a healthy body, which should take precedence over maintaining social civilities.[45]

As the committee's report indicates, physical education programs began to produce a type of disciplined body that was capable of new, orderly movements. Such bodies automatically responded to commands such as "Attention!," "Rest!," or "Turn!" and moved smoothly and rhythmically. They were also able to coordinate the movement of arms and legs while walking and engaging in various games. Furthermore, physical education enabled movement and control of these bodies en masse. In other words, physical education used discursive techniques, as conceptualized by Foucault, to analyze and train subjects to become "docile bodies."

Educators who advocated physical education for young women did not necessarily discourage the development of women's intellects. On the contrary, they wanted to nurture a type of woman who was intelligent and understood her obligations to the nation. In other words, educators used national ideology to transform women into conscious national subjects. On the other hand, physical education was a mechanism through which power worked directly on women's bodies to produce the type of modern subject who would be incorporated in the process of nation building and capitalist production. Formal physical training developed the ability to engage in certain individual movements and participate in organized group activities. The state as well as industrial and commercial enterprises needed such bodies to be healthy mothers and productive workers.[46]

Medicalized Discourse on Women's Clothes and Beauty

Clothing reforms were another mechanism through which women's bodies were examined and regulated by scientific and medical discourses. Health reformers delineated new scientifically legitimized criteria for the size and shape of women's bodies. Reform activists scrutinized conventional kimonos and Japanese hairstyles from a "scientific" perspective and endeavored to devise new styles of clothing and hairdressing for women. These novel fashions resulted in Japanese women adopting new bodily movements and gestures and reshaping their daily routines of dressing and toilet. Dress reforms provided women with scientific and medical knowledge about their body's anatomy and physiology. This education was part of patriotic reformers' efforts to transform women into active, educated subjects who would exert themselves for the nationalist cause.[47]

Clothing reforms were based on the emerging consensus that traditional kimonos were detrimental to women's health. It was not easy to invent appropriate substitutes, though. Western dresses were impractical because of the peculiarities of housing conditions and everyday practices in Japan. As a participant in the ongoing debate about Japanese women's clothing reform, Fukushima Shirō, the owner and chief editor for *Women's Journal*, recommended replacing the kimono's customary hanging sleeves with narrow sleeves. He also recommended wearing *hakamas* (divided skirts), commonly worn by men, to facilitate women's physical mobility. Fukushima praised the practice of wearing *hakamas* among female students and hoped that this style would also be adopted by married women of the upper and middle classes and their servants.[48] (See Fig. 5.1 for the *hakama*-wearing female students participating in an athletic meeting.)

Fig. 5.1 Illustrations of an athletic meeting at Ochanomizu Higher Women's School. Captions (text) in the figure. (Upper row, from left to right) Arena for an obstacle race; the entrance to the arena. (Middle row) A parade. (Bottom row) Hairdressing competition; competition for picking up animals (and insects). Author's note: Many students are wearing *hakamas* and *sokuhatsu* hairstyles; some students are wearing Western dresses (Source: "Ochanomizu kōtō jogakkō undōkai" *Fūzoku gahō*, No. 279, Originally published by Tōyōdō in Tokyo, December 10, 1903; Reprinted in *Fūzoku gahō CD-ROM Version* by Yumani Shobō in Tokyo, 2002)

As part of a spirited *hakama*-wearing campaign launched by *Women's Journal* in 1901, Fukushima aggressively attacked a state official who had prohibited female students from wearing *hakamas* at a welcoming ceremony for the emperor's visit to Miyagi prefecture in northeastern Japan.[49] In response to Fukushima's accusation, prefectural official Yamada Kunihiko justified his order by noting that many girls between ten and 14 were still not in school. The practice of wearing *hakamas* could reduce rates of school attendance because girls who could not afford *hakamas* might stop coming to school out of shame. Moreover, Yamada argued that forcing schoolgirls to wear *hakamas* was tantamount to conducting experiments on them, considering the ongoing debates about the merits and problems of various styles of reformed clothing. Finally, he contended that medical science showed that a wide and bulky sash was the cause of most serious health problems. Wearing *hakamas* would help mitigate this problem, but *hakamas* themselves did not possess innate hygienic merits.

Fukushima fiercely fought back by asserting that the health benefits of *hakamas* had already been verified by medical experts. He emphasized that they should not be spared for the sake of economy because they were critical for students' health, not simply frivolous fashion items. He added that *hakamas* were indeed cost-effective because they could be used as an apron and were often cheaper than a sash for conventional kimonos. Fukushima also considered that encouraging ordinary girls in elementary school to wear *hakamas* was advantageous for spreading this style—which had been adopted mainly by young women of the middle and upper classes—to the lower social strata.

Despite the intensity of this debate, both Fukushima and Yamada adhered to a nationalist and reformist discourse, and shared similar attitudes toward scientific authority and women's education. Both sides based their arguments on an indisputable appeal to medical professionals' views. Both also ardently supported the policy of providing school education for girls. Their desire to increase the rate of school attendance led them to be mindful of the economic burden that the families of schoolgirls had to bear. Yamada at times ridiculed women's *hakama* style as a bizarre fashion, but he did not express the kind of vehement antipathy held by some conservative educators and critics who regarded women wearing *hakamas* as lacking the invaluable womanly decorum that Japanese women had traditionally maintained. This "*hakama* wearing" controversy shows how nationalist discourses on women's health and education framed reformers' discussions on women's clothing. Despite their disagreement on the specific ways clothes should be reformed, many educators and intellectuals shared similar assumptions and goals.

Another fashion that reformers were eager to change was the traditional Japanese chignon, which they believed was highly deleterious to both health and the economy. The movement to urge women to abandon traditional hairstyles was spearheaded by male professionals, who suggested that women should adopt what was called the *sokuhatsu* or "swept-back" style with a bun at the back of the head. The *sokuhatsu* style, which had many variants, was devised during the Meiji period by combining elements of Japanese hairdressing with hairstyles worn by women in Europe and North America. In 1885 under the leadership of military surgeon Watanabe Agata and journalist Ishikawa Eisaku (1858–1886), the Association for *Sokuhatsu*-style Hairdressing for Women was organized in Tokyo (see Figs. 5.2, 5.3, and 5.4 for traditional Japanese hairstyles and Figs. 5.5 and 5.6 for *sokuhatsu* hairstyles).[50]

Fukushima also criticized Japanese hairstyles by adopting pseudo-medical concepts. He argued that chignons hampered the growth of the brain because they were heavy and unhygienic. The generous application of hair oil involved with such hairstyles was particularly problematic

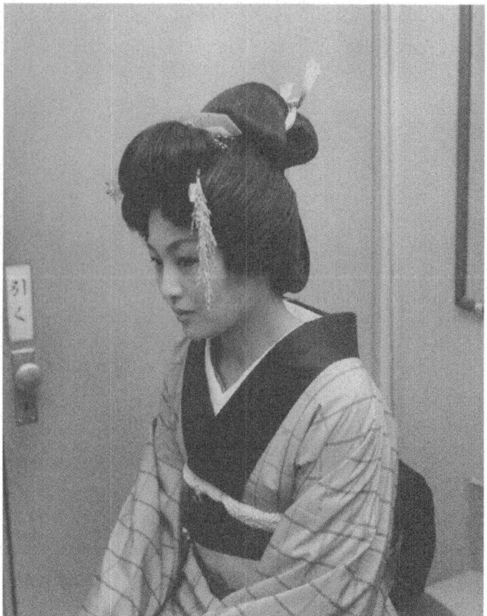

Fig. 5.2 Traditional Japanese "*shimada*" hairstyle, often worn by younger unmarried women (Source: Aflo/ Mainichi Photobank)

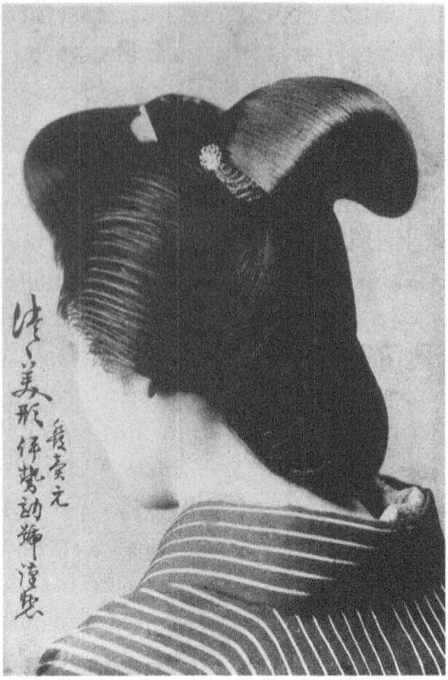

Fig. 5.3 Traditional "*marumage*" Japanese hairstyle, worn by married women (Source: Aflo/Mainichi Photobank)

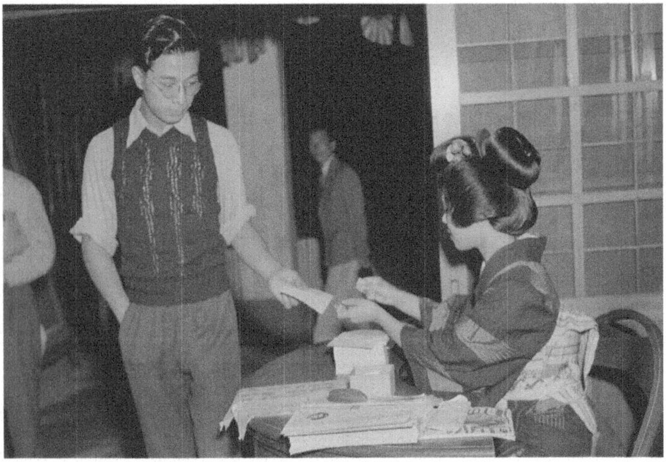

Fig. 5.4 A movie theater ticket collector wearing a "*momoware*" (split peach) hairstyle in 1949. Japanese hairstyles continued to be worn for special occasions even after the end of World War II (Source: Aflo/Mainichi Photobank)

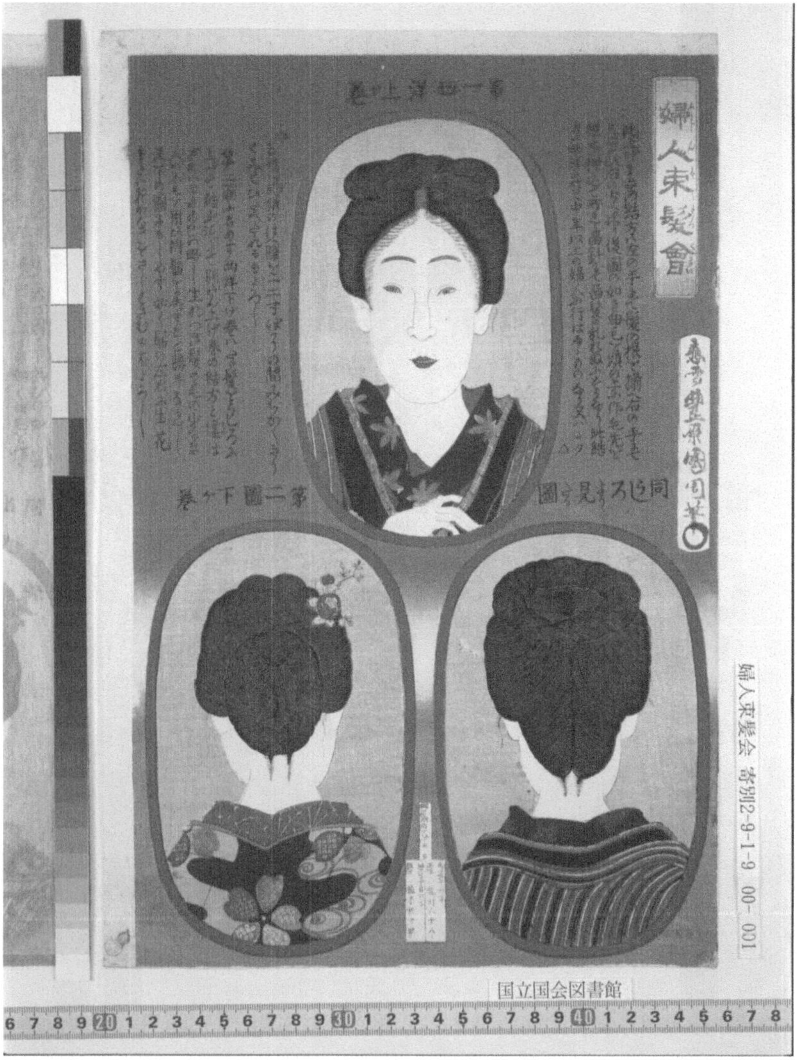

Fig. 5.5 Examples of the *sokuhatsu* hairstyles, presented in a colored woodblock print, *The Association for Women's Sokuhatsu Hairstyles.* Upper row: Western-style topknot chignon. Lower row, right: the rear view of the above. Lower row, left: Western-style low chignon. Author's note: The text that accompanies these illustrations presents instructions for creating the hairstyles (Source: Toyohara Kunichika, *Fujin sokuhatsu kai* [Tokyo: Published by Ueki Rin'nosuke, 1885] from the National Diet Library's Digital Collections)

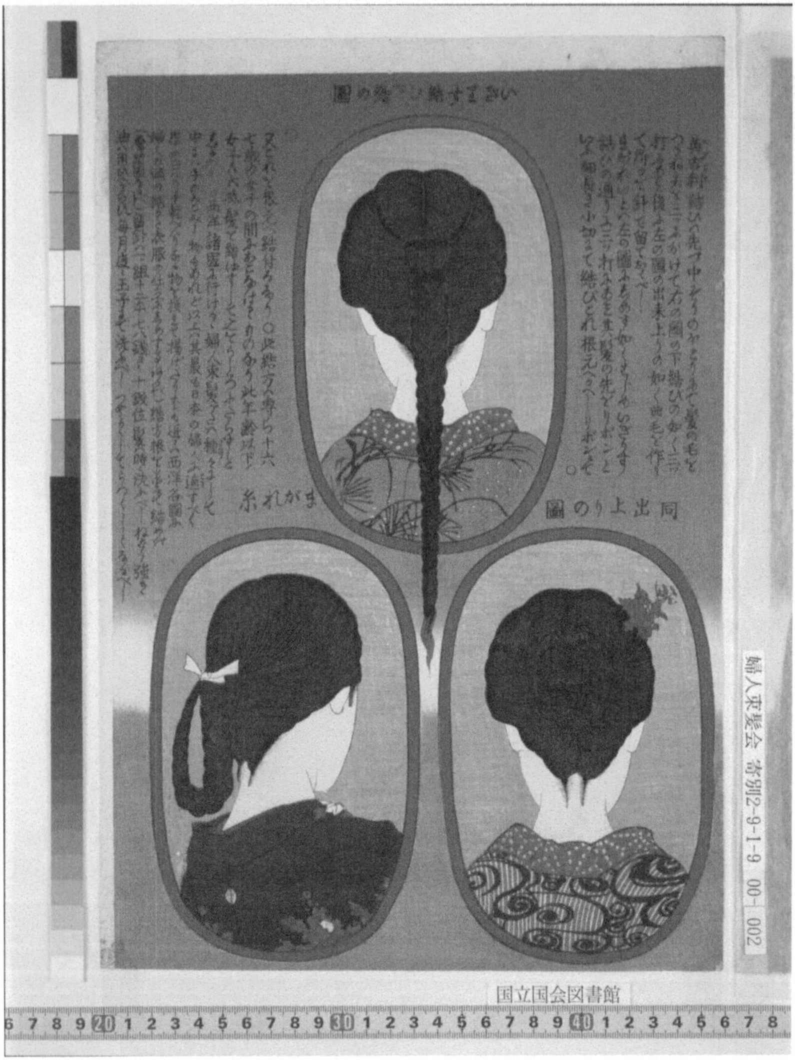

Fig. 5.6 The hairstyles presented in the second page of the woodblock print, *The Association for Women's Sokuhatsu Hairstyles.* Upper row: British hairstyle with a braid hanging down. Lower row, right: the same style as above with a bun. Lower row, left: *Māgareito* [Margaret] hairstyle (Source: Toyohara Kunichika, *Fujin sokuhatsu kai* [Tokyo: Published by Ueki Rin'nosuke, 1885] from the National Diet Library's Digital Collections)

because it blocked pores and attracted dust. Women wearing Japanese-style chignons did not wash their hair often because it was a burdensome chore.[51] Moreover, Japanese hairstyles made women reluctant to engage in physical exercises because vigorous bodily movements would make ornamental hairpins and combs fall out, leaving their hair in an untidy mess.[52]

Fukushima associated the adoption of the *sokuhatsu* hairstyle with new tasks that contemporary Japanese women were required to accomplish for the nation: "Japanese hairstyles may have been adequate during a time when women were treated as decorative dolls. However, we should abolish such elaborate hairstyles now that women need to fulfill their duties as individuals and citizens [*kokumin*]." Specifically, he reasoned that the inactive lifestyle fostered by Japanese hairstyles would eventually result in a degradation of Japanese women's constitution and health. This would prevent many women from performing their primary duty to society and the nation, namely—"producing the next generation of citizens."[53]

In their endeavors to reform women's clothing and hairstyles, health activists were not interested simply in nurturing well developed and healthy female bodies; they also wanted to adorn Japanese women with clothes that would meet certain aesthetic norms. This was important because many prominent physicians, state officials, and educators regarded the physical beauty of Japanese women as a source of national pride. Such concerns were manifested in the warnings of a German-educated physician, Miyamoto Naka, against the hasty adoption of reformed clothes. Miyamoto asserted that many reformed clothes were uncomely and therefore not appropriate. He suggested that relatively young schoolgirls should wear reformed clothes first because they would have a greater effect on their physical growth. Adult women should abandon their traditional kimono style only when more stylish reformed clothes were invented.[54]

For Miyamoto, the aesthetic value of kimonos was indisputable. He claimed that their shape, color, and decorations embodied a refined artistic harmony. He pointed to how a decorative wide sash and fluttering long-sleeves made a superbly balanced outfit, supplemented by the relatively simple shape around the neck, upper body, and lower body. He also praised the application of light-colored neckbands and socks, which added a sense of purity to the total look. He reasoned that the kimono and its various accessories had attained this noble and elegant appearance because it had been continuously improved throughout the Tokugawa period.

Miyamoto also admired the aesthetic quality of Western clothes. He stated that, although Western dresses and kimonos were differently shaped, both were well designed. On the other hand, he viewed the narrow-sleeved garments worn by lower-class Japanese women and the Chinese to be uncomely and comical, and he assumed that existing reformed clothes that looked like such outfits would never be adopted by upper-class Japanese women. Miyamoto's view reflected notions of ethnic and class hierarchy that were being reshaped at the turn of the twentieth century, a period when Japanese intellectuals sought to confer on Japan a status equal to that of Western nations, while looking down on China, which Japan had defeated in the Sino–Japanese War (1894–1895).[55]

Searching for new ideals of feminine beauty, some reform-minded physicians borrowed European scientific concepts. Watanabe Hōkichi and Fujikawa Yū (1865–1940)[56] criticized the Japanese for their habit of only scrutinizing the face, hands, and feet to evaluate a woman's beauty, pointing out the need to take into account the body in its entirety for a scientifically legitimate assessment.[57] They defined the ideal body as perfectly symmetrical and as neither "long and narrow" nor stumpy. For them, the bodies of Japanese women tended to be too short and thin to meet this ideal.

Watanabe further defined a beautiful female body by contending that it should maintain an adequate amount of fat and well-developed muscles. Body fat was an important element in making women's bodies look beautiful because fat would allow women's bodies to appear white, glowing, and round. The level of fat, however, should not be excessive, since this would make the body shapeless. An unhealthy body with excessive fat would lose its subtle curves, particularly those around the arms, legs, under the breasts, and around the jaw. Many lines and bulges would appear with sharp-pointed, white scars, similar to those found in women who had had children. On the other hand, a body that did not possess an adequate amount of fat would lose its round shape and take on an angular form. Such a body was also undesirable. To help women assess whether they were too fat or thin, Watanabe introduced a formula to calculate the ideal body weight for a given height.[58]

$$\frac{\text{Height} \times \text{Bust measurement}}{240} = \text{Ideal weight}$$

Health reformers also sought to reinforce the notion that a body could not be beautiful unless it was in good health. Watanabe deplored the many women who did not concern themselves with health issues but instead concentrated on decorating their bodies. He contended that part of what made the body beautiful was "fiery living energy" [*moyuru seiki*] emanating from within and that only healthy women could possess such beautiful features as a vibrant glint in the eyes, scarlet-colored lips, pink-colored cheeks, and glossy, abundant hair. He insisted that people only feel repugnance toward unhealthy women who applied make-up on "a sickly, pale face" and had "hollow cheeks and black circles under their eyes." Fujikawa emphasized that human beauty would be enhanced by eating nutritious food, breathing fresh air, and getting enough sleep. Good eating habits would help the digestive organs function better. Breathing good air and sleeping soundly would improve the respiratory and nervous systems, respectively.

In addition, Fujikawa introduced scientific techniques to maintain healthy and beautiful skin. He began his lecture by explaining that human skin is covered with tiny wrinkles and hair, just like a piece of velvet cloth. Underneath the skin, there are blood vessels. When they are filled with an adequate amount of blood, the vessels become slightly swollen, tinting the skin with a pink hue. Healthy skin is resilient and produces an adequate amount of oil. To keep one's skin in good condition, Fujikawa deemed it necessary for women to follow proper skin care methods. Noting that the use of soap had become widespread for face washing, Fujikawa recommended basic soaps that did not contain perfumes. Perfumes in soaps, Fujikawa pointed out, would cause skin troubles such as rashes and spots. He also advised against using hard water containing a high quantity of lime for washing. If people had to use hard water, they should put a small amount of boric acid in the water first.

Fujikawa discussed facial massages devised by French physicians to maintain resilient skin and avoid wrinkles, and he even exhibited a German-made massage machine. He described massaging techniques as useful for relaxing tense and contorted facial muscles caused by emotional disturbances. He was particularly concerned about women with hysteric and depressive personalities, who were believed to be prone to anger and anxiety. To avoid damaging their beauty, he recommended that these women "correct" their facial expressions by looking into a mirror once they recovered from emotional agitation. Massaging the face with light cold cream after washing was another effective method.

Health reformers such as Miyamoto, Watanabe, and Fujikawa used scientific authority to define and spread new norms of female beauty and proper beauty regimens. Over time a great deal of knowledge originating in European texts and practices became integrated into Japanese women's own understanding of their bodies and minds. The health-reform movement became a vehicle for the power of science and the desire of nationalistic male experts to reach into women's intimate and daily practices and reshape them according to the experts' needs and models. These new norms induced a heightened level of self-consciousness in Japanese women about their looks and the types of beauty care they practiced, because normalizing discourses constantly reminded them of what their bodies lacked vis-à-vis the new ideal. In effect, they were perpetually encouraged to improve their looks.

Women's increasing self-consciousness about their looks and beauty practices is demonstrated by the questions readers sent to women's magazines. *Women's Journal*, for instance, published a Q & A column concerning women's health and beauty almost every week. Advice columns were an essential component in the women's magazines that began to be published in the 1910s. Japanese scholar Kawamura Kunimitsu discusses the beauty questions subscribers posed to women's magazines. For example, a 17-year-old woman whose face was "always pale" asked if there was medicine to make her face reddish. Another woman asked for a way to put on weight because she could never gain weight even though she was in the habit of engaging in good health practices and eating nutritious food. It was customary for magazines to publish responses from either a physician or a beauty specialist.[59] In this way, women were incorporated into the relentless appropriation of ever-changing beauty norms and techniques that exploited their anxieties and desires, which were "scientifically validated" in new forms of mass media.

The reconstruction of women's beauty regimens occurred in an era when increasing industrialization and commercialization were making new health and beauty products—soap, face lotions, cosmetics, underwear, and menstruation bands—widely available. The dissemination of these new products was closely tied to the development of a commercialized mass culture in which new ideas and products for washing, make-up, and dressing routines emerged from women's magazines, particularly their advertisements, and other media. Guidance from physicians and beauty specialists with professional knowledge was important for the many women who felt lost in the flood of new ideas and commodities.

Increasingly, they were influenced by the cosmetology lectures and magazine articles of medical experts.[60]

Women's Resistances and Collusions

A growing number of women were exposed to new standards and methods for dressing and make-up that had been introduced by medical experts, but how were women themselves responding? In posing this question, I am aware of the serious problems involved in representing women as if they were a unified subject. Differences among women in terms of their class, status, occupation, race, and ethnicity make it impossible for me to discuss the reaction of "women" toward Meiji health reforms. The views and experiences presented in this section are those of women in a specific category: urban, middle-class, educated Japanese women. Although such women represented only a small portion of all Japanese women, it is worthwhile examining their writings about fashion because they reveal the views of those whose thoughts and practices were most directly targeted by discipline and indoctrination. My analysis is focused on ordinary female students and two prominent feminist writer-activists, Yosano Akiko (1878–1942) and Hiratsuka Raichō (1886–1971) (see Fig. 5.7 and Fig. 5.9).

Fig. 5.7 Yosano Akiko (1878–1942) (Source: Aflo/Mainichi Photobank)

Yosano Akiko, a leading female poet of the pre-World War II period, expressed a great interest in young women's clothing.[61] While she was exposed to some of the new scientific knowledge disseminated by health reformers, and even shared with them certain basic notions about emerging modern womanhood and society, she formulated original and radical ideas that challenged their views. All through her life, including the time she attended school, she lacked a solid education in many of the new ideas introduced by health reformers. As the daughter of a well-established pastry shopkeeper, her school education ended upon graduation from a girls' middle school in Sakai City, south of Osaka. Yosano was never part of the women's associations that promoted health reforms, such as the Japan Women's Education Society, the Japan Association for Women's Hygiene, or the Red Cross Society.

Believing that young women should be able to wear attractive clothes of their own choosing, Yosano condemned the reformed clothes and *hakamas* as something that old-fashioned and narrow-minded educators imposed on female students. Her objection to reformed clothing was part of her overall criticism of women's education provided by leading educators. She criticized the prevailing idea that to educate women was to make "good wives and wise mothers." She was convinced that girls should be educated primarily to become "ideal human beings," not just wives and mothers. Yosano, in particular, condemned Japan Women's College and many women's middle schools for teaching housework under the guise of home economics. Women would naturally learn how to sew, cook, and take care of children from experience, she contended; what they needed to learn, either at school or on their own, was the "true knowledge" that would expand their minds and help each become "a perfect human being" and "a perfect woman."[62]

She believed that those responsible for forcing an inferior education on female students were elderly educators incapable of guiding young women. Contemporary young women, whom Yosano called "New Women," were liberated from the cumbersome restrictions that had bound the thinking, emotions, and will of women in previous generations. Young women also enjoyed a much more plentiful material life, which provided clothing, ornaments, and food that were once luxury items. Yosano observed that it was narrow-minded educators with an old-fashioned, Confucian mindset who criticized young women's affluent lifestyles by claiming that such trends promoted their vanity.[63]

She confronted these critics by arguing that the affluent new material culture was an important part of a new civilization, which had brought about freedom and happiness. She asserted that during ancient times, when people had been ignorant, some countries might have fallen into ruin because citizens indulged in material luxuries. The progress of material life in the modern period, however, had been accompanied by advancements in intellectual and cultural life. Because of the improvement in women's education, many young women were intellectually and emotionally mature, and were competent to make their own decisions regarding material goods and extravagance. If many of them still lacked an education that taught them how to handle these newly available material comforts, they would soon be able to obtain one. For Yosano, young women between the ages of 17 and 27 possessed a set of interests that was especially conducive to fostering intellectual and emotional growth. This process made their "inner self" beautiful. Such young women should be provided with the freedom to fully enhance their outward appearance in order to complement their inner development.[64]

Based on this belief, Yosano vehemently defended young women who wore what critics described as extravagant clothing. She contended that their garments were not as luxurious as they seemed by pointing out that, even though they looked ornate in the eyes of men who lacked knowledge of women's clothes, their clothing and accessories failed to constitute a perfectly matching ensemble. Indeed, they were reluctantly wearing kimonos, sashes, and accessories that did not match. Yosano also noted that few students owned kimonos made of silk crepe, a luxurious material commonly used for women's visiting kimonos. Thus, she contended that educators should only begin to issue warnings when students started wearing clothes similar to their wedding gowns. Yosano's criticism was also directed at men who were reluctant to provide decent clothes for their daughters, sisters, and wives; she suggested with a certain amount of cynicism that part of the money they spent at brothels could be set aside for buying new kimonos for women in their own families.[65]

In her eyes, young women wearing *hakamas* and narrow-sleeved kimonos made out of drab-colored cotton or coarse silk looked shabby and unbecoming. In her satirical story about the New Woman, Yosano presented a scene in which the protagonist, a young New Woman, adamantly refused to wear the clothes given to her by her old-fashioned uncle. The clothes were "a strange narrow-sleeved kimono that looked like a hodgepodge between a cooking apron and a nurse's uniform, and hakamas that resembled the ones that men would wear at fencing schools, topped off by a worn-out brown satin sash that her aunt had used for several decades."[66]

The woman turned her back on these clothes saying, "I refuse to wear them. I would rather prefer to remain naked, covered with my young woman's fresh skin rather than wear such moldy, worn-out, and stuffy clothes."[67] She felt a sense of aversion to these clothes, not only because they looked ugly but also because she associated them with obsolete, moralistic Confucian teachings that emphasized the virtue of frugality and simple living.

Although she held a different view of reform-minded physicians and educators, Yosano sometimes adopted their language. She described social phenomena by contrasting the old with the new, glorified the idea of progress, expressed a new form of ethnocentrism, and emphasized the importance of the economy and efficiency. Her positive evaluation of *sokuhatsu* styles, for instance, overlapped with the health reformers' views; they both celebrated the "progress of civilization" and considered the new hairstyles as an indication of women's progress. She observed that if women had adopted the *sokuhatsu* style during the Tokugawa period, they would have been severely condemned. In this new age, however, no one criticized women for wearing the hairstyle because it was the most suitable for contemporary society. Yosano further argued that other impractical and unnecessary customs should be reformed in the way that women's hairstyles had been. Expressing the ethnocentrism that Japanese people harbored during this period, she remarked that if Japanese people had remained obsessed with preserving old customs and neglected to adopt new ways, there would have been no progress and Japan would have remained as primitive as the Ainu or natives in the South Pacific and Africa.[68]

Yosano viewed the fashionable *sokuhatsu* style as evidence of progress in women's intellectual and personal growth. She contended that the reason why a superbly shaped "*hisashi-gami*" (low pompadour) looked somehow unsettling was because many women still lacked the well developed character and intelligence necessary for wearing such a large chignon, even though they were ready to wear other types of *sokuhatsu* (see Fig. 5.8). This is why some women who wore the grand low pompadour looked as if they were walking wigs straight out of the Mitsukoshi department store.[69]

Although her discussions about women's clothes were often shaped by meta-concepts shared by health reformers, Yosano's views were distinct. She emphasized that women should be given the freedom to choose their own clothes and to develop their own intellectual and emotional lives.

Fig. 5.8 Students in science class at Japan Women's College (around the turn of the twentieth century). Author's note: All the students are wearing a *sokuhatsu* hairstyle, called the *hisashi-gami*. This popular hairstyle was also called the "203 Hill," after the Russo–Japanese War, named after a battlefield in the northeast of China where the Russian and Japanese armies fought a fierce battle (Source: Aflo/ Mainichi Photobank)

Yosano encouraged women to take advantage of the attractive new clothing and hairstyles, to enjoy themselves, to nurture their aesthetic sensitivity, and to assert their identity as young women of a new era.[70] This was quite different from health-reformer recommendations, which sought to impose reformed clothing and hairstyles primarily in service to the nation. Health activists were also indifferent to the pleasure women obtained from wearing fashions of their own choosing.

Despite Yosano's disapproval, wearing *hakamas* eventually became the standard dress for schoolgirls. In her autobiographical writings, feminist activist Hiratsuka Raichō, a young woman from an upper-middle-class family attending a primary school and a women's middle school in the Tokyo area in the 1890s and the 1910s, took it for granted that she and her schoolmates would wear *hakamas*. During her last year at primary school and throughout her middle school years, Hiratsuka also wore reformed clothes consisting of *hakamas* and a short, narrow-sleeved kimono that only covered her hips.[71] Although she often wrote about her experience of wearing reformed clothes in a disinterested, matter-of-fact tone, at other times her decision to wear a specific type of clothing had a more emotional significance. One memorable occasion in which Hiratsuka consciously changed her clothes was when she started her studies at Japan Women's College at the age of 17. She stopped wearing the purple *hakamas* and purple kimonos with splashed patterns she had worn in middle school and adopted instead maroon *hakamas* and brown kimonos with vertical stripes. She also gave up wearing her hair down her back and using large ribbons, and began to wear a *sokuhatsu*-style chignon. Hiratsuka deliberately let go of her "childishly dressed appearance" and prepared herself emotionally to engage in college-level study with many older women as her colleagues; the change in dressing style was an important part of her efforts to forge her identity as a grown-up woman.[72] It was clear that Hiratsuka viewed wearing *hakamas* as an indication of being a student. Since she felt that she was a student in spirit throughout her life, she continued to wear *hakamas* even after her marriage. She reluctantly gave them up after the birth of her first child because carrying the baby on her back while wearing *hakamas* looked too strange in the eyes of the public.[73]

When, in 1919, Hiratsuka became involved in the feminist movement through the New Women's Society [Shin Fujin Kyōkai], she lived an active lifestyle and felt acutely the inconvenience of *hakama*-less kimonos. Ichikawa Fusae, Hiratsuka's close ally in the New Women's Society leadership, reminisced in her autobiography that she and Hiratsuka discussed adopting Western clothes. According to Ichikawa, they agreed to wear Western clothes that they thought appropriate to political activism, including visiting the National Assembly (the Diet), meeting with politicians, or organizing public lectures.[74] Finding a tailor, however, who would take Hiratsuka's order was difficult. She wanted not the type of ornate dress that society women wore, but a simple dark-colored suit that would be practical as well as symbolic of her identity as a feminist activist. She eventually found a tailor who had learned his trade in the USA and was teaching Japanese housewives how to sew children's clothes (see Ichikawa and Hiratsuka in Western clothes in Fig. 5.9).[75]

In 1924 Hiratsuka adopted an "ultra-fashionable," bobbed flapper-like hairstyle. She stated in her 1950 memoir that she changed her hairstyle because of the severe headaches she suffered from her stressful work for the New Women's Society, and because she found her long hair cumbersome.[76] She adopted the new style for reasons not so different from the arguments of health reformers who favored promoting women's active lifestyles. To the reformers' chagrin, however, Hiratsuka's style devastated

Fig. 5.9 Photo of a meeting of the New Women's Society (1919). Author's note: The woman front right is Ichikawa Fusae, and the third woman from the left, wearing Western clothes, is Hiratsuka Raichō (Source: Aflo/Mainichi Photobank)

one of their primary goals: preserving the level of womanly decorum and modesty they considered proper.[77] Adopting a new fashion was a way for Hiratsuka to explore a womanhood entirely different from the ideal of a "good wife and wise mother." Advocating the expansion of women's political rights and sexual autonomy as the era's representative "New Woman," she challenged prevailing norms of femininity not only in thinking but also in appearance.[78]

If feminist writers and activists such as Yosano and Hiratsuka challenged the male-dominated health-reform movement in their own ways, what were the responses of ordinary young female students? Did they resist educator control over their fashion? Most schoolgirls at that time could not choose what to wear because many schools established rules regulating color, materials, and styles of kimonos and *hakamas*. At some schools, students were required to wear kimonos made of cotton and not silk, while other schools required that students wear black, maroon, or brown *hakamas*. Many young women willingly adopted such styles because attending girls' middle schools, or other institutions of higher learning, was a privilege that many ambitious women strove for. Thus, they wore *hakamas* with a sense of pride and accomplishment.

Despite the restrictions that their teachers imposed on them, young schoolgirls found opportunities to dress more fashionably. One way was through the adoption of constantly reinvented and elaborate *sokuhatsu* styles. Young women followed the latest hair trend by inserting a core to inflate their hair at the front or by using various ribbons, flowers, and other ornaments to decorate their hair. Thus, the *sokuhatsu* styles were not necessarily "simpler" and "more economical" than old-fashioned Japanese hairstyles. Moreover, among female students in the Tokyo area, it was stylish to wear long *hakamas* with the sash tied as tight as possible.[79] Schoolgirls thus defeated the health reformers' objectives: their long *hakamas* did not promote an active lifestyle or frugality, and the tightly bound waistband functioned much like a constricting *obi* sash.

Schoolgirls succeeded in subverting some of the health reformers' goals, while creating subtle fashion norms of their own. Yet even though their innovative fashions could be interpreted as a form of self-expression, this type of resistance was not necessarily empowering for young women. Many women became excessively concerned with fashion as new styles and trends were promoted through women's magazines, and clothing and cosmetic stores.[80] Many young women spent countless hours modifying details in their clothes, hair, and make-up. In the end, they became enslaved to the constantly and rapidly shifting fashion norms promoted by an increasingly commercialized mass culture.[81] Their intense pursuit of fashion also reinforced a prevailing view of women as being endowed with limited wisdom and intellectual capacity. Thus, young women's efforts to devise new fashions or follow the latest trends often resulted in collaborating with the forces that supported a patriarchal social order.[82]

By comparing the positions that Yosano, Hiratsuka, and ordinary female students took on the issue of young women's clothing, we gain insight into how women could subvert the health reformers' intentions within the constraints of the discursive horizon. Along with health reformers, Yosano celebrated the coming of modernity, but she put utmost emphasis on individual freedom and the appreciation of women's own aesthetic sensitivity. Indeed, the idea of freedom was key to developing her criticisms of clothing reforms. Women should no longer be subject to the control of what she viewed as old-fashioned, narrow-minded, and moralistic educators and patriarchs; rather, they should engage in independent thinking and the creation of arts. For Yosano, young women's choice in clothing—colorful, elegant, and stylish kimonos—went hand in hand with their liberation from feudalistic, patriarchal rule.

Yosano's ideas were formed by observing young women from the upper and middle classes who took for granted their education and material comfort. Women from less privileged social backgrounds adored and coveted the *hakama* style adopted by female students. Many clearly understood that receiving a women's middle school or college education connoted family affluence and respectability, and constituted a reserve of cultural capital available for marrying into a family whose wealth and status surpassed their own. Female student fashion was attractive because it epitomized much celebrated modernity in addition to aristocratic elegance. On a more practical note, students of lesser means would have appreciated the restrictions on the material and color of their kimonos, or would have preferred even a standardized uniform because such regulation would have saved them the embarrassment of being unable to afford the kinds of expensive and colorful kimonos their peers wore. Although many young women found tremendous pleasure in shopping and arranging their clothes and accessories, women from less affluent families could experience humiliation and pain. Being brought up in a well-to-do shopkeeper's family, Yosano lacked empathy. Her thinking was limited by class bias, which ironically was shared by health reformers concerned about the "sedentary lifestyle" of Japanese women—the vast majority of women in Japan, from peasant and working class families, engaged in long hours of strenuous labor.

In the case of Hiratsuka, she, too, developed feminist ideas from her education in health science at Japan Women's College. She remarked that she adopted the *hakama* style, tailored suits, and bobbed hair for health reasons, and because they increased her physical mobility. Her discussion of fashion drew on the language of health and hygiene that she learned by taking health science courses at the college. She bluntly rejected the model of "good wife, wise mother" that her college strongly endorsed and chose fashion that conservatives considered bizarre and outrageous.

Her feminist resistance, in other words, largely depended on her creative use of the education she received at Japan Women's College. Her development of eugenicist ideas from a feminist perspective also shows that she used her knowledge in health science to further her feminism and activism.[83] She appropriated and negotiated ideas of eugenics to formulate policy that she thought would help advance women's interests. For sure, she overlooked the potential that eugenics had for creating new forms of oppression for groups of women who were considered "unfit" to have children. What I want to stress here, however, is not that her eugenics ideas were riddled with problems, but that she launched her challenge to male dominance by relying on expert knowledge she obtained through the modernized education system.

This shows that feminists could take advantage of, modify, and manipulate scientific ideas they had gained from institutions and educators whose interests were aligned not with women themselves, but with the state and experts. Women, however, could only launch such resistance by first having been subjected to the forces of modernity that made them educated, thinking, modern individuals who then increasingly involved themselves in the ever-changing networks of knowledge and power. The more women were educated in modern health science the more extensively modern power colonized them. They could carry out more effective resistance by using the scientific knowledge they had obtained, even forming strategic alliances with state officials and experts. Or the state sometimes co-opted feminists, as happened with Hiratsuka and other Japanese feminists during the Asia–Pacific War from the 1930s through to 1945. In any event, the types and directions of resistance they could pose were fundamentally bound by the discursive horizon of contemporary human sciences and the structure of networks of expert knowledge and power as they related to women's issues.

At the beginning of Chap. 4, I discuss the emergence of a new terrain where the new networks of knowledge and power were formed and operated within the field of reproductive medicine. What I have attempted to show in this chapter is how modern medicalized knowledge of women's bodies began to be circulated outside the medical field. The expansion of women's schools, the establishment of groups to promote women's health, and the rise of print media targeting female readers were key to the dissemination of such knowledge. The Education Ministry, the Home Ministry, the Hygiene Bureau and other organs of the state played an important role in supporting the establishment and development of women's schools and health organizations. Individuals of high status, including members of the imperial family, aided those efforts. Also, the state worked closely with medical experts in the creation of legitimate scientific and medical knowledge about women's bodies, and health and women's magazines popularized that knowledge. In this way, although the state did not exercise total control over the dissemination of new knowledge about women's health, it could promote the circulation of certain forms of it and concurrently curb the spread of other types of information. Modern bio-power spread in this new interspersed terrain of educational institutions, activist groups, and health and women's magazines. In that terrain the state played a unique role in regulating the spread of knowledge about health and bodies.

This basic configuration of key institutions and mechanisms for spreading bio-power has, by and large, endured since around the turn of the twentieth century. For sure, many aspects of this system have changed. Scientific knowledge about women's health and bodies has been revised repeatedly during the past century. Theories of racial and sexual difference from the turn of the twentieth century that I introduce in this chapter have gone through drastic changes. For example, by the late twentieth century, scientists had repudiated "race" as a legitimate scientific category and since the turn of the twenty-first century have started to question the validity of binary models of sexual difference. Also, from the late nineteenth century onward, young women in Japan have attended schools at various levels, while course curricula and extra-curricular programs for health science, physical education, and sports activities have gone through many changes. The expansion of schooling for women has also meant that expert knowledge about health and hygiene gained currency not only among upper- and middle-class women but also among many peasant and working-class women.[84] In other words, an increasing number of women became more deeply incorporated into regimes of bio-power, both as objects whose bodies were trained and drilled through disciplinary practices and as subjects with the will and desire to improve their bodies within the webs of power and knowledge. Further, since the turn of the twentieth century, activist groups have been set up to work on women's health. Included in such organizations were feminist groups that often opposed policies endorsed by the state and organized medicine. They created a complex picture of resistance and collusion on the part of women. Finally, the media through which scientific and medical knowledge about women's health and bodies circulated have also shifted. In addition to print media, television became a powerful tool for educating women on health and beauty in the second half of the twentieth century, and in the twenty-first century, the internet has emerged as another major medium for this.

Despite these changes, the basic institutional arrangements in which expert knowledge was disseminated and trickled down remained more or less constant: the state and organized medicine circulating scientific knowledge from advantageous positions to further their interests. Within this power structure, women unintentionally served the interests of the state and the capitalist economy, which needed healthy and productive bodies, by engaging in physical exercise and health practices through schools, women's magazines, and health manuals. New health practices promoted women's health and hygiene; the interests of women, however, did not always correspond precisely to those of the state, experts, or industrialists, who together sought bodies fit for healthy reproduction or for factory work. Women at times collaborated with the state and experts, and took advantage of new knowledge and practices about health and hygiene, but at other times resisted the imposition of policies and practices they deemed suspicious or found to be against their interests. Such mechanisms are still at work in Japan today, although much of the medical and scientific knowledge has been revised, and new technologies for defining and disciplining bodies have become available in schools, workplaces, and medical institutions that attract women as students, workers, and patients.

In the next chapter, I return to the issues of maternal and infant health that I discuss in Chap. 4. I follow up on issues in this area from the 1910s through to the end of World War II. I explore the development of policies through the war years and the renewed efforts by the state to expand its control over women's health and reproduction. During this period, in addition to trained midwives in private practice, new government and quasi-government programs were created to monitor and improve maternal and infant health, and mothers in rural areas and poor urban households emerged as new targets of health reform. The newly established Ministry of Health and Welfare introduced new systems of reproductive health services, such as the district nurse system and the maternity passbook, furthering the encroachment of bio-power into women's reproductive lives. I trace these developments, paying attention to the ways different political forces within and outside government interacted and converged.

Notes

1. Narita Ryūichi also analyzes the issues of health and hygiene in women between the turn of the twentieth century and the 1920s in "Eisei kankyō no henka no nakano josei to josei kan," in Joseishi Sogō Kenkyūkai, ed., *Nihon josei seikatsu shi*, Vol. 4 (Tokyo: Tokyo Daigaku Shppankai, 1990): 89–124; and "Eisei ishiki no teichaku to 'bi no kusari,' " *Nihonshi kenkyū* 366 (February 1993): 64–89. For an English translation of the latter, see

"Mobilized from Within: Women and Hygiene in Modern Japan" in Hitomi Tonomura, Anne Walthall, and Wakita Haruko, eds., *Women and Class in Japanese History* (Ann Arbor: Center for Japanese Studies, The University of Michigan, 1999): 273.

2. Foucault, *The History of Sexuality, Vol. I*; and *Discipline and Punish*; Susan Bordo, "The Body and Reproduction of Femininity: A Feminist Appropriation of Foucault" in Alison Jaggar and Susan Bordo, eds., *Gender/Body/Knowledge: Feminist Reconstructions of Being and Knowing* (New Brunswick, NJ: Rutgers University Press, 1989); and *Unbearable Weight: Feminism, Western Culture, and the Body* (Berkeley: University of California Press, 1993); Anne Balsamo, *Technologies of the Gendered Body: Reading Cyborg Women* (Durham, NC: Duke University Press, 1996). See also Joan Jacobs Brumberg, *The Body Project: An Intimate History of American Girls* (New York: Random House, 1997).
3. For example, the first president of Japan Women's College (established in 1901), Naruse Jinzō (1858–1919), possessed an extensive political network that was crucial for his fund-raising efforts for establishing Japan Women's College. Sumiko Otsubo Sitcawich, *Eugenics in Imperial Japan: Some Ironies of Modernity* (Ph.D. Dissertation, Ohio State University, 1998): 53–66.
4. The association drew 112 members within a year after was established. Its membership increased continuously: 375 in 1890, 844 in 1896, 1394 in 1900, and 2082 in 1907. Membership declined, however, during the Taishō period, when an increasing number of women interested in health and hygiene became active in other groups and institutions. Kameyama Michiko, "Shiritsu dai nihon eiseikai to *Fujin eisei zasshi* ni tsuite," *Fujin eisei zasshi*, supplementary volume (Tokyo: Ōzorasha, 1992): 13–24.
5. The monthly journal distributed to the members of the Japan Association for Women's Hygiene.
6. Tokyo Higher Women's Normal School was established in 1874 by the Meiji government to provide teacher training for women. In the post-World War II era, it was re-established as Ochanomizu University, one of the foremost publicly funded women's universities in Japan.
7. *Feris jogakuin 110-nen shōshi* (Yokohama: Ferisu jogakuin, 1982): 27, 29.
8. The Ferris Seminary was one of the schools regulated by the Ordinance for Private School. Such schools had a lesser status than those governed by other ordinances, such as the Higher Girls School Ordinance, and faced a decline in student numbers. Ibid., 40–46.
9. Ono Yoshirō also discusses this topic along with the rise of physical education in the late nineteenth and early twentieth centuries. Ono Yoshirō, *"Seiketsu" no kindai: "eisei shōka" kara "kōkin guzzu" e* (Tokyo: Kōdansha, 1997): 198–239.

10. More research into how social Darwinism was introduced and developed in Japan is still needed. The existing literature includes the following. Unoura Hiroshi, "Kindai nihon ni okeru shakai dāwinizumu no juyō to tenkai," in Shibatani Atsuhiro, Nagano Takashi, and Yōro Takeshi, eds., *Kōza: shinka, Vol. 2, shinka shisō to shakai* (Tokyo: Tokyo daigaku Shuppankai, 1991): 119–152; Fujino Yutaka, *Nihon fashizumu to yūsei shisō* (Kyoto: Kamogawa Shuppan, 1998); Suzuki Zenji, *Nihon no yūseigaku* (Tokyo: Sankyō Shuppan, 1983); Watanabe Masao, "Meiji-ki ni okeru shinka ron no juyō," in Watanabe, ed., *Dāwin to shinka ron* (Tokyo: Kyōritsu Shuppan, 1984): 191–210. For English language works, see, for example, Julia Adeney Thomas, *Reconfiguring Modernity: Concepts of Nature in Japanese Political Ideology* (Berkeley: University of California Press, 2002); Ronald Roden, "Baseball and the Quest for National Dignity in Meiji Japan," *American Historical Review*, Vol. 85, No. 3 (June 1980): 511–534.
11. Baelz taught at the School of Medicine of Tokyo Imperial University from 1876 until 1903. For Baelz's life in Japan, see Toku Baelz, ed., Erwin O. E. von Baelz, *Awakening Japan: The Diary of a German Doctor* (New York: The Viking Press, 1932); Toku Berutsu, *Berutsu no nikki* (Tokyo: Iwanami Shoten, 1951–1955); F. Shottorenda, *Eruwin fon Berutsu* (Tokyo: Ōzorasha, 1995, a Japanese translation of *Erwin von Baelz* by Felix Schottlaender, published in 1928); Masumi Shumitto-Muraki, *Hana Berutsu e no tabi* (Tokyo: Kōdansha, 1993).
12. For all citations from Baelz's lecture, see Erwin von Baelz, translated by Mishima Michiyoshi, "Joshi no taiiku," in *Fujin eisei zasshi* 105 (June 1898): 1–32. I have been unable to locate the original German manuscript. During the meeting Mishima either translated Baelz's oral lecture or gave a talk based on Baelz's written manuscript for a Japanese audience.
13. Baelz may have been referring to the regular custom of tying a belt on the lower hip area in order to tuck it into the lower portion of the kimono to prevent it from trailing. This belt was used in addition to the main *obi* sash. Japanese women did not tie the regular *obi* sash "below the pelvis".
14. References by Terada here and in the following pages, unless otherwise noted, are from Terada Yūkichi, "Hōjin no taikaku, Part I," in *Fujin eisei zasshi*, 169 (December 1903): 15–29; and "Hōjin no taikaku, Part II," in *Fujin eisei zasshi*, 170 (January 1904): 23–34.
15. Imamura Yoshio, *Nihon Taiiku shi* (Tokyo: Fumaidō shuppan, 1970): 457–458.
16. For all citations from Inokuchi's lecture, see Inokuchi Aguri, "Beikoku joshi taiiku no genkyō," in *Fujin eisei zasshi* 163 (June 1903): 1–18.
17. Japanese scientists and medical researchers, on the other hand, tended to think that "racial characteristics" of indigenous minority people (e.g., the

Ainu in northern Japan) were fixed and immutable. See, for example, Yuki Terazawa, "Racializing Bodies through Science in Meiji Japan: The Rise of Race-Based Research in Gynecology," in Morris Low, ed., *Building a Modern Japan: Science, Technology, and Medicine in the Meiji Era and Beyond* (New York: Palgrave Macmillan, 2005): 83–102.

18. For women's participation in sports and physical exercise and its link with feminist movements in Europe, see Kathleen E. McCrone, *Sports and the Physical Emancipation of English Women 1870–1914* (London: Routledge, 1988) and J. A. Mangan and Roberta J. Park, *From "Fair Sex" to Feminism: Sport and the Socialization of Women in the Industrial and Post-Industrial Eras* (London: Frank Cass, 1987).
19. Shimoda Jirō, *Joshi Kyōiku* (Tokyo: Kinkōdo shoseki, 1904), reprinted as *Kindai nihon joshi kyōiku bunken-shū*, Vol. 7 (Tokyo: Nihon Tosho Sentā, 1983). For Shimoda's biography, see Tachi Kaoru's commentary in Nakajima Kuni, ed., *Kaisetsu*, in *Kindai nihon joshi kyōiku bunken shu: the first period* (Tokyo: Nihon Tosho Sentā, 1983): 93–104.
20. Henry Herbert Donaldson, *The Growth of the Brain: A Study of the Nervous System in Relation to Education* (London: W. Scott, 1895). Shimoda provides a plausible but erroneous publication date, 1898. Shimoda, *Joshi Kyōiku*, 748.
21. Ibid., 701–702.
22. Havelock Ellis, *Man and Woman: A Study of Human Secondary Sexual Characters*, 3rd edition (London: W. Scott, 1902).
23. Shimoda, *Joshi Kyōiku*, 702–708.
24. Shimoda cited an English translation of Lotze's work entitled *Microcosmus*. Ibid., 749. Shimoda did not give a full reference; however, the following is most likely the version he used. Hermann Lotze, *Microcosmus: An Essay Concerning Man and His Relation to the World*, translated from the German by Elizabeth Hamilton and E. E. Constance Jones, 2nd edition (Edinburgh: T. & T. Clark, 1887).
25. For references to Lotze, see Shimora, *Joshi Kyōiku*, 730–740 here and in the following pages, unless otherwise noted.
26. H. Fehling, *Die Bestimmung der Frau: ihre Stellung zu Familie und Beruf* (Stuttgart: F. Enke, 1892); Max Runge, *Das Weib in seiner geschlechtlichen Eigenart; nach einem in Göttingen gehaltenen Vortrage* (Berlin, Julius Springer, 1904). According to Shimoda's note, this book was published in 1900. Shimoda, *Joshi Kyōiku*, 750.
27. Ibid., 394–399.
28. Nagae served as vice principal of Tokyo Prefectural Women's Higher School. Nakajima, ed., *Kaisetsu*, 33–35.
29. Nagae Masanao, *Joshi kyōiku ron* (Tokyo: Hakubunkan, 1892), reprinted in *Kindai nihon joshi kyōiku bunken-shū*, Vol. 3 (Tokyo: Nihon Tosho Sentā, 1983).

30. Nagae did not include notes for specifying foreign language sources in *Joshi kyōiku ron*. When he did cite foreign scholars' work, in this case Huxley, he simply stated, "According to Mr. Huxley…." It is most likely that Nagae is referring to Thomas Henry Huxley (1825–1895), one of Charles Darwin's (1809–1882) disciples.
31. Ibid., 60–61.
32. Ibid., 27–33.
33. After launching his early career as an educator and a Christian reformer, Naruse Jinzō studied education and Christian theology in the USA between 1890 and 1894. Upon returning to Japan, he served as principal in Baika Women's School and developed a plan to open a women's college. Japan Women's College, which Naruse established in 1901, was one of the few college-level educational institutions for women in Japan. See Nakajima Kuni's commentary on Naruse in Nakajima, ed., *Kaisetsu*, 55–65; Nika Setsu, *Naruse Sensei den* (Tokyo: Ōfūkai, 1928, reprinted as *Denki sōsho*, Vol. 56, Tokyo: Ōzorasha, 1989). Naruse's biography is also given in Sumiko Otsubo Sitcawich, *Eugenics in Imperial Japan*, 53–88.
34. Naruse Jinzō, *Joshi kyōiku* (Tokyo: Aoki Kozando, 1896), reprinted in *Kindai nihon joshi kyōiku bunken-shū*, Vol. 5 (Tokyo: Nihon Tosho Sentā, 1983): 183–185.
35. Ibid., 179–187.
36. Ibid., 182–183, 224–228.
37. Prominent Japanese physical education specialists, Kawase Genkurō and Inokuchi Aguri, emphasized the advantage of Swedish gymnastics over the existing gymnastic methods partly because they were based on up-to-date scientific theories on human physiology and psychology. Moreover, while the previous gymnastics methods required students to memorize the movements in advance, Swedish gymnastics allowed students to engage in movements directed by the instructor's contemporaneous dictation, and it was considered especially suitable for women. Takenoshita Kyūzō and Kishino Yūzō, *Kindai nihon gakkō taiiku shi* (Tokyo: Tōkōkan Shuppansha, 1959): 61–64.
38. The details of the "scarf" game and the *chidori* race are unknown.
39. *Nihon Joshi Daigakkō yonjū-nen shi* (Tokyo: Nihon Joshi Daigakkō, 1942): 102–103. For eugenics theories taught at Japan Women's College and its physical education curriculum, see Otsubo Sitcawich, *Eugenics in Imperial Japan*, 37–52. The Delsartean exercises originated in methods of elocution and bodily movements developed by the French actor François Delsarte (1811–1871). This system was elaborated and widely disseminated in the USA in the 1880s and 1890s by US educators Steel Mackay (1842–1894), Henrietta Hovey (1849–1918), and Genevieve Stebbins (1857–1913). In addition to being used by actors, dancers, and public

speakers, the Delsartean exercises also found enthusiastic followers among middle- and upper-class women who admired the graceful and expressive gestures and movements modeled on those found in statues and pictures from ancient Greece and Rome. For an overview of the influence of Delsartism in the USA, see Nancy Lee Chalfa Ruyter, *The Cultivation of Body and Mind in Nineteenth-Century American Delsartism* (Westport, CT: Greenwood Press, 1999).

40. Takenoshita and Kishino, *Kindai nihon gakkō taiiku shi*, 3–68; Imamura, *Nihon Taiiku shi*, 407.
41. Shirai Kikurō, *Joshi undō to yūgi* (Tokyo: Kodokan, 1909): 38–39.
42. Ibid., 61.
43. Inokuchi Aguri, Kaji Toku, Kawase Motokurō, Takashima Heizaburō, and Tsuboi Gendō, *Taiiku no riron to jissai* (Tokyo: Kokkōsha, 1906): 353.
44. Shirai, *Joshi undō to yūgi*, 27.
45. Inokuchi et al., *Taiiku no riron to jissai*, 405–413. European gymnastics instructed students to keep their eyes focused straight ahead, reversing the instructions of traditional Japanese etiquette that taught women to cast their eyes downward when facing people who were senior in status and age. For the committee's description of the correct standing posture and walking style, see ibid., 89–90, 364–366.
46. Miyake Yoshiko discusses how women were mobilized as mothers and workers for state purposes during the 1930s and 1940s in "Doubling Expectations: Motherhood and Women's Factory Work Under State Management in Japan in the 1930s and 1940s" in Gail Bernstein, ed., *Recreating Japanese Women, 1600–1945* (Berkeley: University of California Press, 1991): 267–295.
47. The issue of clothing reforms for Japanese women generated public discussions throughout the Meiji period. Barbara Molony, "Gender, Citizenship, and Dress in Modernizing Japan," in Mina Roces and Louise Edwards, eds., *The Politics of Dress in Asia and the Americas* (Brighton, GB: Sussex Academic Press, 2007). Sally Hastings discusses Japanese women's adoption of Western clothes by focusing on the empress in "The Empress' New Clothes and Japanese Women, 1868–1912," *The Historian* 55:4 (Summer 1993). In the late 1880s and 1890s, *Jogaku zasshi* [*Women's Magazine*] published numerous articles endorsing clothing reforms for Japanese women. Its authors recognized the necessity for Japanese women to become modern. In contrast, conservative critiques within the state bureaucracy and outside of it opposed women's dress reforms. My research and analysis focus specifically on women's clothing reforms in relation to the health-reform movement at the turn of the twentieth century. For a synopsis of the history of clothing reforms during the Meiji period, see Chūbu katei keieigaku kenkyūkai, *Meijiki katei seikatsu no kenkyū* (Tokyo: Domesu Shuppan, 1972): 144–195.

48. "Joshi Taiiku ron, Part V," in *Fujo shinbun* 116 (July 28, 1902): 1.
49. For Fukushima's response here and in the following pages, unless otherwise noted, see the following articles by Fukushima and Yamada, which appeared in *Fujo shinbun*. "Chakko seishi jiken, Part I," *Fujo shinbun*, 79 (November 11,1901): 1; "Chakko seishi jiken, Part II," *Fujo shinbun*, 80 (November 18,1901): 1; "Chakko mondai ni tsuite," *Fujo shinbun*, 85 (December 23, 1901): 5; "Chakko mondai ni tsuite," *Fujo shinbun*, 86 (January 1, 1902): 6; "Chakko mondai ni tsuite," *Fujo shinbun* 87 (January 6, 1902): 4; "Chakko mondai ni tsukite mata," *Fujo shinbun*, 108 (June 2, 1902): 1; "Chakko mondai ni tsuite no benkai," *Fujo shinbun*, 112 (June 30, 1902): 5. Ishizaki Shōko also discusses the "hakama adoption" debate in "'Fujo shinbun to Fukushima Shirō no joseiron (sono ichi)," in Sōgō Joseishi Kenkyūkai, *"Fujo shinbun" wo yomu* in *Sōgō Joseishi Kenkyūkai, Kindai Shi Bu Kai Kaihō*, 1 (December 1984): 9–12.
50. Honda Masuko, *Jogakusei no keifu: saishoku sareru meiji* (Tokyo: Seidosha, 1990): 33–34.
51. "Joshi Taiiku ron, Part IV," *Fujo shinbun*, 114 (July 14, 1903): 1.
52. Ibid. Also, "Fujin no keppatsu," *Fujo shinbun*, 84 (December 16, 1902): 1, "Sokuhatsu to bōshi," *Fujo shinbun*, 109 (June 9, 1903): 1.
53. "Fujin no keppatsu," *Fujo shinbun*, 84, 1; "Sokuhatsu to bōshi," *Fujo shinbun*, 109, 1.
54. For this and the next page, the following article is the source: Miyamoto Naka, "Nyofuku kairyō ni tsuite rōba-shin wo nobu," in *Fujin eisei zasshi*, 145 (December 1901): 1–21.
55. For the debasement of the images of China by Japanese intellectuals, see Stefan Tanaka, *Japan's Orient: Rendering Pasts into History* (Berkeley: University of California Press, 1993).
56. Fujikawa Yū (1865–1940) was a German-educated medical journalist and historian who introduced new studies in areas including public health, child psychology, sexology, and eugenics from Europe, and published an interdisciplinary journal on "human nature" called in Japanese, *Jinsei* [*Der Mensch*], first published in 1905. Matsumura Noriaki, Hirono Yoshiyuki, and Matsubara Yōko, "Fujikawa Yū, Pioneer of the History of Medicine in Japan," *Historia Scientiarum*, Vol. 8–2 (1998): 157–171.
57. For a discussion of Fujikawa Yū and Watanabe Hōkichi, unless otherwise noted, see Fujikawa Yū, "Bibō no hanashi," *Fujin eisei zasshi*, Vol. 210 (May, 1907): 1–11; Watanabe Hōkichi, "Josei no bi, Part III," *Fujin eisei zasshi*, 209 (April 1907): 26–31; and "Josei no bi, Part II," *Fujin eisei zasshi*, 208 (April 1907): 31.
58. Watanabe introduced this formula as the work of Dr. "Fiyaoruto," who could possibly be the German physician, Karl von Vierordt (1818–1884). Fujikawa also presented a similar formula for evaluating children's physical growth. The measurements above are in centimeters and kilograms.

59. Kawamura Kunimitsu, *Otome no shintai: Onna no kindai to sekushuaritī* (Tokyo: Kinokuniya Shoten, 1994): 28–43.
60. About the increased commercialization of beauty products for young women in the Taishō era, see Kawamura Kunimitsu, *Otome no inori: Kindai josei imēji no tanjō* (Tokyo: Kinokuniya Shoten, 1993): 164–179; and Kawamura, *Otome no shintai*, 22–54. For an English language work on women and consumerism in interwar Japan, see Barbara Sato, *The New Japanese Woman: Modernity, Media, and Women in Interwar Japan* (Durham, NC: Duke University Press, 2003).
61. Yosano Akiko, *Ichigū yori* [*From my Little Corner*] (Tokyo: Hōbunkan Shoten, 1911). This book consisted of a collection of essays that Yosano had previously published in various journals.
62. Ibid., 51–67, 187–189, 201–202, 370–377, 430–438.
63. Ibid., 580–592.
64. Ibid., 51–67, 580–592.
65. Ibid., 485–488.
66. Ibid., 440.
67. Ibid.
68. Ibid., 120–123.
69. Ibid., 413–114. The Mitsukotshi department store located in Nihonbashi, Tokyo, originated in the Echigoya fabric store, founded in 1673. In 1904 the Mitsukoshi store was re-established as a company independent from the Mitsui group companies. It expanded as a department store with a newly constructed three-story Western-style building in 1908, and became a major center of fashionable urban culture. *Sekai daihyakka jiten*, Vol. 29 (Tokyo: Heibonsha, 1981): 396.
70. Ibid., 51–67, 438–442.
71. Hiratsuka Raichō, "Yōsō no omoide," (originally published in 1955), in *Hiratsuka Raichō chosaku-shū*, Vol. 7 (Tokyo: Ōtsuki Shoten, 1984): 303.
72. Hiratsuka Raichō, "Mejiro no omoide," (originally published in 1939), in *Hiratsuka Raichō chosaku-shū*, Vol. 6 (Tokyo: Ōtsuki Shoten, 1984): 311. For more on women's feelings about the clothes young women wore, see also Baroness Shidzue Ishimoto, *Facing Two Ways: The Story of My Life* (Stanford, CA: Stanford University Press, 1984).
73. Hiratsuka, "Yōsō no omoide," 303.
74. Ichikawa Fusae, Ichikawa Fusae jiden, senzen hen (Tokyo: Shinjuku Shobō, 1974): 77.
75. Hiratsuka, "Yōsō no omoide," 303–304.
76. Ibid., 304. For more on the "modern girls," some of whom adopted bobbed hair and flapper dresses, see Miriam Silverberg, "The Modern Girl as Militant," in Bernstein, ed., *Recreating Japanese Women, 1600–1945*, 249–250; and *Erotic, Grotesque, Nonsense: The Mass Culture of Japanese*

Modern Times (Berkeley: University of California Press, 2006). Silverberg also discusses the politics of representing bodies in advertisements in the 1920s and 1930s in "Advertising Every Body: Images from the Japanese Modern Years," in Susan Leigh Foster, ed., *Choreographing History* (Bloomington: Indiana University Press, 1995): 129–148.

77. Barbara Molony presents a useful and informative discussion on the politics of dressing, involving feminists, including Hiratsuka and Ichikawa, and Japanese Modern Girls. Molony, "Gender, Citizenship, and Dress in Modernizing Japan," 93–97.
78. For Hiratsuka's feminist thoughts, see her writings published in the journal *Seitō* [The Bluestockings], whose publication ran between 1911 and 1916. Several of her articles from this period are compiled in Horiba Kiyoko, ed., *"Seitō" josei kaihō ronjyū* (Tokyo: Iwanami Shoten, 1991). For an English language work on the images of the "New Woman" and the feminist movement of the Bluestockings, see Dina Lowy, *The Japanese "New Woman": Images of Gender and Modernity* (New Brunswick, NJ: Rutgers University Press, 2007). Chapter 8 in Sharon Sievers's *Flowers in Salt: The Beginnings of Feminist Consciousness in Modern Japan* (Stanford, CA: Stanford University Press, 1983) is also a good introduction to the Bluestockings movement in the 1910s.

    It should also be noted that when bobbed hair and Western clothes were widespread in the commercialized culture—say, after the mid-1920s—they lost much of the subversive edge that had been present in Hiratsuka's early appropriation. This shows that subversion does not reside within a specific style of clothing; rather, it is generated through the adoption of certain fashions at a particular historical moment and within a specific cultural context.
79. Kohagi, "Ima mukashi no jogakusei, Part I–III," *Fujo shinbun* 332, 333, and 334 (September 17, 24, and December 1, 1906). Sakamaki Hisa, *Otenba saijiki* (Tokyo: Sōshisha, 1979): 105–110, 178–220.
80. Ibid., Kohagi; Honda, *Jogakusei no keifu*, 7–98; Kawamura, *Otome no inori*, 164–179; and *Otome no shintai*, 22–54.
81. See also Yosano's critique of young women's tendency to be excessively concerned about clothes and accessories in Yosano, *Ichigū yori*, 499.
82. As a comparable case of how women's resistance defeats women themselves, see Susan Bordo's discussion of female-specific diseases such as hysteria, agoraphobia, and anorexia that have become prevalent in Europe and the USA during the late twentieth century. Bordo describes how women used these illnesses to resist dominant norms of femininity and how their resistances often resulted in reinforcing a gender ideology that undermined their own interests. Bordo, "The Body and Reproduction of Femininity."

83. Sumiko Otsubo, "Engendering Eugenics: Feminists and Marriage Restriction Legislation in the 1920s," in Barbara Molony and Kathleen Uno, eds., *Gendering Modern Japanese History* (Cambridge, MA: Harvard University Asia Center, 2005): 225–256.
84. Obvious from my discussions in this chapter, Japanese health reformers from the turn of the twentieth century focused mainly on middle and upper class women. Rumi Yasutake in her forthcoming book, tentatively entitled, *Women's Civilizing Mission and Globalization in the Pacific 1820–1837*, notes that in China, the women's education movement from the early twentieth century—where foreign (often missionary) educators from North America and Europe dominated—paid much more attention to the issues of working-class women.

CHAPTER 6

Knowledge, Power, and New Maternal Health Policies (1918–1945)

Chapters 4 and 5 have addressed different aspects of the encroachment of scientific and medical expertise into women's lives. The movements for reforming childbirth customs and women's health practices in the late nineteenth and early twentieth centuries led to the expansion of bio-power and the tightening of control and surveillance of women's bodies by the state and experts. This chapter, covering the 1910s through to the end of World War II and the final chapter of this book, explores the further diffusion of bio-power and the shifting relationships between the state, experts, and women by returning to issues of maternal and infant health policies. During this period, social-work and feminist activists, and progressively minded physicians who exerted their influence from outside the government bureaucracy emerged as new and important players for shaping the policies.

In Chap. 4, I argue that the spread of modern childbirth practices involving trained midwives was central to maternal health policies endorsed by the state and organized medicine in the late nineteenth and early twentieth centuries. Until the 1910s, those who used the services of school-educated midwives were predominantly urban families with sufficient financial resources. From the 1920s onward, however, concerned medical professionals, health activists, and government officials sought to bring the services of trained midwives to women in poor households and rural areas, leading to the establishment of midwifery services run by non-profits

Y. Terazawa, *Knowledge, Power, and Women's Reproductive Health in Japan, 1690–1945*, Genders and Sexualities in History,
https://doi.org/10.1007/978-3-319-73084-4_6

or local municipalities. They organized these services, in part, in reaction to scholarly and journalistic reports that revealed that poor women were receiving substandard maternal and infant health care. The pitiable state of poor mothers and babies was part of a discussion on new "social problems," such as urban poverty, that socialist writers claimed were created by the expansion and deepening of the capitalist economy.

Capitalist development in Japan created dire economic and social problems that led to protests and criticisms—often involving left-wing groups—of the existing regime. Despite suppressing such opposition, the Japanese government pursued progressive labor policy and implemented new relief measures in the 1910s and 1920s, as Sheldon Garon's classic study on this subject shows.[1] The gradual expansion of government health-related programs from the 1910s onward was an important way in which the state attempted to alleviate the problems of poverty and inequality. Maternal and infant health was one of the major areas to which the government paid attention.

The expansion of health programs for poor mothers and children materialized through the intersection of various social and political forces and interests. It occurred in the initiatives of government health officials, but also with the strong advocacy of leftists, social-work activists, feminists, nurses, midwives, and working-class women. Progressive reformers and feminists, for example the social-work activist Kagawa Toyohiko (1888–1960), the leftist physician Majima Yutaka (1893–1969), the radical feminist Hiratsuka Raichō (1886–1971), and a group of socialist feminists—with political ideas very different from the Home Ministry's health bureaucrats—were involved in crafting or operating public or quasi-governmental programs to support needy mothers and children.[2]

The implementation of new policy measures between 1937 and 1942 during the years Japan was pursuing its war in China and against the Allied Forces brought about yet a new phase in maternal health and welfare policy in Japan. In 1937, after a long campaign by feminist and child-welfare activists as well as concerned government officials, the Diet (Japan's national assembly) passed the Maternal and Child Protection Law, which specified welfare provision for poor single mothers. With the law's passage, the government initiated a public-health-nurse system that would provide maternity and infant health care counseling nationwide. The newly established Ministry of Health and Welfare (MHW) introduced a "maternity passbook" registration system.[3] By 1942, mothers were expected to have a passbook in hand that recorded medical data for every newly born child. Additionally, in 1940 the Diet created the National Eugenics Law to make eugenics a legitimate national policy. The new policies fell short of what their advocates intended because the central government provided neither sufficient organizational nor financial support for childcare allowances for single mothers. Also, facing vehement opposition from pronatalist advocates, there were only a small number of Japanese citizens who had sterilization operations under the 1940 Eugenics Law. The influence that public health nurses could exert through maternity counseling was limited, and obviously they were incapable of giving in-depth counseling to the vast majority of pregnant women. The extent to which women complied with maternity registration was limited during the war and immediate postwar period, despite pregnant women being eligible for an additional ration of milk, food, or materials, such as sanitary cotton, that were deemed necessary for "hygienic" childbirth practices. It was not until the 1950s that the majority of expectant mothers in Japan began to register. The significance of these new policies, however, lies in that these measures, which put fertility control under the auspices of government agencies, marked a departure from placing maternal health care in the hands of midwives in private practice. Also, although wartime concerns to nurture future generations of healthy soldiers and workers partly motivated the passage of bills to improve maternal and infant health, these policies were built on a base of reproductive supervision and care that the state and charities had already haphazardly provided from the late 1910s.

The competing interests and ideas that helped to shape maternal health policies were also held by government officials and politicians who worked within state institutions, validating the idea that the state was far from monolithic. As Japan entered the war years in the 1930s, health officials who desired to expand the government's maternal health programs—ultimately for military purposes—competed with militarist bureaucrats who wanted to allocate more funding to military spending. A conflict among generally conservative government officials and politicians also occurred over eugenics legislation. Scientific-minded eugenicists sought to curtail the reproduction of the "unfit," presumably to enhance the quality of the nation's manpower; however, those who championed the ideology of the emperor-centered family state opposed sterilization, arguing that it undermined the morality of citizens, who are obligated to continue future generations of their families.[4] Government policies and legislation on maternal health materialized as different forces and interests that came from within and outside the state.

The implementation of these new maternal health measures and policies expanded the state and organized medicine's networks of bio-power-intensified surveillance of women. As I show in Chap. 5, the reinforcement of control over bodies and reproduction accompanied women into this web of power and was facilitated by their incorporation into it. That is, having been exposed to new scientific and medical knowledge of their bodies and health, women, whether inadvertently or intentionally, became active agents of the expansion of bio-power. They participated in new maternal health programs, and in this way, frequently served the interest of the state in its attempts to reinforce control over them.

Efforts to Provide Free Midwifery Services for Poor Women

Beginning in the 1920s the state and health professionals sought alternatives to relying on trained midwives in private practice to improve maternal and infant health. The concern in this area evolved in a context where urban working-class poverty emerged as an overwhelming social problem.[5] The 1918 Rice Riots, nationwide protests prompted by the inflated price of rice, were pivotal in prompting vociferous public debates on social ills and reforms. A focal point in the debates was the problem of destitute mothers and their children. From across the political spectrum, distinguished government bureaucrats, leading physicians, social-work activists, philanthropists, and feminists of various persuasions called for a new policy to promote the welfare of mothers and infants. In the mid-1910s, prominent feminists launched a famous controversy over the need to protect motherhood [*Bosei hogo ronsō*]. As I discuss later in the chapter, many feminist thinkers supported state subsidies for childbearing and rearing. Advocates included both socialist feminists—who were concerned about the burden working-class mothers had to bear—and radical feminists like Hiratsuka Raichō, among others, who regarded motherhood as an empowering and defining element of feminine identity.[6] From the early 1920s onward, some leftist health activists and feminists challenged state pronatalist policy by endorsing birth control and even abortions. They regarded birth control not only as a means to improve maternal and child health but also to combat poverty by limiting the size of poor families.[7] Ishimoto Shidzue's (1897–2001) efforts to spread birth control drew support from the American birth control advocate Margaret Sanger (1879–1966) and are well known, but it should also be noted that midwives, feminist activists, and physicians, including such figures as Shibahara Urako (1896–1955), Oku Mumeo (1895–1997), and Majima Yutaka (1893–1966), challenged state reproductive policies more overtly to make birth control and abortion services available to working-class women.[8] The foremost concern of more conservative health reformers and medical experts—such as those involved in the Investigative Committee for Health and Hygiene (Hoken Eisei Chōsa-kai, established in 1916), which advised the Home Ministry—was not necessarily the well-being of poor women and children. Instead, they feared that poor maternal health and high infant mortality would negatively affect national interests.

The heightened concerns for maternal and infant health among officials and health experts were prompted in part by statistics on infant mortality. Commenting on data covering the first 16 years of the twentieth century, Namae Takayuki (1867–1957)—an official and social-work activist—noted Japan's high infant mortality rate compared to European nations and New Zealand. Moreover, the rates of many of these countries were on the decline in the mid-1910s, while those of Japan were rising. That Japan's infant mortality rate remained high and, worse, was on the rise, was even more problematic considering that Japan suffered few detrimental effects from World War I.[9] Indeed, the lively discussions officials and medical experts had about international statistics betrayed a sense of national shame. Official discourse associated the nation's health with national strength, and improving the country's health to match the level of the West was an important goal in Japan's quest to become a world power on par with Western nations.[10] One measure for decreasing infant mortality that many health experts and social-work activists endorsed was providing trained midwives for poor women.

Long before the debate on infant mortality rates emerged, leading midwives and obstetrician-gynecologists made charitable efforts to provide the poor with free hospital births and midwife services. It was the prominent reformer and midwife, Muramatsu Shihoko who, in 1889, established the first maternity hospital to offer free childbirth services to poor women. Drawing on the patronage of prominent figures in Meiji political and business circles—the head of the Toki house, a former domainal lord, among others—Muramatsu added philanthropy to her midwifery business and a midwife apprenticeship program she had founded in eastern Tokyo.[11] This early attempt to help destitute mothers was followed by the work of Saeki Riichirō (1862–1953), a leading Kyoto-Osaka area obstetrician-gynecologist who, in 1897, started a fully fledged maternity hospital attached to his privately run Kyoto midwifery school.

By the mid-1920s, local governments and non-governmental groups established new maternity hospitals. In Tokyo, members affiliated with the Young Men's Christian Association who studied at Tokyo Imperial University organized a group called the San'iku-kai [the Association for the Celebration of Childrearing], in 1918, and opened a maternity hospital for working-class mothers in Honjo, a downtown area in eastern Tokyo. Four years later, the Japanese Red Cross established a maternity hospital in central Tokyo that offered free midwifery services to needy women. Prompted by the disastrous 1923 Great Kanto earthquake, the Saiseikai—the imperial endowment that provided health care for needy citizens—initiated services for poor mothers and infants who resided near the hospital it opened in Shiba, a southern section of Tokyo.[12] By the late 1920s, the Tokyo city government was running three maternity hospitals in working-class neighborhoods in eastern Tokyo. Ninety-four births took place in the first year (1924) at two of the hospitals, Tsukiji and Asakusa. The number rose to 1015 the following year, and 1385 in 1926. These clinics drew support from the prestigious St. Luke's Hospital and Keio University Medical School.[13]

The locations of maternity clinics that offered free midwifery services and the number of patients they served were limited. With the idea that trained midwives were crucial to improving the health of working-class mothers, various actors, including charity, health care, and women's groups, along with local governments started the system of so-called "circulating midwives" [*junkai sanba*] or "public midwives" [*kōsetsu sanba*]. Under this system, a local government or a group usually hired one or more midwives to work full- or part-time for a fixed salary. These midwives were supposed to make house visits on a rotating basis to needy expectant mothers in a given community. Midwives provided free of charge both prenatal counseling and midwifery assistance during home deliveries.

A charity group sponsored by the *Osaka Mainichi* newspaper established a different system to provide free midwifery services. Since its inception in 1911, the Mainichi Newspaper Charity Organization had offered free medical care to the poor, and in 1914, it also began to work on maternal health issues. It made trained midwives available by enlisting midwives in private practice to work as staff on an on-call basis. Poor mothers received a reference for one of these midwives from a social worker or a police officer. The designated midwives attended births free of charge but collected fixed fees of 5 yen per birth from the charity. Compensation for midwives proved quite modest, but the program had many volunteers owing to the honor that was associated with working with an eminent newspaper company.[14] In 1923 in the Osaka area, midwives working with the charity organization handled 65 births including five stillbirths. The number of assisted births gradually increased: 153 in 1924, 187 in 1925, and 242 in 1926. In addition to having midwives attend mothers during childbirth, in 1923, the organization's main branch in Osaka established a system in which midwives visited new mothers at home for infant-care counseling.[15]

While Tokyo and Osaka pioneered both the non-profit and publicly run systems, other major cities nationwide followed their examples between 1923 and 1926.[16] Working for city-run visiting-midwife programs, midwives used the office of the city government, a publicly funded clinic, or their own homes as the base from where they embarked on their daily rounds of house visits. While prenatal and postnatal diagnosis along with postnatal counseling were an important part of their job, local sponsoring governments, such as Shizuoka City, limited free midwifery services for home births to the destitute. The reasons for setting strict criteria for the free home-births service were presumably not only budgetary but also out of consideration for the interests of midwives in private practice who would lose potential patients to a publicly funded midwife program.[17]

Historians of midwifery in Japan evaluate the visiting-midwife programs positively. They were a cost-effective and efficient means of bringing modern childbirth practices and proper health education to poor expectant mothers. These publicly funded midwife programs seem especially helpful in bringing the practice of trained midwives to isolated rural communities, many of which lacked both licensed physicians and midwives.[18] On the other hand, the system of circulating midwives indicated the state and medical profession had a deeper and more dynamic intervention into women's reproductive lives. While some expectant mothers voluntarily visited maternity clinics that provided free services, social workers and midwives located and visited pregnant women in poor neighborhoods who did not necessarily desire their intervention and guidance.[19] There is no doubt that many visiting nurses and midwives were devoted to the improvement of the health of their patients, but house visits by school-educated midwives could infringe on the privacy of households and embarrass female patients.

A visiting nurse's report shows the kind of condescension that many health experts probably felt towards their lower-class patients. The unsanitary living conditions of working-class families and what they thought were inadequate childrearing practices among lower-class mothers appalled visiting nurses and midwives, and they often blamed working-class mothers for their living conditions. Yamamoto Toshiko, a visiting nurse, reported from an eastern Tokyo working-class neighborhood where many small factories were located (today's Kinshichō area in the Sumida district). She recognized that unstable employment and tight household finances led to families living in small single tenement rooms with little sunlight or ventilation, an insufficient amount of clean bedding and linens, and inadequate heating. Along with a lack of proper medical care, their inadequate living conditions resulted in high infant mortality and the frequent spread of infectious diseases. Yet, Yamamoto vacillated between understanding and reproaching the mothers. She remarked that because of a lack of exercise and nutrition during pregnancy, patients tended to bear unhealthy children. Overwhelmed with life's difficulties, they failed to give adequate attention to their children's health and hygiene. Even if they were concerned about their children's health, earning a daily living made these mothers seem neglectful. As mothers they followed their loving instinct but did not realize that acting on this often adversely affected their children's health and well-being. Ultimately, Yamamoto strongly censured her patients: "They are ignorant, immoral, incredibly lazy and wasteful. And they do not try to change their ways despite my advice to do so many times. Of course, it should be remembered that we cannot solely blame these women who accept living in poorly ventilated rooms, located in stinking alleys."[20] There is no question that visiting nurses' attention and care benefited needy people, but they also brought a relationship of dominance–subservience between health professionals and their patients. Lower-class patients were inculcated with "proper" health knowledge and practices, even though they may not have necessarily welcomed attempts at indoctrination.

Health reformers shared a strong desire to "enlighten" and "educate" [*kyōka*] mothers from underprivileged backgrounds. Accompanying this urge was an entrenched assumption that working-class women were dimwitted, ignorant and lazy. Inoma Shunichi (1896–1969), a bureaucrat-researcher and an advocate of maternal and infant health, for example, contended that poor people were mentally deficient (*seishin-teki kekkan*), which made them lazy and spendthrifts. To overcome poverty, they should reform themselves. That is, they should be "constantly working, avoid unnecessary consumption, and thus, maintain their mental energy." Based on these beliefs, Inoma spoke highly of various educational events that the

San'iku Association held in addition to operating its maternity hospital and visiting-midwife program. He claimed that it would be ineffective to improve the moral character of lower-class mothers through ordinary lectures, which could not sufficiently attract their attention. Inoma referenced an event where the San'iku-kai organizers brought women to a storyteller's theater [*yosé*] one night. There, after attending a performance of comic stories that were presumably educational, the women declared they would avoid speaking ill of others. Commending the organizers for their strategy of moral inculcation, Inoma noted that the San'iku-kai staff remembered well that an effective education came from "stimulat[ing] dim-witted women."[21]

It should also be noted that social workers, visiting nurses, and public midwives collected detailed data on poor patients and their families. They gathered information—often on index cards—on needy families to identify those who needed help and the specific types of aid individuals and families required, information fundamental to the operation of social or public health services. The Saisei-kai designed different forms for ill patients, expectant mothers, for general health, child health, and infant health.[22] The index-card system was useful for long-term observations as well as classifying patients into different categories. Moreover, cards made sharing client data among health and family counselors, physicians, local government administrators, and the police convenient. Of course, such invasive data collection was done only on potential recipients of aid, namely the poor and underprivileged. From the 1930s, reporters, scholars, social workers, and health professionals commonly used the phrase "people belonging to 'the card class'" when referring to the mostly urban poor who were on the dole.[23] The index card, thus, became a symbol of the stigmatized urban underclass. Receiving free midwifery service was most likely not as stigmatized as accepting government monetary relief, but women who became the patients of public midwives were doubtless not immune from the taint of being associated with the card class.

The history of the visiting-midwife system shows the process of how the state and medical profession came to identify poor expectant mothers as targets of aid, reform, and education. Despite the good intensions of health activists, maternal health care reforms for lower-class women in the 1920s and early 1930s were unmistakably a top-down movement steeped in bigotry. The social distance between reformers and female patients was vast, amplified by the stereotypes of poor women that pervaded both journalistic and scholarly writings from the period. The underlying condescension toward poor women among educated professionals made their involvement intrusive and compulsory, quite different from the help that midwives offered to their middle-class clients. Within the process of expanding midwife services for the poor, mothers' bodies and minds became an object of disciplining and policing by medical and social-work experts as well as the government. While establishing free or low-cost midwifery services for poor women was a concern that officials and activists shared over the nation's health, maternal health care programs may not always have exactly served the interest and welfare of poor women. Although the introduction of trained midwifery services generally benefited poor mothers, increasing surveillance over their reproductive lives entailed women's subjection to new types of state and professional power.

The Aiiku Association and Maternal and Infant Health in Rural Japan

In the 1920s, health activists identified the urban poor as a prime target for free or low-cost maternal and infant health services. A decade later, rural mothers and infants occupied a similar position for social-work activists and health care reformers. The serious nationwide economic downturn that hit Japan from the late 1920s into the early 1930s affected rural areas more acutely than cities. Amplifying the previous concerns of health reformers that the rural poor needed medical and social services, the crisis set off a vigorous debate regarding the requirement for relief programs in the countryside. News and academic reports on severe rural poverty prompted concerned officials and activists to emphasize the urgency of various social and health programs. National statistics also played an important role in invigorating discussions on the deplorable state of public health in rural Japan. Specialists paid particular attention to high infant mortality rates in rural areas in the 1920s and 1930s when rates were gradually decreasing in cities. By the late 1930s, the infant mortality rate in rural areas outpaced the urban rate, reversing the previous trend. News and statistics on poverty and substandard levels of rural health care prompted health officials and reformers to establish community health programs for rural residents, including mothers and infants, from the 1930s onward.

Even during the 1920s, public health authorities noted health problems specific to rural areas, including a lack of licensed physicians and midwives in a large number of villages. Officials at the Home Ministry's

Hygiene Bureau regularly met with local administrators in charge of improving public health. For example, in a 1927 meeting, officials discussed contaminated drinking water, inadequate housing, inferior nutrition, parasitic germs, the frequent occurrence of disease, a lack of professional medical care, and the inferior physique of rural residents. A lack of licensed physicians in rural villages was a prominent issue, and officials understood well that many trained physicians were reluctant to live and practice in isolated rural communities. Historian Fujino Yutaka also points out that the serious economic recession that hit rural areas between 1928 and 1931 most likely made rural medical practices unprofitable. Between 1928 and 1931 rural areas lost 1671 physicians, while cities gained 3287. Recognizing the seriousness of health problems in rural areas, the Home Ministry made efforts to obtain a budget for providing publicly funded physicians to remote villages: an attempt in 1928 failed because the Finance Ministry rejected the proposal. These debates on substandard health care in the countryside prominently featured the issue of maternal and infant health, and health officials considered a shortage of licensed midwives as equally serious to that of physicians.[24]

A significant effort to improve maternal health in rural areas resulted in the founding of the imperial endowment Boshi Aiiku-kai [the Mother–Child Loving Care Association] in 1934. Originally established to commemorate the birth of the crown prince (Heisei Emperor Akihito, b. 1933), the Aiiku Association promoted nationwide attempts to bolster maternal and infant health in farming and fishing villages. The endowment's early efforts concentrated on researching the health conditions of rural mothers and children. The association enlisted a research committee of state officials, prominent physicians, academics, educators and social reformers. Their research examined the characteristics and causes of children's physical and mental development, methods of childrearing, and birth and infant mortality rates.[25]

The Aiiku Association's intervention in rural maternal and infant health involved different types of research. It mobilized not only public health specialists but also folklore scholars, including the prominent Yanagita Kunio. Yoshinaga Naoko asks why Aiiku Association directors, who generally condemned local childbirth and childrearing practices, worked closely with Yanagita and his folklore studies colleagues, who had the tendency to romanticize what they thought were "age-old" local customs.[26] Perhaps the Aiiku Association's collaboration with folklorists followed from its thirst for knowledge of rural childbirth and childrearing cultures. Critical inquiries into the history of anthropology have shown how anthropologists often worked in partnership with the forces of modern imperialism and colonialism worldwide. While much of anthropological scholarship showed a certain fascination and admiration for local cultures, it still provided imperialist adventurers and colonial administrators with invaluable knowledge of local peoples. That is, anthropological knowledge became a basis on which missionaries and colonizers attempted to conquer and reform indigenous peoples. To implement a new nationwide project to improve maternal and infant health in Japan's countryside, educated health reformers also tried to seek and accumulate knowledge on the object of reform—namely rural mothers and infants—and they drew support from folklorists in addition to researchers in other academic fields. In the 1930s, folklore studies inadvertently became a part of the Aiiku-kai led project to intervene in rural women's reproductive lives and to turn them into more disciplined and nationalistic citizens.

The Aiiku Association's project to improve public health in rural Japan partly drew on the previous public health nurse programs. Owing to growing public attention on health problems in the countryside, both the Asahi Newspaper Social Service Agency and St. Luke's Hospital initiated a public health service for rural areas in the Kyoto-Osaka area in 1930, and in the rural Kanagawa prefecture south of Tokyo in 1933. In implementing these pilot programs, health reformers learned of problems almost unique to rural areas, such as a distrust of outsiders, interference with reform efforts from family and relatives, and deeply entrenched local customs and superstitions. In other words, they realized that a public health nurse program fashioned after the US model was more apt for handling urban health issues than health reform in rural Japan.[27] When they launched a series of programs a few years after the first pioneering experiments mentioned above, heath reformers designed programs that were more suited to rural communities. For example, the Northeast Renewal Society [Tōhoku Kōshin-kai] created a public health nurse program that required nurses to live as residents in a rural community. The program assumed that nurses would participate in community events, meetings, and work service. As invaluable, educated members of village communities, nurses also carried out various functions beyond health guidance. They worked as caseworkers, public health administrators, mediators and supervisors for communal agricultural work organized by village agricultural cooperatives, school nurses, and even as substitute teachers. In villages that lacked physicians and midwives, public health nurses provided primary health care and midwifery services.[28] Along with the Northeast Renewal Society and a few other groups, the Aiiku Association developed a unique public health nurse program specifically targeting rural areas.

One of the most prominent projects the Aiiku Association launched was to designate select farming villages as "Aiiku villages," and revise childbirth and childrearing practices under the leadership of a newly hired public health nurse. In the first year of implementation, the association selected ten villages located in northeastern Japan and in the prefectures of Ishikawa and Fukui. Between 1937 and 1943, a few more villages were added to this project with a significant increase in 1939 when 30 villages were enlisted. In 1944–1945, the final years of the Asia–Pacific War, a staggering 1035 villages were added. Criteria for selecting Aiiku villages included high infant mortality rates and the willingness and enthusiasm of village municipal government and local schools for Aiiku Association projects.[29]

For the first few years of this project, public health nurses and municipal officials explored methods that would be effective for improving health issues particular to young mothers and children in Japan's rural villages. The accounts of Kawamura Tsuguyo, who served as a public health nurse in Minamoto village in Yamanashi prefecture located about 100 miles west of Tokyo, reveal the trial-and-error process by which the Aiiku-kai model of public health nursing operated. From her testimony, we can identify basic living and health conditions, and attitudes on health-related issues that were likely to have caused high infant mortality levels.

One factor was the remarkably busy lives of farmers and other adult family members who took care of young children. Economic historians have suggested that Japanese farmers in the early twentieth century increasingly diversified farming operations, growing cash crops besides rice. This change made farming, which had already been labor intensive, even more so than before. One cash crop that many farmers adopted was silkworms. Research into the history of prewar Japan's rural economy shows that women increased their labor to attend to silkworms, which had negative effects on maternal and infant health. In Minamoto village, farmers combined rice production with silkworm farming, charcoal making, and men's engagement in peddling. During busy seasons, farmers and their wives got up at three or four in the morning and worked relentlessly late into night, retiring to bed at around midnight. The nurse, Kawamura, stated that during the season when all family members were busy with silkworm farming, parents left children alone unattended in a corner of a house, which—occupied by silkworm shelves—was dirty. Families provided informal meals lacking in nutrition, such as tea over a bowl of rice and pickles (i.e., *ochazuke*). Under these circumstances, families treated ill patients and pregnant women poorly. Not infrequently, children died from neglect. Under the mounting pressure of work, expectant mothers took little care to cook nutritious meals, and many suffered from exhaustion. Inadequate health care caused maternal health problems, poor prenatal health, and an insufficient supply of breast milk for newborns. In short, increased hours of farm labor led to deterioration in maternal and infant health in rural Japan in the 1920s and 1930s.[30]

Firmly entrenched local customs and beliefs were also a cause of maternal and infant health problems. According to Kawamura, local taboos prohibited pregnant women from eating meat and fish. To restrict the growth of the child and to facilitate easy birth, pregnant women wore a maternity sash, tied very tightly around the belly, and they were encouraged to engage in hard labor. Further, not only did they not use a midwife, but also villagers considered childbirth as extremely dirty and hence used unsanitary rags during deliveries. Families praised women who left the childbed early to restart daily work routines, and they imposed even more strict food taboos after childbirth, allowing women to eat only rice porridge and pickles. While 80% of babies suffered from rashes, believing that curing them with medication would lead to weak infant constitutions, local residents let them heal naturally. They also believed a Shinto god of disease caused measles, and as such, resisted treating children suffering from measles with modern medicine.[31]

Yazaki Kimiyo, a local leader of the Aiiku-kai and colleague of visiting nurse Kawamura, reminds us that expectant mothers initially showed little interest in Aiiku-kai activities. A highly educated woman from a relatively wealthy family Yazaki explained:

> Women do not realize how miserable and lowly they are treated because they take this for granted. They do not listen seriously to our talk about the Aiiku-kai activities; many women seem to think it has nothing to do with their own lives. Some women came close to saying that only women of leisure would participate in such activities for fun…When I was giving a talk warning them about excessive labor looking after silkworms during pregnancy, I was appalled to hear a women comment, "We earn money by spending one month raising silkworms, but it takes twenty years to bring up children and make use of them."[32]

To change such customs and attitudes, Kawamura thought it most urgent to educate mothers in proper maternal and infant health care as well as to create an environment where they could put what they learned into practice. During farmers' off seasons, she organized seminars, neighborhood discussion meetings, and exhibitions, and distributed prenatal and infant care pamphlets. Many women, however, were too busy with farm and housework to attend meetings, and even when they had the time few were willing to invest the time and effort into attending meetings. Having experienced this failure, Kawamura changed tactics and visited individual houses to do counseling and check to see if women were actually putting pamphlet suggestions into practice.[33]

Kawamura and her assistants were well aware that even when mothers were willing to implement new health practices, they could face the opposition of their mothers-in-law. Since it would be offensive to older women to invite them for "educational meetings," Aiiku-kai activists organized meetings to show gratitude to seniors for their long years of service to their families and communities. For these gatherings, activists invited a person of high prestige to give a lecture on health and hygiene or a group of children to sing songs about health and hygiene.[34] The meetings for senior citizens mollified older people who objected that Aiiku Association activities only gave care and attention to young mothers and children, and did not care about older villagers.

The Aiiku Association and their nurses understood that to support expectant mothers and ensure safe deliveries they had to persuade mothers and their families to use the services of trained midwives instead of relying on the help of family members and neighbors. In localities where practicing midwives were available, nurses incorporated midwives into their programs.[35] Even where midwives' services were available, it was difficult for many poor farming families to pay fees for midwifery services. And families that could afford such expenses were, out of frugal habits, reluctant to hire midwives. To address a lack of trained midwives and financial resources, many villages organized mutual help groups, which hired a midwife with funds collected from village residents (josan kumiai).[36] Thus, in one way or another, Aiiku-kai villages secured midwifery services.

Aiiku-kai activists employed a unique technique, organizing select local women into small units that aided nurses with their responsibilities. Each unit was composed of several local women, including young, unmarried and middle-aged, married women, to provide education and counseling for expectant mothers and other villagers.[37] Once an Aiiku-unit identified a pregnant woman in the area for which it was responsible, its members (or the supervising public health nurse) would visit her house, provide a childbirth manual, and educate her and her family in maternal health care.[38] Unit members reported any pregnancy complications to a physician or a midwife. Members had to instruct the family to contact the unit when the mother started labor. After a childbirth check of the health of the mother and the newborn, they had to scrutinize the condition of the birthing room, bedding, and babies' clothes, and advise on nursing and nutrition. The Aiiku unit members visited the mother and child to keep health records. They used a special form to track changes for a year after birth. Unit members reported problems in a child's physical and mental development and sought professional help when necessary.

Like many mutual assistance programs, the intervention of either the nurse or Aiiku unit drew on strong community support but also had coercive and intrusive aspects. The unsolicited and repeated house visits constituted an invasion of privacy of the women and their families, and a forceful interference in their choices of maternal and child nursing care. To be sure, public health nurses or Aiiku unit members subjected expectant mothers to a high degree of monitoring and control. Such surveillance was likely to limit the reproductive choices of women and families to a considerable extent. They could no longer rely on local, untrained midwives to limit the size of their families through abortion and infanticide. Because of the vigilance of the public health nurse and village women active in the Aiiku-kai movement, it was difficult for married couples who had previously been using birth-control measures (even though the spread of birth-control knowledge had been limited in rural Japan) to continue using them. Also, women had to follow meticulous guidelines and radically change childbirth regimens, and these changes brought about a new awareness of bodies tied to knowledge of modern medicine. Aiiku-kai health activists carried out this mundane health education in conjunction with attempts to inculcate rural women in the urban, middle-class ideology of "maternal love" and "the nationalistic mission of motherhood."[39]

The support for the Aiiku Association's programs, including the Aiiku village project, came from the heart of Japan's national government and relied significantly on funding from and the prestige of the imperial family as well as the backing of the Home Ministry, the Education Ministry, the Colonial Ministry [Takumu-shō], and the Imperial Household Agency. Officials in charge of establishing the association recommended Princess Kuninomiya Chikako (wife of Prince Kuninomiya Kuniyoshi, mother of

Empress Kōjun [Hirohito's wife] and grandmother of Prince Akihito or the Heisei emperor) as the association's president.[40] Having accepted the post, she occasionally visited the Aiiku Research Institute in the Azabu area in Tokyo, met with administrators and researchers in the association, and observed kindergarten classes and displays of research results.[41] The Aiiku-kai staff considered such visits a great honor and publicized them in their magazines and newsletters. While the princess may not have visited all Aiiku villages throughout the country, imperial chamberlains serving the Imperial Household Agency occasionally observed Aiiku programs as part of customary local area observation tours. Such visits were likely to encourage villagers—educated elites among them, at least—to participate in association activities.[42]

On the occasion of an imperial chamberlain visit, Wake Ichirō—the mayor and local leader of the Aiiku Association in Kanatsu village in Niigata prefecture—highlighted the importance of Aiiku-kai activities amid the war and the imperial family's support for Aiiku Association programs.

> Now that we are to complete the war in East Asia, having endured the attacks by the Mongols, and strive to build an East Asian (community anew), this (Aiiku) project bears even more significance because it concerns our population policy.
>
> It may be said with due reverence that this program was established to commemorate the birth of the crown prince with a grant provided by Her Majesty, the Empress. As such, day after day, I feel gratitude for the infinite imperial benevolence that made it possible for this program, which used to be merely one of the village reform projects, to expand year after year to flourish as we see today. I was deeply touched anew by having had the honor of [the imperial chamberlain's] observing our program, and thus, I am determined to concentrate on further enriching the content of the program and improving the actual statistics.[43]

Local leaders like Wake attempted to persuade women to direct their devotion to the emperor and the state and be grateful for the grace that the imperial family extended to the emperor's subjects, including women themselves. Thus, the Aiiku village program not only educated rural women in health matters, but it also became a vehicle through which women were exposed to and inculcated in the emperor-centered nationalist ideology.

We find this in a pamphlet that was used to educate Aiiku unit members, for example. Its preface began with a discussion about the importance of departing from the mindset of thinking only about loving one's own children. Instead, villagers must love village children based on a spirit of mutual help. The author asserted that this was the first step for developing "the spirit of the national citizenry" [*kokumin seishin*], which started by nurturing love for one's home province. The next step would be to develop a love for the nation. Further, spreading the knowledge of home medicine and childcare helped Japan remove the shame of being "a civilized nation" that was backward in maternal and infant health. The author concluded that the mission of Aiiku unit members was of grave importance because its work created momentum for the nation's expansion.[44] These ideas were likely to have developed and reinforced a sense of national belonging among some women.[45]

As public health nurses and local supporters engaged in the Aiiku village program, they became aware that to make progress on the issues of maternal and infant health the Aiiku-kai's educational efforts had to go beyond the narrow targeting of mothers and infants and involve the village community as a whole. Existing local health habits appalled well-educated public health nurses who were often from urban, lower-middle-class backgrounds and openly disdained rural residents.

Yoshikumi Ito, a school teacher and a public health nurse, gives us a glimpse of the state of health and hygiene in rural Japan and her reactions to it. In an article she wrote for the journal, *Shakai jigyō* [*Social Welfare*], she deplores farming families' lack of concern for sanitation and cleanliness. For example, she criticized families that wore clothes for five or six days without washing. She complained that farm families did not use sheets or comforters. Taking "whiteness" as a symbol of hygiene, she deplored villagers who did not wash clothes until they became white. They also rarely hung comforters and mattresses in the sun to dry and sanitize them. While several families took turns preparing a bath, 30 to 40 people would wash in the same water, leaving later bathers—typically the baby and young wife, who prepared the bath—the dregs of dirty water. Yoshikumi expressed great concern that the exposure of babies to dirty water could cause disease in infants who were more vulnerable than adults. She was horrified that people from different households washed various things in the same river: She observed people on the banks of a stream washing dishes, pots, and steamers; doing laundry; and cleaning farming instruments including containers that carried night soil.[46]

Moreover, villagers were reluctant to visit a doctor even when they were sick or injured not only because they lacked money to pay the fees but also because they needed to trek steep mountain paths for several hours to see the doctor. One day a villager came to Yoshikumi and asked for help for an injury. The local woman had applied oil distributed by a Buddhist temple and used wild herbs to stanch her wound's bleeding. Despite clear signs of a worsening injury and Yoshikumi's strong suggestion to visit a doctor, the woman insisted Yoshikumi treat the wound so that she did not have to see a doctor. At another time, a mother brought a three-year-old that had fallen into a hearth and was severely burned. The mother had tried to treat the burns by applying old kerosene oil filled with dirt, and in the process she had broken many blisters on the child's face and hands. While Yoshikumi told the mother how inappropriate the treatment was, she managed to treat the burns and prevent permanent disability and disfigurement. By 1940 Yoshikumi had been attempting to reform villagers' health habits for ten years, mainly by means of recurrent house visits. In addition to giving accounts of local health habits, she openly expressed a condescending attitude towards rural women, declaring, "they are incredibly ignorant" and "are far inferior to urban women."[47] Her sentiments were indicative of an attitude of domination that educated health reformers had toward rural women, and it was likely that such attitudes created feelings of humiliation and resentment among villagers.

Such condescension notwithstanding, public health nurses and other local leaders of the Aiiku village program drew up village-wide plans to improve health and hygiene awareness. Their efforts extended beyond individual house visits to include seminars, study group meetings, and tea parties for local residents and mothers, at which Aiiku leaders invited public health nurses, midwives, physicians, or school teachers to speak. Aiiku-kai headquarters suggested holding talks for senior village residents, requesting that a religious leader or influential local figure should give talks to educate the generation of in-laws. Aiiku-kai units worked closely with existing local schools, municipal government offices, charities, and religious and women's groups to deliver these extended educational programs to larger village communities.

Despite its detrimental ideological and practical consequences, the Aiiku village program served the interest of mothers, to some extent, by undermining other adult family members, that is their parents-in-law and husbands. The new childbirth practices demanded that families cut back on mothers' labor time and spend more money on clean linen and cotton, nutritious food, and professional midwifery services. On the other hand, the Aiiku movement promoted pronatalism, which worked against the interests of rural mothers. In general, women were obliged to work long hours regardless of suffering from pregnancy- or childbirth-related sickness—even though the Aiiku program helped give them a little leeway. What would have helped them, indeed, were birth control and abortion services that would help reduce the number of pregnancies and births. A tightened control over abortion practice under the Aiiku program most likely deprived mothers and families of a means to avoid chronic pregnancies and births that often had damaging effects on women's health.

The ideological manipulation of citizens promoted through the Aiiku movement drew on the use of the sweet-sounding words "*aiiku*" (loving childcare) and "*kodakara*" (children as treasure). The Aiiku movement employed these words in conjunction with the ideology of the emperor-centered family state where all Japanese are portrayed as the emperor's children. Children are "treasures" because through the inheritance of the family line they ensure the continuation of individual families, and are indispensable to their development and prosperity. They are also expected to support and carry on the legacy of the nation as a big family with the emperor at its head. Supporters of the Aiiku movement also drew a parallel between parental love for children and the emperor's love for his subjects. The imperial family's involvement in the Aiiku Association highlighted the emperor's continued benevolence to the nation and reassured the Japanese people of his tender and nurturing care for them. Following the example of the emperor, Japanese citizens were urged to support the Aiiku movement, think of children as treasures, and give them loving care.

The irony is that the Aiiku project was closely tied to Japan's ongoing imperialist war. As I show more explicitly in the later sections of this chapter, the impetus for the growing interest of officials and politicians in maternal and infant health in the 1930s and the first half of the 1940s largely derived from the need for young men who could be conscripted into the armed forces. The Aiiku movement extensively played on the imagery of imperial nurturance and tenderness. These representations played an ideological role in obfuscating the government regard for the lives of ordinary citizens, who were in practice expendable for the purposes of war, especially for the perpetuation of the imperial institution.[48]

New Maternal and Infant Health Policies, 1937–1942

While the Aiiku village project reached only select rural communities, the national government passed a series of laws between 1937 and 1942 that led to a new level of state intervention into women's reproductive lives across Japan. The legislation included the 1937 Maternal and Child Protection Law, the 1937 Public Health Centers Act, the 1938 National Health Insurance Act, and the 1940 National Eugenics Act. The establishment of the Ministry of Health and Welfare (MHW) in 1938 bolstered Japan's war efforts by enhancing the health and welfare of its citizens. It also lent momentum to state efforts to promote and regulate reproduction.

Existing studies of the history of state control over reproduction in modern Japan tend to discuss new reproductive policies during the war years in conjunction with the state's totalitarian control over citizens' lives. As I have shown in previous sections, national and local governments along with non-governmental groups had already implemented programs, however limited, to enhance maternal and child health before Japan entered the war in the late 1930s. New maternal health care policy measures introduced in the late 1930s and early 1940s, therefore, should be viewed as a culmination of existing trends rather than a policy shift prompted by the total war regime. That is, the 1937–1942 period was a turning point in the history of reproductive politics in modern Japan, but it is erroneous to view the new momentum in maternal health policy as an anomalous event of Japan's totalitarian regime. Rather, it should be understood as an historic event in the continuous process of expanding state measures for controlling the health and reproduction of the nation that bridged the pre- and postwar periods. Various ideological and political forces were at work in the establishment of these new reproductive policies. For example, the passage of the Maternal and Child Protection Law shows that militarist concerns coincided and merged with feminist and social-work activist agendas.

The 1937 Maternal and Child Protection Law required the provision of financial assistance and other services to needy single mothers. Its passage was largely due to a vigorous movement for the "protection of motherhood" that combined advocacy by women's groups and calls for relief measures for destitute mothers by social-work activists and concerned government officials. The ongoing debates about motherhood and its place in society helped shape the agenda of their campaign. As early as the 1910s—as demonstrated in the aforementioned "motherhood protection controversy"—Japanese feminist thinkers engaged in a vociferous discussion of women's roles as mothers. Renowned female writer Yosano Akiko endorsed a view of what we would today call liberal feminism. She emphasized equality of the sexes and women's economic independence, which would allow women to raise children on their own; she opposed government help for mothers, considering such a policy as replicating women's dependence on men, which was demeaning for the vast majority of mothers who Yosano believed should be able to support single-mother households with the income they earned. Those who opposed Yosano included Hiratsuka Raichō, a radical feminist who thought women's pursuit of motherhood as a central force in developing talents in women and exerting women's influence in larger society. Yamada Waka, a feminist of a more conservative persuasion, considered women's traditional roles as wives and mothers to be the basis on which to build a movement to empower women. While Yosano (who opposed state child support allowances) and many male government officials (who advocated such measures) viewed government allowances as no more than handouts to women in reduced circumstances, Yamada endorsed publicly funded child support as part of women's rights. Socialist feminists were also vehement supporters of government child support allowances based on their observations of the reality of working-class mothers. Many feminists advocating for child support allowances developed the view that childrearing should not remain the sole burden of individual families and women but should be shared by society as a whole. They thought the government had a mandate to provide childcare support.[49] In this way, although feminists, social-work activists, and government officials of diverse persuasions became involved in the movement for motherhood protection, they were not at all philosophically unified.

A feminist organization called the New Women's Society (Shin fujin kyōkai), which Hiratsuka established in 1919, was important for gathering together various feminists to push for the motherhood protection bill and make progress on other issues. It aimed to work on a broad range of feminist issues with the equality of the sexes as its long-term goal. It worked for the expansion of women's political rights, but also took seriously the protection of the rights of women, mothers, and children. It paid special attention to the predicament of working- and lower-class mothers. Some members, such as leading suffragette Ichikawa Fusae (1893–1981), were more interested in the issue of women's suffrage and were not in full agreement with Hiratsuka's idea of motherhood as the defining element of womanhood[50]; however, Ichikawa and others who could be described as liberal feminists joined the New Women's Society

Fig. 6.1 A gathering of the New Women's Society (January, 1920). From left to right (front row): Ichikawa Fusae, Oku Mumeo, and Hiratsuka Raichō (Source: Aflo/Mainichi Photobank)

and eventually became, in the 1930s, an important force advocating for maternal protection legislation.[51]

There was another group of feminists who joined the Society, influenced by Swedish feminist Ellen Key and Hiratsuka's ideas of what was understood as "maternalist feminism." A follower of Hiratsuka and Key was Oku Mumeo.[52] She thought that women with young children should not work outside the home but concentrate on caring for children. She admitted, however, that working-class mothers had no choice but to take up paid jobs for supporting their families. Moreover, due to the low wages earned by working-class women, children and mothers suffered from substandard nutrition, poor hygiene, and a lack of adequate health care. She continued that despite words of admiration and support for pronatalism, members of the ruling class attempted to ban birth control, which would ultimatly help nurture healthy children. Oku's response to the issues of working-class women with children was to endorse government-supported childcare for poor women. Larger society, she believed, should support childrearing by guaranteeing mothers' "right to live." Further, female activists could advance their movement by expanding their personal maternal love, which she found rather limited, by acting as "mothers of society" and loving all of society's children[53] Fukushima Shirō, a prominent male journalist and supporter of the New Women's

Society, also endorsed this line of thought. He emphasized developing maternal love both for one's own children and those of larger society.[54] Using weekly publication *Fujo shinbun* [*Women's Journal*], which he edited, as a forum, Fukushima launched a campaign for motherhood-protection legislation after the New Women's Society dissolved in 1922.[55]

The more conservative Yamada Waka, along with her husband Yamada Kakichi (1865–1934), also backed Hiratsuka in her early efforts to establish a system of publicly funded child support. In endorsing this measure, Yamada Waka remarked that heaven made it women's mission to preserve and improve the [human] species the mission of foremost importance, relegating men's paid employment—producing items necessary for human sustenance—of secondary importance. For women and men to fulfill their missions, it was essential that they maintain healthy and financially stable families. The nation is an aggregate of many families, each being a cell, with the health and financial stability of families of national importance. Families procreate, protect, and nurture the lives of the nation. And the Japanese people are proud of their "beautiful" 3000-year-old family system. For Yamada, providing government childcare support was more than helpful to poor mothers. It prevented poor families from destruction. She argued that reinforcing the importance of motherhood was central to strengthening women's rights. The Maternal Protection Act, therefore, would help single mothers pursue an important mission, nurturing the next generation of citizens. Aid for this purpose was different than a handout. It was a duty of the nation, and from the mother's perspective, an indisputable right as a mother of the nation.[56] Yamada's conservatism is obvious in her emphasis on the importance of the preservation of the family and the nation as well as her belief in separate and fixed gender roles; yet, she was undeniably a feminist who advocated for women's rights and had genuine concerns for women in hardship.

Socialist feminists led by the ideas of Yamakawa Kikue, an important participant in the "motherhood protection controversy," developed their own forms of activism. Challenging the male-dominated socialist movement, female activists involved in establishing an all-encompassing proletarian party in Osaka in the mid-1920s raised the issue of motherhood protection as one of "the six special demands of women." Other demands included the abolition of the legally sanctioned household (*ie*) system, an end to laws that discriminate on the basis of sex, and a minimum wage for all workers regardless of sex or race.[57] In opposition to the radical feminist Hiratsuka, who advocated a state child-support allowance to enable women to be stay-at-home mothers, socialist women promoted women's right to work foremost, and supported government assistance for child-care as complimentary to women's right to work. Working-class women's groups cited the late 1920s economic downturn and the rise in female and male unemployment, destitution, and hunger as reasons for the government to expand maternal and child welfare services, such as free midwifery and daycare services, with free lunches for children.[58] Moreover, as the growth of militarism in the 1930s led the state to censure socialist and communist activism more severely, one way for socialist women to continue their activism was to highlight their demand for motherhood and child protection that was not only in line with the goals of other groups in the Japanese women's movement but also with government officials who saw improving maternal and child welfare as beneficial to rearing a large and healthy population necessary for waging wars.[59]

The Maternal and Child Protection Act derived not only from previous feminists' activism but also from the activist tradition of childcare professionals and social-work reformers. For example, drawing on his long-time experience as a child-welfare activist, Namae Takayuki campaigned for maternal protection laws based on his belief that all children possess the right to a safe delivery and adequate nurturance. To fulfill this right, parents were obligated to provide suitable services for their children. In the event that parents were unable to carry out their obligations, however, government and society must step in.[60] Home Ministry official Moriya Eifu, on the other hand, endorsed the maternal protection bill because children are not entities that parents privately own. They belong to villages, towns, the nation, and the world. As such, it is without question that a local or national government should take the responsibility for raising children if parents lacked resources. Moriya suggested that this law should be named the Child Protection Law and not the Maternal and Child Protection Law.[61] The historian Hayakawa notes that Namae emphasized his strong belief in children's rights while Moriya, failing to mention the rights of either children or women, highlighted the government obligation to provide care for poor mothers and children. Hayakawa also notes that in the views of Namae and Moriya, the government's obligation to protect mothers followed the instrumental value of caring for children. Feminist concerns about women's rights were irrelevant to the passage of the bill.[62] Despite tension between individuals and groups advocating for the bill that would fit their respective political programs, activists felt strongly about passing a bill on maternal and child protection.

Eventually, a unified movement that specifically aimed to bring about the legislation emerged among women's suffrage activists. Their membership later broadened to include reformers from other social-work and women's groups. The initial impetus for the movement came from an urgent speech by a member of the Social People's Women's League [Shakai Minshū Fujin Dōmei], a socialist women's group that called for legislation to bring relief to needy mothers at a 1934 national conference for women's suffrage, sponsored by 16 different women's groups. The conference sought legal reforms to improve the status of and provide better legal protection for women. Conference participants discussed necessary relief for destitute mothers in the context of improving women's social standing. Recognizing the importance of government aid for poor single mothers, feminist activists—including the aforementioned Yamada, Ichikawa, Oku, and Kaneko Shigeri (1899–1977, maiden name: Yamataka), who became an energetic endorser of the bill—from various backgrounds established the League for the Protection of Motherhood [Bosei Hogo Renmei].[63] This movement gained momentum both because of the serious economic problems Japan faced in the late 1920s and early 1930s and growing militarism spurred by the 1931 Manchurian Incident that prompted Japan's invasion of China.[64] During the economic crisis, the media reported sensational stories about single mothers and their children who committed family suicide because of dire economic hardship.

Taking advantage of the widespread public attention to the problem of family suicide, the Maternal Protection League initially intended to give it priority in their petition to the Diet to push the passage of the bill for maternal protection.[65] The group, however, shifted its policy to discuss the need for maternal protection in conjunction with what Diet members considered an urgent issue of providing relief to rural areas that were seriously hit by harvest failure. Unless they shifted tactics, activists felt almost certain their petition would be ignored by the Diet during the 1934 session. The league presented a petition that discussed how harvest failure in northern Japan led to inadequate diets and health care for pregnant women and infants. Children who were born and raised during economic hardship would fail to develop healthy, robust bodies. For example, military conscription physicals in Iwate prefecture demonstrated that prospective conscripts for 1932 and 1933 were inferior to those from previous years because those men had been born during the years of poor harvest around 1913. The Maternal Protection League took advantage of militarist demands for able-bodied soldiers to press their agenda. At the same time, the league dropped from the original bill feminist provisions for the improvement of the legal status of women and the rights of mothers. The law that was finally enacted was more in line with the concerns of state officials and male politicians. The promotion of the welfare of mothers and children was relegated to a secondary concern.[66]

Indeed, feminist and social-work activists made a significant concession to pass the bill. The law stipulated that mothers and grandmothers who met the following qualifications were eligible to apply for a government financial aid package, consisting of an average of 25 sen per day.

1. Mothers or grandmothers obligated to support their child or children (or their grandchild or grandchildren in the case of grandmothers) without support from husbands or the children's fathers.
2. Mothers and grandmothers lacking means to support their children or grandchildren who are under the age of 13 years.
3. Mothers of illegitimate children and/or mothers who had children as a result of common-law marriage.

Mothers engaging in unfit behavior could be denied support, and mayors had the authority to advise mothers on childcare practices.[67]

This bill was a compromise because activists had initially asked the government to offer aid to mothers who had children aged 15 years or younger. They also thought it should be the national government that would provide funding, which would ensure uniform, national distribution and lessen the detrimental effects of fluctuating budgets. The law that passed, however, required local governments to assume financial responsibility for aid. Between April and September 1938, 68,742 women received aid, two-thirds of the applicants qualified to receive assistance. The sum of funds that was allocated for this program was one-quarter of that which was initially projected. Historian Hayakawa argues that the reason that the government allocated a much lesser amount was because it prioritized financial support to families who lost members serving in the military over single mothers.[68]

One aspect of this system was the close surveillance of women by social workers or local government administrators. To receive aid, women had to allow an inspection by government-commissioned social workers and forego family privacy. Social workers not only investigated mothers' financial situations, but they also reviewed their character and behavioral patterns. Such intrusive examinations were stipulated in the act. Aid would not be provided for mothers whose "delinquent behaviors" [*seikō furyō*] made them unfit to raise children. Further, the law authorized mayors to instruct aid recipients on childrearing practices, although it was social workers who actually provided advice on how to be good mothers.[69] In other words, in addition to offering material assistance to poor mothers, the act subjected poor women to government vigilance, requiring them to behave and conform to the norms of the "good mother." The new surveillance over poor women's morals and behavior was one of the ways through which the state and middle-class social welfare experts attempted to reform sexual and family practices of lower-class women whose customs widely diverged from the middle-class norms of a stable single-partner family system. Poor women might have multiple husbands or change common-law husbands over time.[70] The mechanisms of discrimination and surveillance of poor women continued to remain an important aspect of public welfare provisions for poor single mothers in the post-World War II years as the government created a new welfare system and publicly funded dormitories for single mothers.

Declining birth rates in the late 1930s along with militarist demands for more manpower led the government to pursue a more energetic pronatalist policy. To increase population, the government issued The Guidelines for Establishing Population Policy [Jinkō seisaku kakuritsu yōkō] in 1941. Not all officials endorsed a pronatalist policy and some medical experts working for the MHW strongly opposed it, arguing that producing more children would increase the dependent population, a useless burden, which would hinder Japan from concentrating on its war efforts. They argued that it would take at least 15 years for newborns to make a contribution to such endeavors.[71] While the military's endorsement of pronatalism prevailed, it is to be noted that for both the proponents and opponents of pronatalism, the value of these various "populations" lay in their utility for achieving the national goal of fighting a successful war. The well-being or sacrifices of these "populations" was secondary to the national interest.

As a strongly nationalistic, military-led policy, the guidelines foregrounded an awareness of the state of the country and advancing the Japanese people as a nation. More specifically, the document suggests the Japanese people should develop the Greater East Asia Co-prosperity Sphere, which would encompass many nations in Asia. People should realize that, to achieve this goal, an increase in the Japanese population and the improvement of its quality were essential. The guidelines explicitly condemned the political philosophy of individualism. Instead, it called for reinforcing a worldview that gave primacy to the protection and interests of the state and family.[72]

Within this ideological construct, women would have to deny individual interests that conflicted with the state or their family interests. It should be noted that this was not necessarily a new development in the history of reproduction and maternal health care in Japan. As I discuss in this chapter and Chap. 4 and 5, the government and the medical establishment almost always formulated reproductive and maternal health policy to assure the interest of the state rather than that of individuals. The deepening authoritarianism that occurred as the war went on, tightened reproductive control both by coercion and by education, or, at least, prompted the government to "attempt" to control reproduction more stringently even if such endeavors were of limited success.

To increase the population, the government had either to increase the number of births or decrease the number of stillbirths and early deaths of children and young adults. The guidelines adopted both goals. But to decrease the number of early deaths, the government needed to allocate resources for maternal and infant health and tuberculosis prevention—infant deaths and tuberculosis were two of the most important factors that increased the mortality rate. As scholar Hino Shūitsu points out, simply spreading propaganda to decrease deaths without actually providing resources was ineffective, and the government—pressured to meet many different wartime needs—had been unable to provide significant funds for public health programs.[73] Under these circumstances, the only viable policy seemed to be to encourage people to have more children, even though many would die from a lack of effective public health measures. In Hino's words, because infant mortality rates were unlikely to improve dramatically, this policy amounted to having women "bear more children and letting more of them die" [*umeyo shineyo*] instead of having them "bear more children (simply) to increase the population" [*umeyo fuyaseyo*]. The health bureaucrats who authored a guide for understanding the new policy asserted that it is preferable to have women bear more children and let a significant number of them die (*tasan tashi*)—to increase the dynamic "life's energy" of the Japanese people—rather than having them bear fewer and help more to survive (*shōsan shōshi*).[74] Obviously they were indifferent to the risks and burdens for women of repeated pregnancies and births, which frequently resulted in mothers' illness or death.

Actual measures to increase the number of births included: encouraging early marriages; providing favorable treatment for families with many children; improving maternal and infant health care; and more stringently controlling abortions and the use of contraceptives. To encourage earlier marriages, the guidelines suggested promoting less expensive weddings, lending money for marriage expenses, and having private groups and publicly funded agencies conduct match-making services. Another recommendation was to curb the employment of women over the age of 20, and to encourage employers to improve working conditions that would hinder women from getting married. In reality, however, pushing women to have children would prove difficult when they were in war industry jobs. Many of those jobs required long hours, strenuous labor, and were detrimental to women's reproductive health.[75] To support families with many children, the guidelines recommended family allowances or child benefits and tax provisions that favored large families while penalizing single people. Although it never actually implemented policies to relieve large families of financial burdens, the government awarded couples who had many children. Beginning in September 1940, the MHW awarded families with more than ten healthy children who were at least six years old. This continued as a yearly event until 1944, and from 1941, award-winning families with children who attended middle or high schools received a scholarship. Those mothers, however, were often not happy with so many children because of the onerous childcare responsibilities and the heavy physical burdens that frequent pregnancies and childrearing incurred.[76]

With rising concerns about population and reproduction, officials and experts considered improving the quality of the population as important as increasing its size. This belief eventually led to the 1940 National Eugenics Law. Scientists and non-governmental groups had long been exploring the implementation of eugenics policies in Japan. Indeed, not only government officials and medical professionals of conservative, nationalist persuasion, but also some feminists and activists on the left sought to implement eugenics policies, believing they would improve people's quality of life.[77] In 1930 the Home Ministry—through its Investigative Committee for Health and Hygiene—began to plan for the implementation of eugenics policies by creating the Special Committee for National Hygiene [Minzoku Eisei Tokubetsu Iinkai]. This official eugenicist group collaborated with individual scholars and non-governmental groups to write a sterilization bill, the National Eugenics Protection Bill, and presented it to the Diet several times in the mid-1930s. Although these attempts failed, they paved the way for the passage of the 1940 National Eugenics Law.

Within the ideological context of Japan in the mid-1930s, eugenics policy for sterilizing certain groups was controversial, not because law-makers and activists were concerned about the human rights of the targeted people but because many thought the sterilization of any Japanese citizen denies the obligation to continue the family lineage. Carrying on the family line ultimately helped the nation perpetuate itself, prosper, and increase the number of able-bodied males for military service. Pronatalism was ideologically justified because individual families constituted the Japanese nation and the nation itself was a family presided over by the people's father, the emperor. Activists and Diet members who were in favor of eugenics policies succeeded in passing a bill in 1940 because the bill that was passed was highly compromised. It strongly emphasized the surveillance of abortions as a eugenics measure.[78] Thus, the outcome of the National Eugenics Law was twofold: the sterilization of those who were considered "unfit" for biological reproduction, and the prevention of abortion by those who were presumably fit to reproduce. While the law resulted in a small number of sterilizations, it blocked abortions for mothers who needed them for medical reasons.

In other words, an inadvertent and serious consequence of the 1940 Eugenics Law was that its enforcement prevented doctors from providing abortions to women whose lives were threatened by pregnancy, which, prior to the legislation, they had been able (in their professional judgment) to terminate without the interference of state authorities. Under the National Eugenics Law, all abortions were theoretically subject to government scrutiny. Afraid that they would be punished for providing an illegal abortion, many obstetrician-gynecologist were hesitant to perform abortions.[79] Expectant mothers with tuberculosis faced grave risks during and after childbirth. Medical researcher Segi Mitsuo, who later became a leading MHW official, published an article criticizing the National Eugenics Act for tightening abortion prohibition. He noted that the promulgation of the law led tuberculosis-infected mothers to cancel abortions, thus causing many maternal childbirth deaths. Because of the article, Segi incurred serious censure from the National Police Department along with the Eugenics Section of the MHW.[80] In any event, the National Eugenics Act increased maternal mortality by restricting access to abortion services.[81]

The eugenics movement in pre-World War II Japan culminated in the passing of the National Eugenics Law and brought about another unexpected outcome. In light of the major objectives of eugenics policy, it was not surprising that eugenicists endorsed limiting the biological reproduction of people thought to have hereditary diseases, including mental disability.[82] What was unique in twentieth-century Japan was that people with non-hereditary diseases, for example, Hansen's disease (leprosy), became one of the major targets for forced sterilization or abortions. Those who endorsed such policies often reasoned that people with Hansen's disease would face financial difficulties raising children, either within or outside the sanatoriums in which they were unjustly segregated. They would end up producing children who would receive inadequate care and education. In the Epilogue I return to the issue that the state and medical experts denied certain groups of people their reproductive rights in pre-World War II Japan despite support for pronatalist policies. In fact, government reproductive policy bridged pre-World War II pronatalism and postwar anti-natalism.

To reinforce its pronatalist policy, the government gave priority to renewed efforts to support maternal and infant health. A major project that it pushed was establishing community health centers nationwide to employ district nurses for health counseling. The push for community health centers led to the passing of the Health Center Law in 1936—it took effect in 1937.[83] As I indicate earlier, groups formed by public-minded health experts and social-work activists had provided counseling and education to expectant mothers in limited ways before the establishment of the system of public health centers. The new government initiative reinforced these services mostly through local governments or non-governmental groups and renewed its emphasis on increasing the population.[84] The government's initial plan was ambitious: it aimed to establish 500 health centers and an additional 1100 branches within ten years. The process of executing this project went slowly in the beginning, but as public awareness of the benefits of health centers increased, their number rose. By 1941, 187 health centers had been established nationwide. Centers increased to 306 in 1943 and to 770 the following year.[85] An expert strongly supportive of the district nurse system remarked that it would be desirable to provide one nurse per rural village of 300–500 households or 2000–3500 people.[86] As I discuss later, the number of public health nurses was insufficient to meet the goal.

As a part of the plan to establish community health centers, the MHW issued an ordinance in 1941 stipulating professional qualifications for public health nurses who would staff the health centers. The edict required that district nurses pass an exam or enroll in an MHW-approved training program that usually took several months to complete. The examination covered anatomy, physiology, infectious diseases, maternal and child health, emergency medicine, nursing, nutrition, public health laws, and social work. The law allowed those who had studied either midwifery or nursing for longer than a year to take the examination.[87] Training programs, designed for women who had a background either in midwifery or nursing, in the first place, emphasized subjects that public health nurses would find useful but to which regular midwifery and nursing programs paid sporadic attention. These subjects included childcare and early education; public health administration and laws; social work; and health-related problems in rural areas.[88]

While health authorities and scholars expected district nurses to implement public health measures, the prevention and treatment of infectious diseases (e.g., tuberculosis, trachoma, diphtheria, and venereal diseases), the promotion of maternal and infant health, and the introduction of social-work programs were stressed.[89] Health professionals designed policies so that district nurses—unlike medical doctors and nurses who interacted with patients only after they had already become ill—intervened actively into community members' daily lives and advised to change problematic health practices. In rural communities that lacked other medical professionals, however, the roles nurses played went far beyond health guidance and preventive medicine support.

Indeed, district nurses took up different responsibilities serving different communities. Typically, in urban areas where they could send their patients to medical doctors and midwives in private practice or charity hospitals, public health nurses did not provide medical care or childbirth services. In isolated rural areas where there were neither licensed medical doctors nor midwives, public health nurses handled illnesses, injuries, and births to the extent they could. Nurses serving villages without a physician were busy treating illnesses and injuries and had little time for preventive medicine. Also, in many rural areas, there was the widespread practice of sharing a single nurse between the village and its elementary school. In rural communities, public health nurses rented a farmhouse or set up an office in a room in the local elementary school as a makeshift health center.[90]

Health officials and specialists acknowledged public health nurses working in rural areas faced demanding and daunting responsibilities, and had to work with limited resources. Their salaries were modest. They typically worked for local governments as low-level clerks, and they seemed not to receive remuneration that properly reflected their medical expertise. For example, Fujima Asayo, who served as a district nurse for the Tokyo city government in 1930, received a monthly salary of 50 yen.[91] In contrast, in 1929, a police officer in Tokyo with a wife earned 63 yen per month. The 50 yen monthly salary, however, was sufficient for a couple to make a decent living in Tokyo at the time.[92] I have not been able to find records indicating the salaries of district nurses working in rural Japan, but their earnings most likely were much less than what Fujima made in Tokyo—although nurses working in rural areas probably received many modest gifts from their patients, such as home grown vegetables. And nurses regularly worked well into the evening carrying out house visits.[93] Budgetary constraints meant that district nurses lacked clerical assistance. One nurse related that the local government asked her to take care primarily of paperwork; she could only conduct house visits during her lunch break or after work.[94]

With the allocation of a modest budget, the public health nurse project was, from the government's perspective, a cost-effective alternative to more expensive measures, such as providing publicly funded physicians and hospitals, or free medications. From a practicing district nurse's point of view, lacking a nearby licensed physician made making progress on local health issues difficult.[95] Because since the Meiji period, nurses were, by definition, women, and women usually had fewer obligations from a financial viewpoint, nurses subsidized government public health services through their low wages.

The implementation of a nation-wide public nurse system led to increased state surveillance on citizens' health and reproduction. Public health nurses collected health-related data more systematically and meticulously than had been done before. This was not at all a new undertaking as licensed midwives in private practice had been keeping records of individual clients. Non-governmental groups and local governments that created visiting nurse programs had developed multiple forms to record different types of client. Quasi-governmental group Saisei-kai created different types of index cards for ill patients, expectant mothers, children, and general health.[96] These previous developments of record-keeping became the basis from which public health nurses created their data system. In addition to keeping records, public health nurses also developed forms that their assistants (recruited from local residents) could use to report clients' conditions when nurses conducted house visits on their own. They could share knowledge about patients, if they thought it necessary, with medical doctors, public health administrators, and the police. It is true that local midwives in private practices had already been collaborating with public health officers and the police; with the introduction of the public health nurse system, however, the government could obtain knowledge on illnesses, pregnancies, and births more directly and with greater accuracy. Also, unlike visiting nurses working for charity groups, government-appointed district nurses began to collect data on not only poor women, but also women from privileged backgrounds. The government could potentially use this expanded and presumably more accurate knowledge of the reproductively active female population to control their health and sex practices or to "improve" their health care in general.

The district nurse system meant that individual midwives in private practices, or a small number of publicly funded maternity clinics and midwives, no longer guided expectant mothers. Instead, a district nurse—a government employee—kept the national interest in mind and could monitor and supervise women's reproductive lives. The numbers of health centers and district nurses were still too small to cover the nation as a whole, but as of 1940 there were 18,447 public health nurses working in either local government or non-profit groups in Japan (this statistic omits the facilities in Japan's colonies, such as Korea and Taiwan). In some prefectures, including Iwate, Shiga, and Okinawa, there were a small number of district nurses, less than 20 for each prefecture, while other prefectures, such as Aichi, Niigata, Osaka, Hyōgo, and Fukuoka, employed more than 1000 nurses. Aichi had close to 5000 public health nurses, including many "maternal counselors" [Bosei hogo-in].[97] Despite the limited number of district nurses available for many geographic areas, it is plausible that they had a significant influence on maternal and infant health. They actively sought out pregnant women and mothers with infants regardless of the mothers and their families' readiness or willingness to talk to the nurses. Most likely, this form of intrusive intervention helped to improve the health of many mothers and children, but the system had a coercive aspect with its investigation and counseling of mothers, and given that the nurse was armed with professional knowledge and backed by both local and national governments, the power relationship between the nurse and her clients was lopsided.

In 1942, the MHW introduced another measure, reinforcing state surveillance of pregnancies and births: maternity registration, the system of supplying each mother with a maternity handbook [ninsanpu techō]. This program strongly encouraged expectant mothers to report their pregnancies to the local municipal office to receive a maternity book—although those who failed to report received no penalty. Pregnant women were required to see a midwife or a physician under this system, and the medical professional was supposed to fill out the forms included in the maternity book to keep records of each pregnancy and birth. The book also served as a tool for education. It listed a set of instructions for expectant mothers that would enhance maternal and infant health.[98] The maternity handbook system allowed the government to monitor and support expectant mothers more efficiently than the previously existing system that relied on the mediation of midwives, physicians, and public health nurses. In other words, the state health bureaucracy sought to expand both the scale and rigor of reproductive surveillance by initiating this new pregnancy registration system.[99]

Segi Mitsuo, the MHW official who was the main architect of the system, declared its foremost aim was to expose expectant mothers to professional medical care at an early stage of pregnancy, which he and other health officials believed would improve maternal and infant health. Accordingly, recommendations for expectant mothers listed in the maternity handbook included a monthly check-up by a midwife or physician. If the mother was unable to make monthly visits to a midwife or physician, she should at least have three prenatal check-ups, including an initial visit after the discovery of pregnancy, another one either in the fifth or sixth month, and the last prenatal check-up in the eighth or ninth month.[100]

Segi asserted with a condescending tone that expectant mothers who used the book and the general public lacked an understanding of the significance of the system, namely, that it was the mother's exposure to early prenatal care that paid public health dividends. Interestingly, he remarked that public ignorance was not a problem because, in the end, medical experts were the only ones who could appreciate the benefits of early prenatal counseling.[101] It should also be remembered that the data that were recorded in the book were largely used for medical and administrative purposes, and they were primarily for the review of medical experts and municipal office workers. Experts and the state were the ones who interpreted, understood, and utilized knowledge of women's bodies to manage

Fig. 6.2 An example of the maternity book (56 pages in total). The maternity book includes general guidance on taking care of a pregnancy and a child's health care, along with pages where either a health provider or the mother enters information on the health of the mother and child. A book is issued to each child and is kept by the mother. This particular book was issued by Kanagawa prefecture around the mid-1980s. Photograph by the author (2017)

and control women's health. Women themselves were not the main interpreters of their bodies, nor did they use the knowledge of their bodies that was documented in the books.

As of the late 1930s and early 1940s, it was quite unusual for underprivileged women either in cities or the countryside to seek prenatal counseling from a trained midwife or a doctor. A push for pregnant women to see a midwife and obtain a maternity handbook partly came from local municipal offices and public health nurses. It was common practice that the head of the neighborhood association or a municipal officer urged and assisted the women to report their pregnancies to the municipal office.[102] Little consideration was given to the issue of the privacy of pregnant women or their families.[103] Health bureaucrats were eager to gather as accurate information as possible about pregnancies, and for this reason they had newspaper reporters tell their readers that women should report their pregnancies regardless of their marital status.[104] Ordinary women,

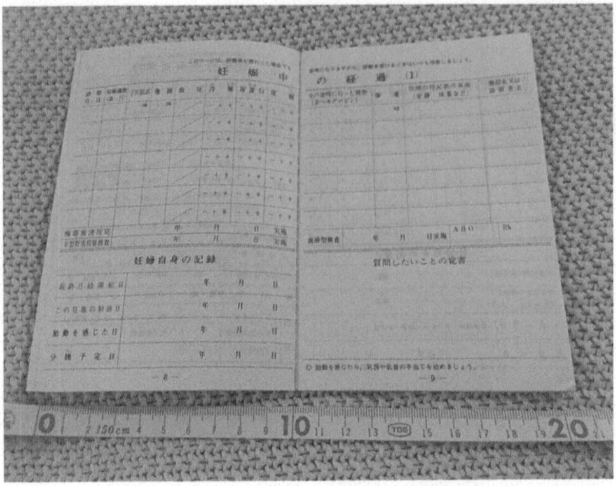

Fig. 6.3 Inside the maternity book (pages 8–9). These are the pages where the obstetrician-gynecologist and the midwife keep the mother's health data collected during pregnancy visits. In the bottom section on page 8, the expectant mother makes notes of the start date of her last menstruation, the first pregnancy visit to the obstetrician-gynecologist, the first time the mother feels fetal movement, and the expected date of delivery. Photograph by the author (2017)

however, seem to have lacked an incentive to cooperate voluntarily with the system.[105]

One of the reasons, often the only one, that women were willing to participate in the scheme was because upon reporting their pregnancies to the government they received special rations, such as absorbent cotton, bleached cotton cloth, milk, other dairy products, or an extra amount of rice and sugar.[106] Such rations allotted specifically to expectant mothers would help women and their families significantly in the scarcity-ridden wartime economy, and the incentive served as a major means to expand the program.

Despite women's general lack of interest, MHW administrators intended to educate mothers in maternal and infant health, and thereby, increase the nation's birth rate. Accordingly, the maternity handbook listed recommendations and admonitions for expectant mothers. It included a discussion about diet, a recommendation of moderate exercise, and warnings against physically demanding work during pregnancy. In suggesting mothers eat nutritiously during the postpartum period, the book urged pregnant women and their families to reject superstition. The maternity book also discussed specific medical problems—severe morning

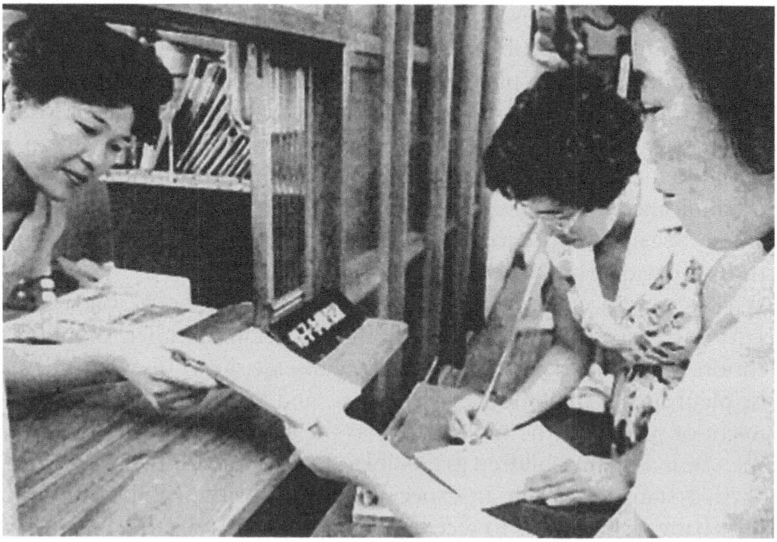

Fig. 6.4 Issuing a maternity book at the window of a municipal office (*c*. 1960) (Source: *Window of Government* [Monthly Magazine, published by the Government of Japan, August, 1960] as cited in Japanese Wikipedia entry on "Boshi kenkō techō" https://ja.wikipedia.org/wiki/%E6%AF%8D%E5%AD%90%E5%81%A5%E5%BA%B7%E6%89%8B%E5%B8%B3. Photograph in public domain)

sickness, fever, vaginal bleeding, pain in the hips and belly, swelling, numbness, and excessive discharge from the womb—that might arise during pregnancy and the postpartum period.[107]

In addition to presenting recommendations for mothers, the maternity book included forms to record the health of the expectant mother and the newborn. A midwife or physician who performed a check-up on a pregnant patient was required to note the dates of medical visits followed by a brief report on the status of their health. There were also sections where the mother or a medical professional would record the date of the last menstrual period before pregnancy, existing medical problems, the number of children who were alive, and whether or not the woman had had stillbirths. Another page was assigned to the topic of childbirth. The date and time of delivery and the duration of pregnancy were followed by the child's characteristics: sex, weight, and whether born alive or dead. Moreover, the midwife or physician who attended the birth was obligated to give their name, whether a surgical procedure to deliver the child was used, the reason for a miscarriage or premature birth, and any symptoms that would indicate a problem with the child's growth.[108] As I discuss in Chap. 4, midwives and physicians in private practice had already been using clinical charts (often in the form of index card) similar to those in the maternity book. And medical professionals were required to make patients' medical charts available to the local state authorities upon request. The significance of the new maternity handbook system was that the government began to procure knowledge on individual pregnancies and births more widely and directly from mothers.

What was noticeable on examining the maternity book was a list of specific medical problems to which the expectant mother should pay attention. These included beriberi, kidney disease, heart disease, tuberculosis, pleurogenic pneumonia, peritonitis, and syphilis. The widespread problem of syphilis at the time, as well as the government's desire to have women bear as many children as possible, was indicated by a passage in the book that suggested even an expectant mother with syphilis can bear a healthy baby if she begins to receive proper treatment for the disease early during her pregnancy. The book repeatedly noted that mothers should consult with a midwife or a physician if they detected any health problem found on the list.[109] With these instructions, pregnancy and childbirth were represented as disease-ridden events. They were defined as "medical" events as opposed to the existing notion of childbirth as an everyday event embedded in the religio-cultural structure of rural communities.

The maternity book system, thus, expanded medicalized modern childbirth education to women in rural and working-class communities through the direct involvement of the government. Most of the advice that the handbook gave did not go beyond the ordinary, but its recommendations set the standard for maternal and infant health care, and thereby discredited many local customs. Because local practices often compelled mothers to engage in practices that were either unhealthy or physically demanding, some of these recommendations, if they were put into practice, would benefit the mother. For example, the guidelines emphasized the importance of eating nutritious food, getting sufficient rest and sleep, and maintaining good bodily hygiene, all practices of which age-old local customs had disapproved.

While the growing influence of medicalized advice freed women from the oppression of local folk childbirth customs, they instead became subject to another type of power urging them to conform to new norms of maternal health set by the government and the medical profession. Already by the 1920s in urban Japan, the medical profession in collaboration with the media and, to a lesser extent, the government had helped to shape a modern childbirth culture with a set of standard practices and required products. Many women and their families, especially those from urban middle-class backgrounds, desired to follow the "proper" childbirth practices suggested by medical professionals and childbirth manuals. Pressure for women to obey medical authorities' recommendations was not simply exercised through force, but women engaged in activities because power also worked from the inside out, by making women desire to conform. Discourses on proper childbirth practices began to incite feelings of obligation and anxiety, prompting women to follow childbirth practices that medical authorities approved for women of respectable backgrounds. In other words, women became self-conscious agents who willingly reinforced the power of medical authorities despite their interests possibly conflicting with those of their female patients.

All the maternity book advice to be applied during pregnancy and the postpartum period no doubt heightened mothers' self-consciousness about their bodies and health practices. Caring for oneself during pregnancy became a mother's moral obligation. The first article in the government-issued maternity handbook read: "Strong children are born from strong mothers. Mothers ought to take good care of their health during pregnancy to give birth to good babies and, thus, to serve the nation."[110] Within this construct, women were perceived as objects of reform, but they also became subjects that willingly obey such instructions. The type of moral inculcation that pregnancy and childbirth guidelines fostered helped women become active agents of health reform and nationalism. In this way, they became enmeshed in a power dynamic through which they often inadvertently collaborated with the government in its pursuit of pronatalist policies. Mothers were thus created with a new sense of body and self that could be understood through medical language. They came to think that taking care of their bodies during pregnancy and childbirth made them better citizens.

How universal the implementation of the maternity handbook system was is debatable. Midwife journals, published in many localities, publicized the system, and licensed midwives along with public health nurses spread the word about the program among expectant mothers. Historian Owada Michiko shows that in Shizuoka prefecture expectant mothers who obtained a maternity book received a small additional ration of rice and a special ticket to exchange for pieces of cloth that were large enough to make a maternity sash as well as a baby's kimono and underwear. Also, once they obtained a certificate from a physician or midwife to verify that they could not produce sufficient breast milk for the newborn (with the percentage of the shortfall recorded), they could receive a special ration of milk and sugar from the relevant municipal office. The public health nurse who gave this testimony in an oral history interview cautioned that given the low status of young wives in rural farming families, there was no way that expectant or new mothers could eat the small additional rations by themselves; they were obligated to offer them to their in-laws, husbands, and children first.[111] While she acknowledged that the details of the maternity passbook rationing system differed by location, anthropologist Nishikawa Mugiko indicates that in the northern Ishikawa prefecture where she conducted her research, many women asked midwives and physicians to issue the pregnancy certificates to obtain extra rations. An interviewee, retired midwife Takejima Mii related that with increasing competition among midwives to acquire more patients, midwives in other towns used their connections with municipal officers to deny rations to mothers who hired Takejima.[112]

In Tokyo women were reported to have pretended they had a newborn or that they could not produce sufficient breast milk for their child. According to the Asahi newspaper, this prompted the Tokyo branch of the Aiiku Association to inspect the production of breast milk in each mother.[113] Women's testimonies corroborated the report. Ishinabe Hideko, a northern Tokyo resident, recounted that she and other mothers had to visit a community health center and feed their babies breast milk for about 20 minutes. Then, health officers weighed each baby. When they ascertained that women could not produce a sufficient amount of breast milk, the officers distributed a fixed ration of milk to each woman.[114] The news of women cheating, alongside Ishinabe's testimony, suggests that there was a somewhat functional system to distributing additional rations of milk, and most likely other items as well, to malnourished mothers in the Tokyo area.

Ishinabe indicates that she had access to rationed milk until 1944, even if its portion was not large enough. This somewhat corresponds with Segi's recollection of his experience helping Tokyo's prefectural government officers distribute milk and other rationed items. He described the aftermath of a Tokyo air raid on March 10, 1945, when he and other health officers could not deliver rationed items in badly bombed areas of eastern Tokyo. Segi noted that he and others continued their efforts to distribute rationed goods, traveling all over the country. Because prefectural governments directly handled the distribution of rationed articles, the nationwide bombing of Japanese cities most likely hampered such efforts by severely damaging local government administrative functions.

The foods that could be rationed became less and less available, even with Segi and others' efforts to deliver rations even on a small scale and develop substitute foods, such as powdered soybean to replace milk.[115] The historian Ishizaki Shōko calls attention to the sporadic implementation of the rationing plan in conjunction with the maternity book system, giving the example of her own mother who went to the local health center to discover that in her village the maternity book and rationing system did not exist.[116] This may be because later during the war, some local governments became dysfunctional owing to bombings and the absence of resources (i.e., both rationed goods and available municipal officers).

The scarcity of sources makes it difficult to assess the extent to which expectant mothers participated in the maternity book system. Immediately after it was initiated, even though supplementary rations motivated women to sign up with the system, the project was in its incipient stage, and many women were uninformed and unfamiliar with it. From 1944 to 1945, the shortage of food and other necessities became so acute that even the government began to experience difficulties in procuring basic necessities, such as milk, for mothers with infants. Moreover, many mothers moved from cities to the countryside to avoid air raids, and such migration most likely hampered the maternity registration and rationing systems. We know from Segi's testimony, however, that MHW officials were traveling to many parts of the war-ravaged nation by train. Despite air raid damage to the railway system, rations for mothers and infants did get distributed, if unevenly and in miniscule amounts, throughout the war years.[117]

More important to remember, perhaps, was that Segi, who continued to serve the MHW after the end of the war, succeeded in convincing health officers working for the postwar occupation administration (the General Headquarters Supreme Commander of Allied Powers, GHQ) to maintain the maternity book system.[118] This system was expanded during the postwar years and by the 1970s almost every mother who delivered a baby in Japan came to be incorporated into it.[119] It has served as an important mechanism for improving maternal and infant health care as well as supervising citizens' reproduction to this day.

The maternity book system is not the only reproductive policy program that continued after the end of World War II. Indeed, Japan's postwar government continued with policies from other reproduction-related legislation, including the 1937 Maternal and Child Protection Law, the 1937 Public Health Centers Act, and the 1941 National Eugenics Act, that is, the postwar state used a system of reproductive surveillance and control that it had established during the 1937–1942 period as a basis for developing new systems.[120] The state's vigilance and control over citizens' biological reproduction were not relaxed during the postwar years but became more subtle and meticulous with the help of an expanded health bureaucracy that could amass and utilize increasingly and presumably more accurate data on the health of citizens. My point is that there is an underlying continuity between the prewar and postwar policies on reproduction and maternal health.

For sure, certain significant changes occurred in reproductive policy practices in postwar Japan. One change involved the decline of midwife-assisted births alongside the rise of physician-attended hospital births. Scholar Ōbayashi Michiko argues that US health officers working at GHQ were ignorant of the skill and knowledge level of licensed Japanese midwives because midwives in the USA had been historically excluded from the modern medical establishment and remained untrained—many during their heyday were illiterate. With the assumption that midwives were uneducated and unreliable, US health officers drew up the 1948 Public Health Nurse, Midwife, and Nurse Law, which deprived midwives of the authority to provide a range of medical treatments, such as administering injections, using surgical instruments, and prescribing and providing medication. In other words, the law affirmed the dependent and subordinate status of midwives that the prewar midwife law had stipulated. With this 1948 law, midwives in the postwar era failed to gain the freedom and autonomy necessary to expand their businesses, nor could they resist the trend from the 1960s onward that expectant mothers increasingly preferred a hospital birth.[121]

Further, dramatically reversing its previous pronatalist policy, the Japanese government in the postwar years pursued an aggressive antinatalist policy prompted by anticipated high birth rates after veterans returned from the battlefields, fear of overpopulation, and a presumed shortage of food and resources. While the government strongly endorsed this movement to limit the size of the family, major corporations and their labor unions espoused and promoted it because they deemed that this "family planning" movement would "raise employees' standard of living" and reduce the cost of health benefits.[122] Local governments and agricultural corporations in rural areas also firmly supported the movement and disseminated the ideas and information associated with it through local public health centers, district nurses, and women's groups.[123]

In the postwar era, theories of eugenics continued to exert a strong influence in shaping ideas and policy on reproduction.[124] The postwar family-planning movement wholeheartedly championed the rhetoric of eugenics. Once the fear of overpopulation was replaced in the 1960s with a concern that a shortage of workers would dampen Japan's high economic growth, the anti-natalist movement, instead of simply calling for a reduction in the number of children, began to emphasize the importance of birth control for bearing "a good child" and preventing the birth of "an unfortunate child," that is, a child with a birth defect. It is also well known that the 1948 Eugenic Protection Law significantly liberalized abortion practices. Under this law, Japanese women could obtain abortions with approval from an obstetrician-gynecologist who possessed a license for providing abortions. Yet, as its name indicates, one of the important aims of the law was to "curb the birth of unfit offspring" (including cases where the child's parents could not provide adequate economic support to raise a "fit" child), and it relegated the protection of the mother's health to a secondary concern. Relatively less well known is that while only a limited number of sterilization operations were performed under the prewar National Eugenics Law, a remarkable number of such operations were carried out under the postwar Eugenic Protection Law. Official statistics account for 538 sterilization cases between 1941 and 1947. Such operations for the first decade after the enactment of the Eugenic Protection Law amounted to over 300,000 with more than 7600 cases of compulsory sterilizations.[125] Some of the victims of compulsory sterilization, such as Hansen's disease patients, sought compensation and an apology from the government, beginning in the 1990s.

Acknowledging these changes—expanded eugenics schemes and a reversal of pronatalist policy—which occurred in the postwar years, I emphasize the importance of paying attention to the consistent trend of the state and medical experts intervening in women's reproductive lives. Located in this larger trend of increasing state reproductive surveillance, policies implemented from the 1930s to the end of World War II constituted the basis for expanding bio-power not only in the war years but also the postwar years.

Before the 1930s, trained midwives played pivotal roles supervising women's reproduction. They diagnosed and managed pregnancies and births drawing on their knowledge of modern medicine as well as data on their patients. They were crucial players in the workings of bio-power at the micro-level. The enactment of important legislation in the late 1930s and early 1940s reduced somewhat the importance of midwives and relegated some of their functions to new players and systems. The promulgation of the Health Center Law and the Maternal and Child Protection Law led to the mobilization of district nurses and social workers for identifying mothers in need, gathering data on them, and providing reproductive and childrearing counseling. The maternity handbook system allowed the government to collect data on its citizens' pregnancies and births without the mediation of private health professionals.

With the establishment of various programs for maternal and infant health, data on an increasing number of women and children became incorporated into national medical statistics at the macro-level to make women and children the object of bio-power. They were classified into different groups according to their body characteristics. While many were considered fit for reproduction, some were categorized as unfit, from a eugenics standpoint. Obtaining knowledge of women's reproductive health both on an individual level and collectively via statistical methods was central to the state being able to implement its reproductive policies effectively. The laws and programs on maternal and infant health that were put in place from the 1930s through to the early 1940s reinforced the mechanisms of the workings of bio-power. The new regime of reproductive surveillance foreshadowed the more refined and less overtly forcible, but not necessarily more lax, reproductive control of the state and organized medicine that developed after the end of World War II.[126]

During the postwar years, another turning point in modifying maternal and child health and welfare occurred in the mid-1960s with the promulgation of the 1964 Maternal and Child Welfare Law and the 1965 Maternal and Child Health Law. While the creation of these laws aimed to revise maternal health and welfare policies more fitting to the contemporary needs of mothers and children, policy measures stipulated by these acts were congruent with the purposes of those policies that had been initiated in the late 1930s and early 1940s. In particular, the Maternal and Child Welfare Law shared a problem with the 1937 Maternal and Child Protection Law. Both laws treated the mother and child as one unit, and women's rights to their own welfare were rendered subordinate to the well-being of their children. Also, the 1937 Health Center Law foreshadowed the 1965 Maternal and Child Health Law (revised in 1994), a central stipulation of which concerns the obligation of local municipalities to provide mothers, their spouses, and children with guidance that they see physicians, midwives, or district nurses, regarding pregnancy, childbirth, and childrearing. Finally, the Maternal and Child Health Law and its revised version affirmed the central role played by the maternity book system for government policies in this area. In this way, the new institutions and laws established during the war years offered a new administrative structure and have had the long-lasting effect of shaping Japan's maternal and infant health policies.

Within such a structure, modern bio-power has expanded, but with the rise of the Japanese women's liberation movement in the late 1960s, an invigorated feminist reproductive rights movement became critical of government policies. Yet it, too, became a part of the expanding networks of bio-power, even though, at times, it was successful in taking advantage of modern medical knowledge for its cause. That is, feminists who fought against the government were not ignorant of modern medicine. Rather, many of them were highly educated, and to exert their influence on reproductive health they had to work within the networks of knowledge and power, not outside it. Investigating the development of reproductive health policies in postwar and late-twentieth-century Japan, however, is certainly beyond the scope of this book.

Notes

1. Sheldon Garon, *The State and Labor in Modern Japan* (Berkeley: University of California Press, 1990).
2. For a useful English language source on the history of state reproductive policy in both prewar and postwar Japan, see Kozy Kazuko Amemiya, *The Road to Pro-Choice Ideology in Japan: A Social History of the Contest between the State and Individuals over Abortion* (Ph.D. Dissertation, University of California, San Diego, 1993); Shōko Ishizaki, "Principles of Procreation and the Family in Modern Japan: Factors behind Decisions on Family Size," in Hiroko Tomida and Gordon Daniels, eds., *Japanese Women: Emerging from Subservience, 1868–1945* (Folkestone, Kent: Global Oriental, 2005).
3. For a history of the Ministry of Health and Welfare as an entity separate and independent from the Home Ministry, see Soeda Yoshiya, *Naimu-shō no shakai-shi* (Tokyo: Tokyo Daigaku Shuppan-kai, 2007): 546–559.
4. Matsubara Yoko, "Dai go-shō: Sengo no yūsei hogo hō to iu na no danshu hō" in Yonemoto Shōhei, Nudeshima Jirō, Matsubara Yoko, and Ichinokawa Yasutaka, eds., *Yūsei gaku to ningen shakai* (Tokyo: Kōdansha, 2000): 181–182.
5. Enami Shigeyuki and Mihashi Toshiaki, drawing on a Foucauldian perspective, published a unique and fascinating study on changing represen-

tations of the poor in Japan's modernity. Enami Shigeyuki and Mihashi Toshiaki, *Saiminkutsu to hakurankai* (Tokyo: JICC Shppankyoku, 1989). Note also that during the interwar period, large cities with large urban poor populations, such as Tokyo and Osaka, conducted various studies on the poor that made the social problem visible and concrete.

6. Kouchi Nobuko, *Shiryō boshi hogo ronsō* (Tokyo: Domesu Shuppan, 1984).
7. Note that the rampant problem of poverty, exacerbated by the births of many children, led some government health officials to endorse birth control. Eugenics played an important role in convincing them of the usefulness of birth control. Before the government began to pursue a more stringent pronatalist policy in the late 1930s, counselors working at health centers run by the city of Tokyo offered birth control information, and the national government overlooked such practices. Thus, birth control activism was generally a challenge to government policy, but not totally subversive of it.
8. For birth control activism in Japan during the early twentieth century, see Helen Hopper, "'Motherhood in the Interest of the Sate': Baroness Ishimoto (Katō) Shidzue Confronts Expansionist Policies Against Birth Control, 1930–1940" in Tomida and Daniels, eds., *Japanese Women, Emerging from Subservience, 1868–1945*; Ishizaki Shōko, "Principles of procreation and the family in modern Japan: Factors Behind Decisions on Family Size"; "Seishoku no jiyū to sanji chōsetsu undō," *Rekishi hyōron*, March 1992; "Kindai nihon no sanji chōsetsu to kokka seisaku," *Sōgo joseishi kenkyū*, 15 (1998); Ogino Miho, *"Kazoku keikaku" e no michi: kindai nihon no seishoku wo meguru seiji* (Tokyo: Iwanami Shoten, 2008): 2–85. On Shibahara, see Fujime Yuki, "Aru sanba no kiseki: Shibahara Urako to sanji chōsetsu," *Nihonshi kenkyū*, 366 (February 1993): 90–112; and Yuki Fujime, "One Midwife's Life: Shibahara Urako, Birth Control, and Early Shōwa Reproductive Activism" in *Wakita Haruko, Anne Bouchy, and Ueno Chizuko, eds., Gender and Japanese History, Vol. 1 (Osaka: Osaka University Press, 1999)*: 297–325; Shirai Chiaki, ed., *Umi sodate to josan no rekishi* (Tokyo: Igaku Shoin, 2016). For a translation of the autobiography of Katō, a leading birth control activist, see Kato Shidzue, Baroness Shidzue Ishimoto, *Facing Two Ways: The Story of My Life* (Stanford, CA: Stanford University Press, 1984), and for her biography, Helen M. Hopper, *Kato Shidzue: A Japanese Feminist* (Hoboken, NJ: Pearson, 2003).
9. Namae Takayuki, "Sanpu oyobi nyūji no hogo," *Shakai jigyō*, 6:7 (October 1922): 23.
10. Ōkuni Michiko, *Hoken-fu no rekishi* (Tokyo: Igaku Shoin, 1973): 16.
11. Muramatsu Shihoko had been largely unknown until the local historian Harashima Sachiko brought her work to life. See Harashima Sachiko,

Meiji no joi Muramatsu Shihoko, Ro-bungaku no chichi Yasugi Tadatoshi (a private publication by Harashima Sachiko, Tokyo, March 2001); *Shihoko no ayunda michi* (a private publication by Harashima Sachiko, Tokyo, May 2007). See also Ishihara Tsutomu and Harashima Sachiko, "Honpō kōya no san'in setsuritsusha Muramatsu Shihoko no anseidō to sono jizen jigyō," *Nihon ishigaku zasshi*, 52:1 (March 2006); Ishihara Tsutomu, "Josanpu no rekishi," *Perineitaru kea*, 25:12–27:4 (December 2006–April 2008): 280–296.

12. The Saisei-kai units visited 75,223 households between July 1924 and June 1925 and attended 12,018 ill or injured patients. On the Saisei-kai [the Imperial Relief Association], see also Sheldon Garon, *Molding Japanese Minds: The State in Everyday Life* (Princeton, NJ: Princeton University Press, 1997): 48–49.
13. Tokyōshi Shisei Chōsakai, *Toshi ni okeru ninsanpu hogo jigyō ni kansuru chōsa*, originally published in 1928, reprinted in Ichibangase Yasuko, ed., *Senkanki shuyō toshi shakai chōsa hōkokushū josei-hen daīki, 6* (Tokyo: Kindai Shiryō Kankō kai, 1998): 85–108.
14. As of 1921, the Osaka City Midwife Association set the regular midwifery fees for childbirth at more than 5 yen. This charge only covered the delivery of the baby. In addition to this fee, mothers had to pay at least 1 yen for other services, such as prenatal diagnosis and help with the ritual of putting on the maternity sash, or for bathing the baby in the days after childbirth. Aoki Hidetora, *Ōsaka-shi sanba dantai shi* (Osaka: Osaka-shi Sanba Kai, 1935): 168–169.
15. Tokyo Shisei Chōsakai, *Toshi ni okeru ninsanpu hogo jigyō ni kansuru chōsa*: 108, 121–122.
16. For discussions of the vibrant visiting midwife nurse programs both established by city the of Osaka and non-profit groups, see Tokyo Shisei Chōsakai, *Toshi ni okeru ninsanpu hogo jigyō ni kansuru chōsa*: 108–123; Ōkuni, *Hoken-fu no rekishi*: 6–10, 16–40.
17. By the mid-1930s there were many licensed midwives, enough to create an oversupply in large cities such as Tokyo. For example, midwife Yoneyama Matsu failed to procure even one client over a year around 1935 in eastern Tokyo, and this prompted her to move to her hometown in Yamanashi prefecture where she had no problem finding clients. Fujita, *Osan kakumei*, 151.
18. For the spread of the public midwife system in rural Japan, see, for example, Yumoto Atsuko, "Nagano-ken ni okeru junkai sanba no seiritsu to fukyū," *Shinano*, 53:7 (July 2001). Having faced the difficulties in prompting trained midwives to open their private practice, a number of villages in Nagano prefecture hired a public midwife in 1925. By 1935, 97 villages were running a public midwife system.

19. In early twentieth-century Japan, many who took up the responsibility were social workers, or local government-appointed "hōmen iin" [district commissioner]. Others were local notables without social-worker training. They often worked without receiving remuneration for their services out of a willingness to help the poor and unfortunate in their neighborhoods. See Garon, *Molding Japanese Minds*: 52–57.
20. Yamamoto Toshiko, "Hōmon kangofu no me ni eijitaru suramu no nyūyōji," *Shakai jigyō*, 17:1 (April 1933).
21. Tokyo Shisei Chōsakai, *Toshi ni okeru ninsanpu hogo jigyō ni kansuru chōsa*: 98. Inoma's view of the poor was shared by many government officials and middle-class reformers. See Garon, *Molding Japanese Minds*: 40–58.
22. For examples of the index card systems from the Saisei-kai and Osaka prefecture relief services for deceased or injured veterans' families, see Maki Tetsuo, "Onshi zaidan saisei-kai junkai kangofu jigyō no zenkokuteki soshiki-ka e no zenntei, sono ni," *Shakai jigyō*, 23:6 (September 1939): 76–86.
23. See, for example, "Hoken-fu no kutō wo kiku zadankai 3," *Fujo shinbun*, 2134 (May 4, 1941): 9; "Hoken-fu no kutō wo kiku zadankai 4," *Fujo shinbun*, 2135 (May 11, 1941): 8; "Junkai sanba no jigyō kyōka: Arata ni ninsanpu kādo sakusei," 13.
24. Fujino Yutaka, "'Mui-son' mondai no tōjō," *Toyama kokusai daigaku jinbun shakai gakubu kiyō*, 2 (2002): 103–108.
25. Yoshinaga Naoko, "Shōwa zenki ni okeru shussan no henyō 'bosei no kyōka,'" *Tokyo Daigaku daigakuin kyōikugaku kenkyūka kiyō*, 37 (1997): 2, 9; Onshi Zaidan Boshi Aiiku-kai Gojū-nen Shi Henshū Iinkai, *Boshi aiikukai gojū-nen shi* (Tokyo: Onshi Zaidan Boshi Aiiku-kai, 1988): 23–25.
26. Yoshinaga Naoko, "Nōson ni okeru san'iku no 'mondai ka'," in Kawagoe Osamu and Tomobe Ken'ichi, eds., *Seimei to iu risuku* (Tokyo: Hōsei Daigaku Shuppan-kai, 2008): 131–138. Yanagita's contributions include, for example, Yanagita, *San'iku shūzoku goi*; "Shōni seizon-ken no rekishi," *Aiiku*, 3 (September 1935).
27. Ōkuni, *Hoken-fu no rekishi*, 97–103.
28. Ōkuni, *Hoken-fu no rekishi*, 104–119.
29. Yoshinaga, "Shōwa senzen ki ni okeru shussan no henyō to 'bosei no kyōka'," 23–24. In 1940, there were 178 cities [*shi*], 1706 towns [*machi*], and 9614 villages [*mura*] in total nationwide. By 1945, there was an increase in the number of cities and towns while the number of villages decreased. The numbers of cities, towns and villages were 205, 1797, and 8518, respectively.
30. Yoshinaga, "Nōson ni okeru san'iku no 'mondai ka'," 119–121.
31. Ibid., 127.
32. Ibid., 125–126.

33. Yamanashi-ken Shakai-ka, "Minamoto mura aiiku jigyō nit suite" (1941), cited in ibid., 128.
34. Kawamura Tsuguyo, "Yamanashi-ken Minamoto-mura ni okeru nyūyōji no hoken jōtai'," *Aiiku shinbun*, 3:2 (January 15, 1940): 4.
35. Shibuya Kon'ichi, "Waga mura no aiiku jigyō wo kataru," *Aiiku shinbun*, 2:8 (July 15, 1939): 7.
36. "Aiiku-han to shussan fujo kumiai," *Aiiku shinbun*, 1:8 (November 15, 1938): 7.
37. "Nōson wo meirō ni suru aiiku-han no soshiki to jigyō: dai-ikkō, seisei no seishin," *Aiiku shinbun*, 1:1 (April 15, 1938): 6; "Nōson wo meirō ni suru aiiku-han no soshiki to jigyō: dai-nikō, hanin to yakuin," *Aiiku shinbun* 1:2 (May 15, 1938): 7.
38. "Nōson wo meirō ni suru aiiku-han no soshiki to jigyō: dai-ikkō, seisei no seishin," *Aiiku shinbun*, 1:1 (April 15, 1938): 6.
39. "Jinkō seisaku kakuritsu kōyō," reprinted in Ichikawa Fusae, ed., *Nihon fujin mondai shūsei*, 2, (Tokyo: Domesu Shuppan, 1977): 521.
40. Onshi Zaidan Boshi Aiiku-kai Gojū-nen Shi Henshū Iinkai, *Boshi aiiku-kai gojū-nen shi*, 21.
41. "Sōsai miya denka aiiku kenkyūjo ni onari," *Aiiku shinbun*, 2:1 (August 15, 1938): 7; "Sōsai kuninomiya taihi denka honkai ni rairin asobasaru," *Aiiku shinbun*, 5:1 (December 15, 1941): 1.
42. "Jijyū aiikumura wo goshisatsu," *Aiiku shinbun*, 5:8 (July 15, 1942): 4; "Jijyū goshisatsu no kōei ni yokushite," ibid., 4–5.
43. Wake Ichirō, "Jijyūkyō saken no kōei ni yokushite," *Aiiku shinbun* 5:8 (July 15, 1942): 3.
44. Matsuyama Teruo, *Aiiku han'in techō* (Tokyo: Onshi Zaidan Aiiku -kai, 1936), 3, introduced in Yoshinaga Naoko, "Onshi zaidan aiiku-kai ni yoru aiiku-mura jigyō no sōsetsu to tenkai," *Kenkyūshitsu kiyō*, 32 (2006): 4.
45. The nationalist ideology portrayed women not only as mothers of individual children but as mothers of the nation (*minzokuno haha*). Women's bodies were, by extension, "maternal bodies of the nation" (*minzoku no botai*). See, for example, Kawamura Hyōtarō, *Nōson hoken-fu* (Tokyo: Sangabō, 1942): 32–44.
46. Yoshikumi Ito, "Shisetu hōmonfu to shite no jukkanen wo kaerimite," *Shakai jigyō*, 24:10 (October 1940): 53–58.
47. Ibid.
48. In his study of Japan's surrender toward the end of World War II, Tsuyoshi Hasegawa makes the point that the Japanese state was clearly less concerned about the massive loss of civilian lives from the repeated air raids on Japanese cities and the dropping of the atomic bombs than by the prospect of the discontinuity of the imperial house that an unconditional surrender would incur. Tsuyoshi Hasegawa, "Were the Atomic Bombings of Hiroshima and Nagasaki Justified?," in Marilyn Young and

Yuki Tanaka, eds., *Bombing Civilians: A Twentieth Century History* (New York: New Press, 2009).

49. Kouchi, *Shiryō boshi hogo ronsō*. See also Hayakawa Noriyo, "Taishō-ki fujin kaihō shisō no kakki-ten wo saguru," *Jinmin no rekishi-gaku*, 61 (November 1979); Sawayama Mikako, "Hiratsuka Raichō ni okeru haha no kenri to kodomo no kenri no tōitsu no shisō wo megutte," in Ningen bunka kenkyūkai hen, *Josei to bunka: shakai, bosei, rekishi* (Tokyo: Hakuba Shuppen 1979).
50. Hayakawa suggests that Hiratsuka's ideal that women with young children stay home to raise their children could be misinterpreted to support conservative thinkers' endorsement of separate gender roles, as for example, in the "good wife, wise mother" [*ryōsai kenbo*] ideology. Hayakawa, "Taishō-ki fujin kaihō shisō no kakki-ten wo saguru," 2.
51. Imai Konomi, *Shakai fukushi shisō to shite no bosei hogo ronsō* (Tokyo: Domesu Shuppan, 2005): 113–132; and "Oboegaki: Boshi hogo-hō seiritsu made no kiseki," *Osaka Taiiku Daigaku Kenkō Fukushi Gakubu Kenkyū Kiyō*, 1 (March 2004): 69–73.
52. It is worth noting that Hiratsuka and Oku were a part of the settlement house movement that the Christian reformer Kagawa Toyohiko promoted from the early 1920s onward. By establishing settlement houses, middle-class reformers lived close to and worked with their lower-class neighbors in the working-class areas of large cities. Through their involvement in this movement, Hiratsuka and Oku witnessed the dire problems of poor mothers first hand, which eventually shaped their ideas for the Maternal Protection Law. Imai, *Shakai fukushi shisō to shite no boshi hogo ronsō*, 102–132; and "Shakai undō to shite no shakai fukushi: Oku Mumeo no katsudō wo tōshite," *Kirisutokyō shakai mondai kenkyū*, 55 (December 2006).
53. Oku Mumeo, "Josei no shakai-teki shinshutsu to nyūyōji," *Shakai jigyō*, 18:1 (April 1934): 28–31; "Ika ni bosei wo aigo subekika," *Shakai jigyō kenkyū*, 22:6 (June 1934): 107–111. See also Hayakawa's interpretation of Oku's ideas on childcare support. Hayakawa, "Senji-ki no bosei-ron," in Tokyo Rekishi Kagaku Kenkyū-kai Fujin Undō-shi Bukai, ed., *Onna to sensō: Sensō wa onna no seikatsu wo dou kaetaka* (Tokyo: Shōwa Shuppan, 1991): 253.
54. "Boseiai no kakuchō," *Fujo shinbun*, 1335 (January 10, 1926): 1. See also Hayakawa, "Senji-ki no bosei-ron," 252–253.
55. Fukushima published numerous editorials endorsing maternal protection legislation in *Fujo shinbun*. Among his major pieces are: "Bosei hogo no kyūmu," *Fujo shinbun* 1339 (February 7, 1926): 1; "Boshi fujo undō," *Fujo shinbun*, 1569 (July 1, 1929): 1; "Boshi fujo-hō no kyūyō," *Fujo shinbun*, 1583 (October 12, 1930): 1; "Boshi hogo no seishin," *Fujo*

shinbun 1760 (March 4, 1934): 1; "Boshi fujo-hō undō," *Fujo shinbun*, 1785 (August 26, 1934): 1; "Boshi hogo-hō gikai tsūka," *Fujo shinbun*, 1920 (March 28, 1937): 1. For Fukushima's campaign to push maternal and child protection legislation, see also Imai, *Shakai fukushi shisō to shite no boshi hogo ronsō*, 140–155.

56. Yamada Waka, "Nyūyōji no shakaiteki kankyō to sono kaizō," *Shakai jigyō*, 18:1 (April, 1934): 33–34; "Bosei hogo undō to boshi fujo-hō," *Shakai jigyō*, 20:7 (Oct, 1936): 65–67.
57. Imai Konomi, "Shakai minshū fujin dōmei no boshi fujo-hō seitei undō," *Kirisutokyō shakai mondai kenkyū*, 49 (December 2000): 183–185.
58. Ibid., 186–190.
59. Ibid., 192, 195.
60. Namae Takayuki, "Boshi fujo-hō no kigen," *Fujo shinbun*, 1354 (May 23, 1926): 6–7.
61. "Kōza, Jidō fujo-hō no naiyō: Moriya Shakai-kyoku Dai-ni Buchō," *Fujo shinbun*, 1379 (November, 14, 1926): 4–5.
62. Hayakawa, "Senji-ki no bosei ron," 252–252.
63. Imai, "Oboegaki: Boshi hogo-hō seiritsu made no kiseki," 69–72; Hayakawa, "Senji-ki no bosei ron," 247–251.
64. Even before devastating economic recessions hit Japan in the late 1920s, politicians, government officials and concerned activists—witnessing various social problems, unemployment, and rural poverty—had sought to reform Japan's poor laws. Their efforts to revise the 1874 poor law (Jutsukyū kisoku) led to the establishment of the new Relief Law (Kyūgo-hō) in 1929. As is obvious from my discussion, the movement for maternal protection legislation had a long history, having originated before the issue of the revision of the poor law system attracted public attention around 1922. See Yokoyama Kazuhiko and Tada Hidenori, eds., *Nihon shakai hoshō no rekishi*(Tokyo: Gakubun-sha, 1991): 30–33. Terawaki Takao examines how officials and social-work activists who supported child protection legislation instead of general poor law legislation failed in their attempts at reform. Terawaki Takao, "Showa shotō ni okeru kyūhin rippō seitei hōshin no kakutei to jidō fujo hōan no kisū (jō)," *Nagano daigaku kiyō*, 17:4 (March 1996); "Showa shotō ni okeru kyūhin rippō seitei hōshin no kakutei to jidō fujo hōan no kisū (ge)," *Nagano daigaku kiyō*, 18:2 (September 1996). For an English language source on this law, see Garon, *The State and Labor in Modern Japan*, 143–156, 179.
65. Imai, "Oboegaki: Boshi hogo-hō seiritsu made no kiseki," 73–74.
66. Ibid., 74–75; Hayakawa, "Senji-ki no bosei ron," 247–251. Hayakawa draws attention to how officials and legal scholars relied on family-based concepts to view the mother and child as an inseparable entity for legal purposes. She discusses the repercussions of this emphasis on the impor-

tance of the family relationship and maternal love for shaping the details of the Maternal and Child Protection Law and revising motherhood ideology to suit militarist purposes.

67. Shakai-kyoku Shakai-bu (Naimu-shō Shakai-kyoku), ed., *Boshi hogo-hō tō no setsumei* (Tokyo: Shakai-bu Shakai-kyoku, 1937).
68. Hayakawa, "Senji-ki no bosei ron," 246–251. As early as 1917, the Diet passed the military relief law [Gunji kyūgo-hō]. The military had already been offering various forms of aid for veterans and their families before the promulgation of this law, but this legislation certainly helped improve relief measures for veterans and military families. The government revised this law in 1937 (renamed "Gunji fujo-hō") to further expand relief for ex-servicemen and their families. Nihon shakai jigyō daigaku kyūhin seido kenkyū-kai, ed., *Nihon no kyūhin seido* (Tokyo: Keisō Shobō, 1960): 163–167, 284–293.
69. Shakai-kyoku Shakai-bu, *Boshi hogo-hō tō no setsumei*: 15–16.
70. See Akamatsu Keisuke, *Hijōmin no minzoku bunka* (Kyoto: Akashi Shoten, 1986); and *Hijōmin no sei minzoku* (Kyoto: Akashi Shoten, 1991).
71. Hino Shūitsu, "Senji-ka no kenkō to hoken seisaku," in Tokyo Rekishi Kagaku Kenkyū-kai Fujin Undō-shi Bukai, ed., *Onna to sensō*: 29–32.
72. "Jinkō seisaku kakuritsu kōyō" reprinted in Ichikawa, ed., *Nihon fujin mondai shūsei*, 2, 520.
73. Through her research on local women's history in Shizuoka prefecture, historian Owade Michiko shows how government attempts to increase the population during the war years failed. Owada Michiko, "Sanji seigen kara tasan shōrei e: bosei ni miru shizuoka-ken kindai josei-shi no tenkai," *Shizuoka-ken kindai-shi kenkyū*, 14 (October 1988): 61–63.
74. Ibid. See also Owada's discussion on Hino's points. Ibid., 53–54.
75. It is true that there was reluctance on the part of the state to use women's labor for heavy industry jobs out of concern for maternal protection. The government began to forcibly conscript women into wartime production only in 1944. By then, however, many women had already been working in physically demanding jobs. The government relaxed labor regulations to address labor shortages, which led to deteriorating working conditions for many women. Hayakawa, "Senji-ki no bosei-ron," 264–271; Miyake Yoshiko, "Doubling Expectations: Motherhood and Women's Factory Work Under State Management in Japan in the 1930s and 1940s" in Gail Bernstein, ed., *Recreating Japanese Women, 1600–1945* (Berkeley: University of California Press, 1991): 281–292.
76. See, for example, Owada, "Sanji seigen kara tasan shōrei e," 59. For a useful discussion in English on Japan's wartime policy of population

increase, see Miyake, "Doubling Expectations," 277–281. For more on this topic, see Hayakawa, "Senji-ki no bosei-ron"; Ogino, *"Kazoku keikaku" e no michi*, 112–140; Takahashi Hiroyuki, "Senji 'jinkō seisaku' no sai-kentō," in Kawagoe et al., eds., *Seimei to iu risuku*, 141–176.

77. See, for example, Sumiko Otsubo, "Engendering Eugenics: Feminists and Marriage Restriction Legislation in the 1920s," in Barbara Molony and Kathleen Uno, eds., *Gendering Modern Japanese History* (Cambridge, Mass: Harvard University Asia Center, 2005): 225–226.
78. Hayakawa, "Senji-ki no bosei-ron," 254–257; Matsubara Yoko, "Dai go-shō: Sengo no yūsei hogo hō to iu na no danshu hō," in Yonemoto Shōhei, Nudeshima Jirō, Matsubara Yoko, and Ichinokawa Yasutaka, eds., *Yūsei gaku to ningen shakai* (Tokyo: Kōdansha, 2000): 176–183.
79. Ibid., 182.
80. Segi Mitsuo, "Dai-ikkai: Boshi eisei gyōsei no taisei-ki," *Sanfujin-ka no sekai* 29:4 (April 1977): 520. For a discussion of how the National Eugenics Law and more stringent restrictions on abortions affected maternal deaths from tuberculosis complications, see Owada, "Sanji seigen kara tasan shōrei e," 53–54.
81. Nakagawa Yoneko, a public health nurse in Kōchi prefecture, related how the government policy of *umeyo fuyaseyo* (give birth to more children and increase the population) strictly prohibited abortion services even when the expectant mother was suffering from toxaemia and on the verge of death. The new policy with the issuance of the Guidelines for Establishing Population Policy set in 1941 brought a new kind of distress to practicing public health nurses and midwives, who could no longer suggest saving the lives of mothers by way of abortion. Kimura Tetsuya, *Chūzai hoken-fu no jidai, 1942–1997* (Tokyo: Igaku Shoin): 37.
82. Carolyn Stevens's anthropological study introduces major issues in the field of disability studies in Japan, including a brief history of the Japanese government's disability policy. Carolyn S. Stevens, *Disability in Japan* (London and New York: Routledge, 2015). See also Reiko Hayashi and Masako Okuhira, "The Disability Rights Movement in Japan: Past, Present, and Future," in *Disability and Society* (Abingdon-on-Thames, United Kingdom: Taylor and Francis Publishing, 2001): 855–869; and Karen Nakamura, *A Disability of the Soul: An Ethnography of Schizophrenia and Mental Illness in Contemporary Japan* (Ithaca: Cornell University Press, 2013). Mark Bookman's new historical studies in this area are also worth noting. Mark Bookman, "Remarkable Opportunity and Untold Hardship: Impairment in Interwar Japan," Paper presented at the 46th Annual Meeting of the Mid-Atlantic Region Association for Asian Studies, Philadelphia (October 2017).

83. Along with maternal health, the government had also recognized tuberculosis as one of the most urgent public health issues, and this concern was central to the implementation of the Health Center Law. Ōkuni, *Hoken-fu no rekishi*, 76.
84. Another health care measure that the national government pushed in the 1920s and 1930s was to expand health insurance to more people. The passing of the health insurance bill in 1922 led to the creation of a health insurance plan mostly targeted at manual workers in heavy industries that were potentially hazardous and prone to severe accidents. The 1938 National Insurance Act primarily helped the rural population with health insurance coverage provided through health insurance associations established by local mutual aid or trade associations. Yokoyama and Tada, eds., *Nihon shakai hoshō no rekishi*, 42–56. For classical studies of the history of the Japanese health insurance system, see Saguchi Takashi, *Nihon shakai hoken seido-shi* (Tokyo: Keisō Shobō, 1977); *Kokumin kenkō hoken* (Tokyo: Kōseikan, 1995). For an English language source, see Garon, *The State and Labor in Modern Japan*, 7, 112, 198, 205. Recent studies have paid attention to the push by local groups for health insurance coverage for rural residents. They challenge the view that the 1938 legislation was a result of a top-down process, originating in the wartime health movement. Takashima Yūko, "Kokumin kenkō hoken seido keisei katei ni okeru iryō riyō kumiai no reikishi-teki ichi," *Ōhara shakai mondai kenkyūjo zasshi*, 564 (November 2005). Expectant mothers who had employee health insurance were entitled to receive 30 yen to pay for midwifery fees and other necessities for childbirth. Komiyama Shin'ichi, *Hoken-fu dokuhon* (Tokyo: Kyōdō Kōsha, 1943): 239.
85. Nakayama Makiko, *Shintai wo meguru seisaku to kojin: Boshi kenkō sentā jigyō no kenkyū* (Tokyo: Keisō Shobō, 2001): 38–40; Akagi Asaji, *Nihon no hokenfu* (Tokyo: Tokuwa Shobō, 1943).
86. "Mondai: shakai hoken-fu aruiwa hokenfu jigyzō ni tuite," *Shakai jigyō*, 25:2 (February 1941): 30.
87. Akagi, *Nihon no hoken-fu*, 34–38.
88. Kawashima Hyōtarō, *Nōson hoken-fu* (Tokyo: Sengabō, 1942), 80.
89. Akagi, *Nihon no hoken-fu*, 13–16.
90. Koshimizu Sōkichi, "Waga mura no hoken-fu wo kataru: Yamanashi-ken Minamoto-mura no maki," *Aiiku shinbun* 3:2 (January 15, 1940): 5.
91. Ishizaki Shōko, "Hoken-fu e no michi wo kirihiraita Fujima Asayo," in Shinjuku-ku Chiiki Josei-shi Hensan Iinkai, ed., *Shinjuku ni ikita josei tachi*, 3 (Tokyo: Shinjuku-ku Chīki Josei-shi Hensan Īnkai, 1996): 101–106.

92. Iwase Akira, *"Gekkyū hyakuen" saraīman: senzen nihon no heiwa na seikatsu* (Tokyo: Kōdansha, 2006).
93. "Hoken-fu no kutō wo kiku zadankai 2," *Fujo shinbun*, 2133 (April 27, 1941): 8–9.
94. Ibid.
95. "Hoken-fu no kutō wo kiku zadankai 5," *Fujo shinbun*, 2136 (May 18, 1941): 8.
96. Maki Tetsuo, "Onshi zaidan saisei-kai junkai kangofu jigyō no zenkokuteki soshiki-ka e no zenntei, sono ni," *Shakai jigyō*, 23:6 (September 1939): 79–81.
97. Ōkuni, *Hoken-fu no rekishi*, 132–133.
98. "Ninsanpu techō kitei no seitei," *Jinkō mondai kenkyū*, 3:8 (August 1942): 32–34. Note also that the 1940 National Physical Fitness Law (Kokumin tairyoku-hō) created a similar physical fitness handbook (tairyoku techō) for monitoring and guiding children's health. Tomono Kiyofumi, "'Umu' koto e no kokkateki manazashi no seiritsu: sensō ga umi otoshita 'boshi techō'," *Nihon fujin mondai konwa-kai kaihō*, 52 (1992): 59.
99. Nakayama Makiko surmises that Segi's observation of the German maternal health system in 1939 helped him create the maternity book system. Segi himself also credited the previous maternity registration system, which relied on the cooperation of practicing physicians and midwives (but not necessarily the government). Segi gave credit to MHW official Itō Hajime who originally suggested using a handbook (or notebook) that was handy to carry and keep for the maternity registration system. Nakayama, *Shintai wo meguru kojin to seisaku*, 47–48. Segi Mitsuo, "Dai-nikai: Boshi eisei gyōsei no taisei-ki," *Sanfujin-ka no sekai* 29:5 (May 1977): 661–662. See also Ōbayashi Michiko, *Josanpu no sengo* (Tokyo: Keisō Shobō, 1989), 105–114.
100. "Ninsanpu techō kitei no seitei," 33.
101. Segi, "Dai-nikai: Boshi eisei gyōsei no taisei-ki," 662.
102. Nakayama, *Shintai wo meguru kojin to seisaku*, 47–49.
103. The maternity book system fascinated US health officials working to reconstruct Japan's health administration under General MacArthur's occupation authority. Some even discussed the possibility of implementing a similar system in the USA, but they generally thought that it would not work due to American women's desire to keep their pregnancies private. Segi, "Dai-nikai: Boshi eisei gyōsei no taisei-ki," 661.
104. Tomono, "'Umu' koto e no kokkateki manazashi no seiritsu," 59.
105. Hayakawa, "Senji-ki no bosei-ron," 261.

106. Owada, "Sanji seigen kara tasan shōrei e," 62.
107. "Ninsanpu techō kitei no seitei," 33.
108. Ibid., 33–34.
109. "Ninsanpu techō kitei no seitei," 33.
110. "Ninsanpu techō kitei no seitei," 33.
111. Owada, "Sanji seigen kara tasan shōrei e," 62.
112. Nishikawa Mugiko, *Aru kindai sanba no monogatari: Noto, Takejima Mii no katariyori* (Toyama: Katsura Shobō, 1997): 214–217.
113. "Bonyū wo kensa: Gyūnyū no haikyū wo tekisei-ka," *Asahi shinbun* (September 26, 1944): 2.
114. Kitaku Sōmubu Josei Seisaku-ka, ed., *Senjika ni kurashita josei tachi* (Tokyo: Domesu Shuppan, 1997): 132.
115. Segi, "Dai-nikai: Boshi eisei gyōsei no taisei-ki," 662–663.
116. Personal conversation, June 2011.
117. Segi, "Dai-nikai: Boshi eisei gyōsei no taisei-ki," 662–663.
118. Ōbayashi Michiko, *Josanpu no sengo*, 114–116; Segi Mitsuo, "Dai-sankai: Boshi eisei gyōsei no taisei-ki," *Sanfujin-ka no seikai* 29:6 (June 1977): 785.
119. See, for example, Ujiya Yoshio, "Jidō fukushi no tachiba kara mita boshi techō ni tsuite," *Fukushi kenkyū*, 9 (September 1958). Social worker Ujiya disapproved that many mothers and staff working for infant homes were remiss in entering information in maternity books and that rural mothers tended to obtain their books, and thus register their pregnancies, at a later stage (i.e., after sixth months). He was not, however, concerned that expectant mothers failed to acquire the book. This does not mean that all pregnant women registered their pregnancy by obtaining the book, but it reasonable to assume that the majority of pregnant women in Japan participated in the maternity book system by the 1960s.
120. The Maternal and Child Protection Law along with the 1929 Relief Law were an important basis for establishing the postwar welfare system, with the 1951 Social Welfare Law being key to its development. The government expanded health and welfare provisions for mothers and children through legislation such as the 1947 Child Welfare Law [Jidō fukushi-hō], the 1950 Public Assistance Law [Seikatsu hogo-hō], the 1964 Mother and Child and Widows Welfare Law [Boshi oyobi kafu fukushi-hō], and the 1965 Maternal and Child Health Law [Boshi hoken-hō].
121. Ōbayashi, *Josanpu no sengo*; Nakayama Makiko, "Shussan no risuku kaihi wo meguru poritikkusu," in Kawagoe and Tomobe, eds., *Seimei to iu risuku*.

122. Tama Yasuko, *"Kindai kazoku" to bodi poritikkusu* (Kyoto: Sekai Shisō-sha, 2006); Ogino, *"Kazoku Keikaku" e no michi*, 141–254; Ōbayashi, *Josanpu no sengo*, 166–224.
123. For the family planning movement in rural Japan, see Tsujimura Teruo, *Sengo shinshū josei shi* (Tokyo: Kasei Kyōiku-sha, 1966), 274–287.
124. For an excellent discussion on Japan's postwar eugenics policy, see Matsubara, "Dai go-shō, Nihon: yūsei hogo hō to iu na no danshu hō."
125. Yoko Matsubara, "The Making of the Eugenic Protection Law of 1948: Reinforcing Eugenic Policy after WWII," a paper presented at the 51st Annual Meeting of the Association for Asian Studies, Boston (March 1999), 5.
126. For a new oral history study on public health nurses from the mid- to late twentieth century, see Kimura, *Chūzai hoken-fu no jidai, 1942–1997*.

CHAPTER 7

Epilogue

Having traced the history of reproductive medicine in Japan though 1945, it is apparent that both the state and medical professionals specializing in reproductive medicine have tended to endorse pronatalism. Early modern and modern regimes in Japan by and large deemed population growth to be a positive thing. Population expansion would boost the strength of the regime and the tax base. They perceived abortion and infanticide contrary to Confucian morality, which preached human benevolence and celebrated family prosperity, and served as the ideological base of the early modern and modern state. The modern bio-power that arose and expanded in the nineteenth century and beyond often colluded with pronatalist ideology, working against the interests of women and families for whom frequent pregnancies and many children were liabilities rather than assets. In this epilogue, however, I would like to discuss how bio-power and anti-natalist ideologies interacted to prevent licensed prostitutes and people with disabilities and diseases such as Hansen's disease (leprosy) from having children.

As I have shown in the latter half of this book, pre-World War II bio-power and the accumulation of knowledge about citizens' bodies and reproduction often worked toward the improvement of reproductive health and pronatalist purposes. Bio-power functioned in the area of reproductive medicine, however, to classify human beings into different categories according to their desirability for biological reproduction. Combined with the classifications that arose with the creation of laws and

Y. Terazawa, *Knowledge, Power, and Women's Reproductive Health in Japan, 1690–1945*, Genders and Sexualities in History,
https://doi.org/10.1007/978-3-319-73084-4_7

government regulations, medical science, steeped in eugenics ideas, increasingly labeled certain people unfit for biological reproduction in light of the interests of the state, society, and the very persons who were to become the targets of anti-natalist policy. Within the networks of power and knowledge, the reproductive rights of certain groups were ignored. Even medical experts and health activists were remiss in helping licensed prostitutes and people with disabilities or chronic diseases to have decent and humane sexual and reproductive lives.

Although the Meiji state pursued pronatalist policies for the majority of women—centering its efforts on middle- and upper-class women—licensed prostitutes, were the target of different policies. The state exerted control over the health of prostitutes through licensing and mandatory bi-weekly pelvic exams. Gynecological exams were central to the government's efforts to control and prevent the spread of venereal disease (VD), and thus an important measure among state public health policies.[1]

The system of public prostitution identified the bodies of licensed prostitutes and grouped them separately from other Japanese women. Although unprotected sex was likely to lead to pregnancies and the transmission of VD, officials and health activists paid almost no attention to these problems. To prevent pregnancies, prostitutes used homemade contraceptives, but if they became pregnant they used the root of ground cherry or other traditional folk abortifacients—many of these remedies were passed on by brothel madams. Modern medical products advertised in print media began to circulate by the early twentieth century, adding to their pharmacopeia. Underground doctors who illegally carried out surgical abortions ended many pregnancies. Prostitutes' "reproductive" health did not matter. Their exposure to dangerous and ineffective abortifacients or crude abortion procedures by unqualified physicians was a problem that few writers or activists discussed when discourses on pronatalism and women's reproductive health were proliferating.

Many Japanese feminist historians see continuity between Japan's systems of licensed prostitution and so-called "comfort women"—a system of sexual slavery perpetuated by the Japanese military during the Asia–Pacific War (1931–1945). It was the government and military's attempt at controlling sex and the spread of VD, particularly in the use of bi-weekly pelvic examinations. The vast majority of women, either from the Japanese colonies or occupied territories, were mobilized as military sex slaves by deception and force. They were frequently brutally beaten and at times served 40 to 50 soldiers per day. Japanese military bases in Southeast Asia that included a "comfort station" were targets of Allied bombing, resulting in the death of sex slaves incarcerated at the facilities. Although the situation under which "comfort women" worked was unique and particularly severe and brutal, they were subject to bi-weekly gynecological examinations, a practice that was inherited from Japan's licensed prostitution system. Soldiers who used comfort women facilities were told to use condoms, and in some cases, they did, but often they ignored orders, which resulted in pregnancies.

A Korean comfort woman, Song Shindo (1922–2017), testified that she became pregnant several times. One of the pregnancies resulted in a stillbirth. Her babies were adopted by local families.[2] After the end of the war, Song migrated to Japan, where she has been living ever since. In the 1980s and 1990s, after Japan normalized diplomatic relations with the People's Republic of China, the Japanese government investigated Japanese children (and adults) who had been left in China (especially in the northeast, where many Japanese settlers resided) after the war. Over 2400 children (they were already adults by the 1980s) and more than 3500 Japanese women returned to Japan. Whenever Song watched the news about the children who had been left behind in China, she felt tremendous guilt for abandoning her children and wondered if they were among those Japanese left in China who she saw on the television.[3] In Chap. 6, I discuss pronatalist government policies during the war years that celebrated childbirth and educated women to be good mothers. During the same period, comfort-women pregnancies and births were cursed rather than welcomed. Song, who did not bear or adopt a child after working as a comfort woman, felt intense guilt that she left her children in China, even though this was the only choice she had. It led her to strongly desire another child—longing for motherhood—later in life.

Others whose biological reproduction was suppressed included the diseased and disabled. In her informative article, Yoko Matsubara notes that eugenics emerged when scientists and the general public began to cast doubts on the basic social Darwinist assumption that human society would continue progressing without artificial interventions if society simply allowed its weaker members, in a natural selection-like process, to perish. Social Darwinists and eugenicists agreed that the theory of evolution applied to social progress. Eugenicists, however, emphasized that artificial selection could improve human society. In Japan, Herbert Spencer's ideas made social Darwinism highly influential in the 1880s and 1890s. By the turn of the twentieth century, European-educated Japanese scientists had already introduced eugenics in *Jinsei* [*Human Nature*] and other journals, and in the 1920s and 1930s, concerned officials and scientists formed an influential eugenics movement, leading to the establishment of the National Eugenics Law (1940) and Eugenic Protection Law (1948). The Japanese government continued to mention eugenics in its policies to improve the quality of human resources through the mid-1970s. Under eugenics-oriented policies, many sufferers of Hansen's disease and people with disabilities faced forced sterilization and abortions.[4]

Japan uniquely subjected Hansen's disease patients to voluntary or involuntary sterilization and abortions despite the Norwegian physician Gerhard Armauer Hansen's discovery of the leprosy bacteria, in 1873, which confirmed that the disease is an infectious and not a hereditary one. The encouragement of leprosy patient sterilization and abortions continued until the mid-twentieth century. The government recognized leprosy as a serious public health threat and began to pursue a policy of segregating leprosy patients in sanatoriums in the first decade of the twentieth century, after the West criticized Japan for leaving Hansen's disease patients begging in streets without government assistance. At one sanatorium, Director Mitsuda Kensuke (1876–1964), a passionate advocate of segregation, began sterilization operations on volunteers in 1915. He cited the financial and practical difficulties of raising children at the facility. At many leprosy sanatoriums, sterilization was required before residents could marry, and eventually became a compulsory procedure for all inmates.[5]

To the disappointment of Mitsuda and others who endorsed sterilization, medical experts asserted leprosy's infectious etiology, and the 1940 National Eugenics Law rejected mandatory sterilization. Leprosy patients, however, became targets of sterilization under the 1948 Eugenic Protection Law, which stipulated sterilization and abortion were allowed in cases of disease, poverty, or large families in which an unfavorable environment would cause children to grow up weak, sickly, or inferior. Sterilization and abortion were approved for leprosy patients and their spouses to prevent the transmission of the disease to potential children. Between 1949 and 1979, 301 male leprosy patients had vasectomies, far fewer than the 1003 who had been sterilized before the promulgation of the Eugenic Protection Law. On the other hand, because the Eugenic Protection Law permitted such procedures for female patients, the number of sterilizations and abortions among female patients skyrocketed to 1171 and 7629, respectively. In 1998, Hansen's disease patients in Japan began filing lawsuits against the government, challenging the constitutionality of the 1953 Leprosy

Protection Law that legislated compulsory segregation of leprosy patients. In 2001, the plaintiffs were vindicated in all aspects of the law.[6] With the ruling and other trends that recognize the importance of human rights, increasing attention has been paid to the violation of the reproductive rights of those who had experienced discrimination, including leprosy patients.

For the same reasons that the Eugenic Protection Law discouraged the biological reproduction of leprosy patients, it supported sterilization and abortions for people with disabilities. A 1951 Eugenic Protection Law revision targeted people with mental illness and diminished mental ability for sterilization and abortions at a time when anti-natalist policies were aggressively pursued. In 1952, the regulation was extended to include spouses of those who suffered from mental illness or disability, including those whose illness or disability were not hereditary. The following year the Ministry of Health and Welfare issued an edict that physicians could carry out sterilizations and abortion procedures against the will of patients with mental illness or disabilities. For compulsory operations, physicians and hospital workers could use physical restraint, anesthesia, and deception.[7] Each year between 1949 and the early 1970s, several hundred abortions were carried out without consent. During the peak years between 1954 and 1960, more than 1000 women went through involuntary abortions each year. Even though compulsory abortions declined dramatically in the 1980s, the Eugenic Protection Law remained in effect until 1996, when it was replaced by the Maternal Protection Law.[8]

A focal point of contemporary abortion debates in Japan is aborting fetuses with defects, based on prenatal testing. Disability rights advocates have raised concerns that the rise in abortions of fetuses with defects will ultimately lead to heightened discrimination against people with disabilities. As a result, they endorse policies for tighter government regulations on abortion. On the other hand, feminists fear that increased state intervention in the issue will impinge on women's right to abortion. The debate is being carried out at a time when the Japanese government wishes to pursue renewed pronatalist policies to address the nation's declining population and when right-wing ideologues, working closely with the Liberal Democratic Party leadership, have launched a backlash against feminism, hoping to bring back the idea of women's primary roles as mother and wife.

In this short epilogue, I want to show how the government of modern Japan denied particular groups the right to have children and proper reproductive health care despite favoring pronatalism and otherwise seeking improved maternal and infant health care. Fundamental to denying reproductive rights was classifying people into different categories: some women's reproduction was highly valued and privileged while the reproduction of sex workers or people with diseases and disabilities was not. It is not that the state ignored these marginal women. Rather, it attempted to control them by mandatory pelvic examinations for licensed prostitutes and the institutionalization of lepers and people with mental illness. This was achieved through the state and medical professionals working closely together—a distinctly "modern" phenomenon, as I argue in this book. The modern bio-power that is central to the classifying of citizens and the medical management of their bodies often works in a way to benefit the interests of the state and organized medicine but not those of women, at times creating egregious oppression and discrimination. However, there were many instances where women resisted, and women activists gained victories—even if they were partial and limited—utilizing their medical or social scientific knowledge. I hope this book will help feminist activists strategize and navigate within the networks of bio-power in innovative and flexible ways, finding new allies and approaches.

Notes

1. Hayashi Yoko discusses licensed prostitution in light of the modern Japanese government's public health efforts. Hayashi Yoko, *Sei wo kanri suru teikoku: kōshōseido ka no "eisei" mondai to haishō undō* (Osaka: Osaka Daigaku Shuppankai, 2017).
2. Zainichi no Ianfu Saiban wo Sasaeru Kai, ed., *Ore no kokoro wa makete nai* (Tokyo: Kinohana sha, 2007): 160–165.
3. Ibid.
4. Matsubara Yoko, "Shōgaisha kara umu koto wo ubatta kyōsei funin shujutsu," *Otokomo onnamo*, Fall and Winter Issue, No. 128 (2016): 29–34.
5. Yamashita Tomoko, "Hansen byō wo meguru danshu ni tsuite," *Seimei no rinri Vol. 3: Yūsei seisaku no keifu* (Fukuoka: Kyūshū Daigaku Shuppankai, 2013): 323–345.
6. Ibid., 345–351.
7. Ishimura Kumiko, "Chūzetsu kisei to yūsei shisō," *Ningen bunka gaku kenkyū shūroku* (November 2001): 29.
8. Matsubara, "Shōgaisha kara umu koto wo ubatta kyōsei funin shujutsu," 30–31.

Select Bibliography

Select Primary Sources in Japanese Language

Journals

Aiiku shinbun (1938–1944).
Josan no shiori (1897–1904).
Fujin eisei zasshi (1898–1907).
Fujin kōron (1921).
Fujokai (1928–1933).
Fujo shinbun (1901–1903, 1926–1941).
Shakai jigyō kenkyū (1934).

Books

Akagi, Asaji. *Nihon no hoken-fu*. Tokyo: Tokuwa Shobō, 1943.

Fukui, Shigeko. *Ninpu no barazono*. Osaka: Ogata Byōin Josanpu Gakkai, marketed by Maruzen, 1898.

Fujikawa, Yū, Kure Shūzō, and Masuda Tomomasa, eds. *Nihon sanka sōsho*. Tokyo: Matsuzaki Tomekichi, 1895. Reprinted Version, Kyoto: Shibunkaku Shuppan, 1971.

Fujimura, Takahide, ed. *Ōsaka sanbakai kaisoku*. Osaka: Ōsaka Sanbakai Honbu, 1899.

Y. Terazawa, *Knowledge, Power, and Women's Reproductive Health in Japan, 1690–1945*, Genders and Sexualities in History,
https://doi.org/10.1007/978-3-319-73084-4

Inokuchi, Aguri, Kaji Toku, Kawase Motokurō, Takashima Heizaburō, and Tsuboi Gendō. *Taiiku no riron to jissai*. Tokyo: Kokkōsha, 1906.

Kagawa, Genteki. *Sanron yoku*. 1775. Reprinted in Sanka Bunken Dokushokai, ed., *Heiseiban sanron, sanron-yoku*. Tokyo: Iwata Shoin, 2008.

Kagawa, Gen'etsu. *Sanron*. 1765. Reprinted in Sanka Bunken Dokushokai, ed., *Heiseiban sanron, sanro- yoku*, op. cit.

Kagawa, Mansai. *Sanka kikai yōhō-sho*. 1891. Reprinted in Fujikawa et al. eds., *Nihon sanka sōsho*, op. cit.

Katsuki, Gyūzan. 1692. *Fujin kotobukigusa*. Reprinted in Fujikawa et al. eds., *Nihon sanka sōsho*, op. cit.

Kawamura, Hyōtarō. *Nōson hoken-fu*. Tokyo: Sangabō, 1942.

Komiyama, Shin'ichi. *Hoken-fu dokuhon*. Tokyo: Kyōdō Kōsha, 1943.

Kondō, Shōichi. *Joshi shokugyō annai*. Tokyo: Hakubunkan, 1906.

Matsushita, Sekijin. *Sanshū okugōri san'iku fūzoku zue*. Nagoya: Seibunkan Shoten, 1937.

Naruse, Jinzō. *Joshi kyōiku*. Tokyo: Aoki Kozandō, 1896. Reprinted in Nakajima Kuni, ed., *Kindai nihon joshi kyōiku bunken shū, dai 1 ki, Vol. 5*. Tokyo: Nihon Tosho Sentā, 1983.

Oku, Ressai. *Kaisei kōhō hiketsu*. Reprinted in Fujikawa et al. eds., *Nihon sanka sōsho*, op. cit.

Sanka Bunken Dokushokai, ed. *Heiseiban sanron, sanron-yoku*. Tokyo: Iwata Shoin, 2008.

Sasakawa, Misu. *Sanba jūsan kai*. Niigata: Sasakawa Misu, 1892.

Shakai-kyoku Shakai-bu (Naimu-shō Shakai-kyoku), ed. *Boshi hogo-hō tō no setsumei*. Tokyo: Shakai-bu Shakai-kyoku, 1937.

Shimoda, Jirō. *Joshi kyōiku*. Tokyo: Kinkōdō Shoseki, 1904. Reprinted in Nakajima Kuni, ed., *Kindai nihon joshi kyōiku bunken shū, dai 1 ki, Vol. 7*. Tokyo: Nihon Tosho Sentā, 1983.

Shirai, Kikurō. *Joshi undō to yūgi*. Tokyo: Kōdōkan, 1909.

Terajima, Ryōan. *Wa-kan sansai zue*. Translated into modern Japanese by Shimada Isao, Takeshita Atsuo, Higuchi Motomi. Tokyo: Heibonsha, Tōyō bunko series, 1985–1991.

Tōjō, Ryōtarō and Doi Mamoru. *Shinsen sanba gaku*. Tokyo: Maruzen, 1906.

Yamada, Hisao. *Uka hatsumō zukai*. 1852. Compiled in Fujikawa et al. eds., *Nihon sanka Sōsho*, op. cit.

Yosano, Akiko. *Ichigū yori*. Tokyo: Kanao Buneidō, 1911.

Yoshioka, Yayoi. *Watashi no jikken shitaru anzan to ikuji*. Tokyo: Teibi Shuppansha, 1921.

Select General Sources

Achiwa, Gorō. *Kindai nihon no igaku: seiyō igaku juyō no kiseki*. Kyoto: Shibunkaku Shuppan, 1982.

Akamatsu, Keisuke. *Hijōmin no minzoku bunka*. Tokyo: Akashi Shoten, 1986.

———. *Hijōmin no sei minzoku*. Tokyo: Akashi Shoten, 1991.

Aoki, Hidetora. *Ōsaka-shi sanba dantai shi*. Osaka: Ōsaka-shi Sanba Kai, 1935.

Aoyagi, Seiichi. *Shinryō hōshū no rekishi*. Kyoto: Shibunkaku Shuppan, 1996.

———. *Kindai iryō no akebono: Bakumatsu meiji no iji seido*. Kyoto: Shibunkaku Shuppan, 2011.

Bartholomew, James R. *The Formation of Science in Japan*. New Heaven and London: Yale University Press, 1989.

Bartholomew, James R. and Sitcawitch, Sumiko Otsubo. "Eugenics in Japan: Some Ironies of Modernity, 1883–1945," *Science in Context* 11.3 and 4, 1998.

Befu, Harumi. "Corporate Emphasis and Patterns of Descent in the Japanese Family," in Robert J. Smith and Richard K. Beardsley, eds., *Japanese Culture: Its Development and Characteristics*. Chicago: Aldine, 1962: 34–41.

Bernstein, Gail, ed. *Recreating Japanese Women, 1600–1945*. Berkeley: University of California Press, 1991.

Berry, Mary Elizabeth. *Japan in Print: Information and Nation in the Early Modern Period*. Berkeley: University of California Press, 2006.

Bordo, Susan. "The Body and Reproduction of Femininity: A Feminist Appropriation of Foucault" in Jaggar, Alison and Bordo, Susan, eds., *Gender/Body/Knowledge: Feminist Reconstructions of Being and Knowing*. New Brunswick, NJ: Rutgers University Press, 1989.

Bourdieu, Pierre. *Distinction: A Social Critique of the Judgement of Taste*. Cambridge, MA: Harvard University Press, 1984.

———. *The Logic of Practice*. Stanford, CA: Stanford University Press, 1990.

Bowers, John Z. *Medical Education in Japan*. New York: Harper & Row, 1965.

Bray, Francesca. *Technology and Gender: Fabrics of Power in Late Imperial China*. Berkeley: University of California Press, 1997.

Burns, Susan. "Kenryoku, chi, saiseisuru shintai," *Misuzu*, No. 368 (November 1991): 1–21.

———. "Contemplating Places: The Hospital as Modern Experience in Meiji Japan," in Helen Hardacre and Adam L. Kern, eds. *New Directions in the Study of Meiji Japan*. Leiden, New York, and Köln: Brill, 1997: 702–718.

———. "Bodies and Borders: Syphilis, Prostitution and the Nation in Japan, 1860–1890," *U.S. Japan Women's Journal, English Supplement*, No. 15 (1998): 3–30.

———. "Constructing the National Body: Public Health and the Nation in Nineteenth-Century Japan," in Brook, Brook and Schmid, Andre, eds., *Nation*

Work: Asian Elites and National Identities. Ann Arbor: University of Michigan Press, 2000: 17–49.

———. "The body as text: Confucianism, reproduction, and gender in Tokugawa Japan," in Elman, Benjamin, Duncan, John, Ooms, Herman, eds., *Rethinking Confucianism: Past and Present in China, Japan, Korea, and Vietnam*. Los Angeles: University of California, UCLA Asian Pacific monograph series, 2002: 178–219.

———. "When Abortion Became a Crime: Abortion, Infanticide, and the Law in Early Meiji Japan," in Howell, David and Baxter, James, eds., *History of Folklore Studies in Japan*. Vol. 127. Kyoto: International Research Center for Japanese Studies, 2006: 37–55.

Butler, Judith. *Gender Trouble: Feminism and the Subversion of Identity*. New York and London: Routledge, 1990.

———. *Bodies that Matter: On the Discursive Limits of "Sex."* New York and London: Routledge, 1993.

Chiba, Tokuji and Ōtsu Tadao. *Mabiki to mizuko: kosodate no fōkuroa*. Tokyo: Nōsanson Bunka Kyōkai, 1983.

Chung, Yuehtsen. *Struggle for National Survival: Eugenics in Sino-Japanese Contexts, 1896–1945*. London and New York: Routledge, 2002.

Cornell, Laurel L. "Retirement, Inheritance, and Integrational Conflict in Preindustrial Japan," *Journal of Family History* 8 (Spring 1983): 55–69.

———. "Infanticide in Early Modern Japan? Demography, Culture, and Population Growth," *Journal of Asian Studies* 55:1 (February 1996): 22–50.

Cornell, L. L. and Hayami, Akira. "The *Shūmon Aratame Chō*: Japan's Population Registers," *Journal of Family History* 11:4 (1986): 311–328.

Davis, Ann-Marie Lynne. *Bodies, Numbers, and Empires: Representing "the Prostitute" in Modern Japan, 1850–1912*. Ph.D. Dissertation, University of California, Los Angeles, 2009.

Dreyfus, Hubert S. and Rabinow, Paul. *Michel Foucault: Beyond Structuralism and Hermeneutics*. Chicago: The University of Chicago Press, 1982.

Drixler, Fabian. *Mabiki: Infanticide and Population Growth in Eastern Japan, 1660–1950*. Berkeley: University of California Press, 2013.

Duden, Barbara. *The Woman Beneath the Skin: A Doctor's Patients in Eighteenth-Century Germany*. Cambridge, Mass.: Harvard University Press, 1991.

Dureau, Christine and Low, Morris. "The Politics of Knowledge: Science, Race and Evolution in Asia and the Pacific," *History and Anthropology* 11 (2–3), 1999: 131–156.

Ellis, Havelock. *Man and Woman: A Study of Human Secondary Sexual Characters*. 3rd ed. London: W. Scott Pub. co.; New York: Scribner, 1902.

Fehling, H. *Die Bestimmung der Frau: ihre Stellung zu Familie und Beruf*. Stuttgart: F. Enke, 1892.

Foucault, Michel. *Archeology of Knowledge*. New York: Harper Colophon, 1972.

———. *The Birth of the Clinic: An Archaeology of Medical Perception*. New York: Pantheon Books, 1973.
———. *The Order of Things: An Archaeology of the Human Sciences*. New York: Vintage/Random House, 1973.
———. Gordon, Colin, ed. *Power/Knowledge: Selected Interviews and Other Writings, 1972–1977*. New York: Pantheon Books, 1977.
———. *The History of Sexuality, Vol. I*. New York: Pantheon, 1979.
———. *Discipline and Punish*. New York: Vintage, 1979.
Frederick, Sarah. *Turning Pages: Reading and Writing Women's Magazines in Interwar Japan*. Honolulu: University of Hawai'i Press, 2006.
Frühstück, Sabine. *Colonizing Sex: Sexology and Social Control in Modern Japan*. Berkeley: University of California Press, 2003.
Fuess, Harald. "Informal Imperialism and the 1879 'Hesperia' Incident: Containing Cholera and Challenging Extraterritoriality in Japan," *Japan Review*, No. 27 (2014): 103–140.
Fujikawa, Yū. *Nihon igakushi*. Tokyo: Shinrisha, 1941.
Fujime, Yuki. "Aru sanba no kiseki: Shibahara Urako to sanji chōsetsu," *Nihonshi kenkyū*, No. 366 (February 1992): 90–112.
———. "One Midwife's Life: Shibahara Urako, Birth Control, and Early Shôwa Reproductive Activism," in Wakita et al., eds., *Gender and Japanese History, Vol. 1*: 297–325.
———. *Sei no rekishigaku*. Tokyo: Fuji Shuppan, 1997.
Fujino, Yutaka. *Nihon fashizumu to yūsei shisō*. Kyoto: Kamogawa Shuppan, 1998.
———. "'Mui-son' mondai no tōjō," *Toyama kokusai daigaku jinnbun shakai gakubu kiyō* 2 (2002): 103–114.
Fujita, Makoto. *Osan Kakumei*. Tokyo: Asahi Shinbunsha, 1988.
Furth, Charlotte. "Concepts of Pregnancy, Childbirth, and Infancy in Qing Dynasty China," *Journal of Asian Studies*, Vol. 46, No. 1 (February 1987): 7–35.
———. "From Birth to Birth: the Growing Body in Chinese Medicine," in Kinney, Anne Behnke, ed., *Chinese Views of Childhood*. Honolulu: University of Hawai'i Press, 1995: 157–191.
———. *A Flourishing Yin: Gender in China's Medical History, 960–1665*. Berkeley: University of California Press, 1999.
Fushimi, Yūko. *Kindai nihon ni okeru shussan to ubuya*. Tokyo: Keisō Shobō, 2016.
Gallagher, Catherine and Thomas Laqueur, eds. *The Making of the Modern Body: Sexuality and Society in the Nineteenth Century*. Berkeley: University of California Press, 1987.
Garon, Sheldon. *Molding Japanese Minds: The State in Everyday Life*. Princeton, NJ: Princeton University Press, 1997.
Gould, Stephen Jay. *The Mismeasure of Man*. New York: Norton, 1996.

Gramlich-Oka, Bettina. "The Body Economic: Japan's Cholera Epidemic of 1858 in Popular Discourse," *East Asian Science, Technology, and Medicine*, No. 30, (2009): 32–73.
Harada, Tomohiko and Tanaka Yoshio, eds. *Tōhoku hokuetsu hisabetsu burakushi kenkyū*. Tokyo: Akashi Shoten, 1981.
Hardacre, Helen. *Marketing the Menacing Fetus in Japan*. Berkeley: University of California Press, 1977.
Hashimoto, Akihiko. *Edo bakufu shiken seidoshi no kenkyū*. Tokyo: Kazama Shobō, 1993.
Hastings, Sally. "The Empress' New Clothes and Japanese Women, 1868–1912," *The Historian*, Vol. 55, No. 4 (Summer 1993): 677–692.
Hattori, Toshirō. *Edo jidai igakushi no kenkyū*. Tokyo: Yoshikawa Kōbunkan, 1978.
Hayakawa, Noriyo. *Kindai tennōsei kokka to jendā*. Tokyo: Aoki Shoten, 1998.
———. "Taishō-ki fujin kaihō shisō no kakki-ten wo saguru," *Jinmin no rekishigaku*, 61 (November 1979): 1–10.
Hayami, Tōru. *Rekishi jinkō gaku no sekai*. Tokyo: Iwanami Shoten, 1997.
Hayashi, Reiko, ed. *Nihon no kinsei, Vol. 15: Josei no kinsei*. Tokyo: Chūōkōron, 1993.
Hayashi, Yōko. *Sei wo kanri suru teikoku: Kōshōseido ka no "eisei" mondai to haishō undō*. Osaka: Ōsaka Daigaku Shuppankai, 2017.
Hino, Shūitsu. "Senji-ka no kenkō to hoken seisaku," in Tōkyō Rekishi Kagaku Kenkyū-kai Fujin Undō-shi Bukai, ed., *On'na to sensō*. Tokyo: Shōwa Shuppan, 1991: 29–32.
Homei, Aya. "'Seijō san' to 'Kindai eisei': Kindai sanba no senmon bunya wo meguru poritikkusu," *Seibutsugaku shi kenkyū*, No. 70, (December 2002): 1–16.
———. "Birth Attendants in Meiji Japan: The Rise of the Biomedical Birth Model and a New Division of Labour," *Social History of Medicine* 19, No. 3 (2006): 407–424.
———. "Midwives and Medical Marketplace in Modern Japan," *Japanese Studies* 32, No. 2 (2012): 275–293.
Honda, Masuko. *Jogakusei no keifu: saishoku sareru meiji*. Tokyo: Seidosha, 1990.
Howell, David L. *Geographies of Identity in Nineteenth-Century Japan*. Berkeley: University of California Press, 2005.
Ichikawa, Fusae, ed. *Nihon fujin mondai shūsei 2*. Tokyo: Domesu Shuppan, 1977.
Imai, Konomi. *Shakai fukushi shisō to shite no bosei hogo ronsō*. Tokyo: Domesu Shuppan, 2005.
Imamura, Yoshio. *Nihon Taiiku shi*. Tokyo: Fumaidō Shuppan, 1970.
Ishimoto, Shidzue. *Facing Two Ways: The Story of My Life*. Stanford, CA: Stanford University Press, 1984.

Ishizaki, Shōko. "Principles of Procreation and the Family in Modern Japan: Factors behind Decisions on Family Size," in Tomida, Hiroko and Daniels, Gordon, eds., *Japanese Women: Emerging from Subservience, 1868–1945*, Folkestone, Kent, GB: Global Oriental, 2005: 278–306.

———. "Nihon no dataizai no seiritsu," *Rekishi hyōron*, No. 571 (November 1997): 53–70.

———. "Meiji-ki no seishoku wo meguru kokka seisaku," *Rekishi hyōron*, No. 600 (April 2000): 39–53.

———. "Hoken-fu e no michi wo kirihiraita Fujima Asayo," in Shinjuku-ku Chiiki Josei-shi Hensan Iinkai, ed., *Shinjuku ni ikita josei tachi* 3. Tokyo: Shinjuku-ku Chiiki Josei-shi Hensan Iinkai, 1996: 101–106.

Jannetta, Ann Bowman. *Epidemics and Mortality in Early Modern Japan*. Princeton, NJ: Princeton University Press, 1987.

Johnston, William. "Jūhasseiki Nihon no igaku ni okeru kagaku kakumei: ranpō no hatten no tame no shisōteki na zentei," *Nihon igaku zasshi* 27, No. 1 (1981): 6–20.

Jordanova, Ludmilla. *Sexual Visions: Images of Gender in Science and Medicine between the Eighteenth and Twentieth Centuries*. Madison: The University of Wisconsin Press, 1989.

Joseishi Sōgō Kenkyūkai, ed. *Nihon joseishi*, vol. 3–4. Tokyo: Tōkyō Daigaku Shuppankai, 1982.

———, ed. *Nihon josei seikatsu shi*, Vol. 3–4. Tokyo: Tōkyō Daigaku Shppankai, 1990.

Kaiho, Yōko. "Kinsei ishikari basho hōsō ryūkōshi nōto: Bunka 14-nen wo chūshin ni," *Sapporo no reikishi*, No. 53 (August 2007): 22–33.

Kamata, Hisako, Miyasato Kazuko, Suganuma Hiroko, Furukawa Yūko, and Sakakura Yoshio. *Nihon-jin no koumi, kosodate*. Tokyo: Keisō Shobō, 1990.

Kanazu, Hidemi. "'Datai mabiki' kō," in Kinugasa Yasunobu, ed., *Kinsei shisō shi no genzai*. Kyoto: Shibunkaku Shuppan, 1995: 401–421.

———. "'Nihon sankagaku' no seiritsu," *Edo no shisō, Vol. 6*. Tokyo: Perikansha, 1997: 63–78.

Kanzaki, Nobutake. *Kikigaki yūkaku Narikomaya: Fushigina basho no fōkuroa*. Tokyo: Kōdansha, 1989.

Kaptchuk, Ted J. *The Web That Has No Weaver*. New York: Congdon & Weed, 1983.

Kasahara, Hidehiko and Kojima Kazutaka. *Meiji ki iryō eisei gyōsei no kenkyū*. Kyoto: Mineruva Shobō, 2011.

Kawagoe, Osamu and Tomobe Ken'ichi, eds. *Seimei to iu risuku*. Tokyo: Hōsei Daigaku Shuppankai, 2008.

Kawamura, Kunimitsu. *Otome no inori: Kindai josei imêji no tanjō*. Tokyo: Kinokuniya Shoten, 1993.

———. *Otome no shintai: On'na no kindai to sekushuaritī*. Tokyo: Kinokuniya Shoten, 1994.

Khun, Thomas. *The Structure of Scientific Revolutions*. Chicago: The University of Chicago Press, 1962.

Kikuchi, Isao. "Hōsō ryūkō to ainu shakai: Jūkyūseiki zenki no jinmei sōshitsu to ezochi kaihatsu," *Rekishi kagaku*, No. 171 (December 2002): 1–11.

Kimura, Naoko. *Shussan to seishoku wo meguru kōbō: sanba josanpu dantai to sankai no Hyakunen*. Tokyo: Ōtsuki Shoten, 2013.

Kimura, Tetsuya. *Chūzai hoken-fu no jidai, 1942–1997*. Tokyo: Igaku Shoin, 2012.

Kitaku Sōmubu Josei Seisaku-ka, ed. *Senjika ni kurashita josei tachi*. Tokyo: Domesu Shuppan, 1997.

Kizu, Yuzuru. *Nyonin kinsei: gendai kegare kiyome kō*. Osaka: Kaihō Shuppansha, 1993.

Koizumi, Eiichi. *Dataizai kenkyū*. Tokyo: Ganshōdō Shoten, 1934.

Kōseishō Imukyoku. *Isei hachijū-nen shi*. Tokyo: Insatsukyoku Chōyōkai, 1955.

Koshiba, Masako. "Boshi kenkō techō," *Joseishi kenkyū*, No. 23, '88-XII (December 1988): 36–37.

Kouchi, Nobuko. *Shiryō boshi hogo ronsō*. Tokyo: Domesu Shuppan, 1984.

Kōzai, Toyoko. "Ainu wa naze 'Yamani ni nigeta' ka?: Bakumatsu ezochi ni okeru 'waga kuni saisho no kyōsei shutō' no okuyuki," *Shisō*, No. 1017 (January, 2009): 78–101.

———. "Yobō sesshu to iu 'eisei': shutō no rekishi no hanshō kara," *Gendai shiso*, 38:3 (March 2010): 216–231.

Kuriyama, Shigehisa. "Between Mind and Eye: Japanese Anatomy in the Eighteenth Century," in Charles Leslie and Young, Allan, eds., *Paths to Asian Medical Knowledge*. Berkeley: University of California Press, 1992: 21–43.

Kyōtofu Ishikai, ed., *Kyōto no igakushi*. Kyoto: Shibunkaku Shuppan, 1980.

LaFleur, William R. *Liquid Life: Abortion and Buddhism in Japan*. Princeton: Princeton University Press, 1992.

Laqueur, Thomas. *Making Sex: Body and Gender from the Greeks to Freud*. Cambridge, MA: Harvard University Press, 1990.

Leavitt, Judith Walzer, ed. *Women and Health in America*. Madison: University of Wisconsin Press, 1984.

Liu, Yanchi. *The Essential Book of Traditional Medicine, Vol. 1*. New York: Columbia University Press, 1988.

Lotze, Hermann. *Microcosmus: An Essay Concerning Man and his Relation to the World*. 2nd ed. Translated from the German by Elizabeth Hamilton and E. E. Constance Jones. Edinburgh: T. & T. Clark, 1887.

Low, Morris. "Medical Representations of the Body in Japan: Gender, Class, and Discourse in the Eighteenth Century," *Annals of Science*, Vol. 53, No. 4 (July 1996): 345–359.

———, ed. *Building a Modern Japan: Science, Technology, and Medicine in the Meiji Era and Beyond.* New York: Palgrave Macmillan, 2005.

Lowy, Dena. *The Japanese "New Woman": Images of Gender and Modernity.* New Brunswick, NJ: Rutgers University Press, 2007.

Maruyama, Masao. *Nihon seiji shisōshi kenkyū.* Tokyo: Tōkyō Daigaku Shuppankai, 1952.

———. *Studies in the Intellectual History of Tokugawa Japan.* Translated by Hane, Mikiso. Princeton: Princeton University Press, 1974.

Matsubara, Yōko. "Minzoku yūsei hogo to Nihon no yūsei-hō no keifu," *Kagakushi kenkyū*, No. 201 (1997): 42–50.

———. "Dai go-shō, nihon: sengo no yūsei hogo hō to iu na no danshu hō" in Yonemoto Shōhei et al., eds., *Yūsei gaku to ningen shakai: seimei kagaku no seiki wa doko e mukau no ka.* Tokyo: Kōdansha, 2000: 181–182.

Mitchell, Timothy P. *Colonising Egypt.* Cambridge, UK: Cambridge University Press, 1988.

———, ed. *Questions of Modernity.* Minneapolis: University of Minnesota Press, 2000.

Miyaji, Naoko T., and Lock, Margaret. "Monitoring Motherhood: Sociocultural and Historical Aspects of Maternal and Child Health in Japan," *Daedalus*, 123:4, Health and Wealth (Fall 1994): 87–112.

Miyake, Yoshiko. "Doubling Expectations: Motherhood and Women's Factory Work Under State Management in Japan in the 1930s and 1940s" in Bernstein, ed., *Recreating Japanese Women, 1600–1945*: 267–295.

Miyasaka, Yasuko. "Aru sanba no seikatsushi: kekkon, kosodate to hatarakukoto," *Ningen Bunka Kenkyu Nenpo* 13. Tokyo: Ochanomizu Joshi Daigaku Ningen Bunka Kenkyū Sentā, 1989.

Molony, Barbara. "Gender, Citizenship, and Dress in Modernizing Japan," in Roces, Mina and Edwards, Louise, eds., *The Politics of Dress in Asia and the Americas.* Brighton, GB: Sussex Academic Press, 2007: 81–100.

Murakami, Nobuhiko. *Meiji joseishi: On'na no shokugyō, chūkan-kōhen.* Tokyo: Rironsha, 1971.

Nagatomo, Chiyoharu. *Kinsei kushihonya no kenkyū.* Tokyo: Tōkyodō Shuppan, 1982.

Nakamura, Ellen Gardner. *Practical Pursuits: Takano Chōei, Takahashi Keisaku, and Western Medicine in Nineteenth-Century Japan.* Cambridge, MA: Harvard University Asia Center, distributed by Harvard University Press, 2005.

Nakano, Yūko. "Women and Buddhism: Blood Impurity and Motherhood," in Okuda Akiko and Okano Haruko, eds., Watts, Alison, trans., *Women and Religion in Japan.* Göttingen: Harrassowitz Verlag, Wiesbaden, 1998.

Nakayama, Makiko. *Shintai wo meguru seisaku to kojin: Boshi hoken sentā jigyō no kenkyū.* Tokyo: Keisō Shobō, 2001.

Namihira, Emiko. *Byōki to chiryō no bunka jinruigaku*. Tokyo: Kaimeisha, 1984.

———. *Kegare*. Tokyo: Tōkyōdō Shuppan, 1985.

———. "Pollution in the Folk Belief System," *Current Anthropology*, 28:4 (1987): 65–74.

Nankin chūigakuin, ed., and Chūigaku gairon hōyaku iinkai, trans. *Chūgoku kanpō igaku gairon*. Tokyo: Chūgoku Kanpō, 1965.

Narita, Ryūichi. "Eisei kankyō no henka no nakano josei to josei kan," in Joseishi Sogō Kenkyūkai, ed., *Nihon josei seikatsu shi*, Vol. 4. Tokyo: Tōkyō Daigaku Shppankai, 1990: 89–124.

———. "Mobilized from Within: Women and Hygiene in Modern Japan" in Tonomura, Walthall, and Wakita, eds., *Women and Class in Japanese History*: 257–273.

Nihon Gakushi-in, ed. *Meiji-zen nihon igakushi, Vol. 1–5*. Tokyo: Nihon Gakujutsu Shinkōkai, 1955–1964.

Nishikawa, Mugiko. *Aru kindai sanba no monogatari*. Toyama: Katsura Shobō, 1997.

———. "The Modern Midwife (sanba) and the Transformation of Childbirth at the Local Level in Japan," *U.S. Japan Women's Journal*, No. 24 (2003): 82–101.

Norgren, Tiana. *Abortion Before Birth Control: The Politics of Reproduction in Postwar Japan*. Princeton, NJ: Princeton University Press, 2001.

Ōbayashi, Michiko. *Josanpu no sengo*. Tokyo: Keisō Shobō, 1989.

Oberländer, Christian. "The Rise of Western 'Scientific Medicine' in Japan: Bacteriology and Beriberi," in Low, ed., *Building a Modern Japan*: 13–36.

Ochiai, Emiko. "Edo jidai no shussan kakumei," in Ochiai, Emiko, *Kindai kazoku to feminizumu*. Tokyo: Keisō Shobō, 1989: 56–78.

———. "Aru sanba no nihon kindai: Raifu hisutorī kara shakaishi e" in Ogino et al., *Seido to shiteno on'na*: 258–322.

———. "Kinsei matsu ni okeru mabiki to shussan," in Wakita Haruko and Susan Hanley, eds., *Jendā no nihonshi, Vol. 1*. Tokyo: Tokyō Daigaku Shuppankai, 1994: 425–459.

———. "The Reproductive Revolution at the end of the Tokugawa period," in Tonomura, Walthall, and Wakita, eds., *Women and Class in Japanese History*: 187–215.

Ogata, Masakiyo. *Nihon sankagaku shi*. Tokyo: Maruzen, 1918, reprinted, Tokyo: Kagaku Shoin, 1980.

Ogawa, Keiko. "Seiyō kindai igaku no dōnyū to sanba no yōsei," in Shirai, ed., *Umi sodate to josan no rekishi*: 19–67.

Ogino, Miho, Tanabe Rekiko, Himeoka Toshiko, Chimoto Akiko, Hasegawa Hiroko, and Ochiai Emiko. *Seido to shiteno on'na*. Tokyo: Heibonsha, 1990.

Ogino, Miho. *"Kazoku keikaku" e no michi: kindai nihon no seishoku wo meguru seiji*. Tokyo: Iwanami Shoten, 2008.

Okitsu, Kaname. *Edo senryū on'na hyakkei*. Tokyo: Jiji Tsūshin sha, 1994.

Ōkuni, Michiko. *Hoken-fu no rekishi*. Tokyo: Igaku Shoin, 1973.

Onshi Zaidan Boshi Aiikukai, ed. *Nihon saniku shūzoku shiryō shūsei*. Tokyo: Daiichi Hōki Shuppan, 1975.

Ooms, Herman. "A Structural Analysis of Ancestral Rites and Beliefs," in Newell, William H., ed., *Ancestors*. The Hague: Mouton Publishers, 1976: 61–90.

———. *Tokugawa Ideology: Early Constructs, 1570–1680*. Princeton: Princeton University Press, 1985.

Ōta, Motoko, ed. *Kinsei nihon mabiki kankōshi shiryō shusei*. Tokyo: Tōsui Shobō, 1997.

Otsubo, Sumiko. "Feminist Maternal Eugenics in Wartime Japan," *U.S.-Japan Women's Journal. English Supplement*, No. 17 (1999): 39–76.

———. "Between Two Worlds: Yamanouchi Shigeo and Eugenics in Early Twentieth-Century Japan," *Annals of Science*, Vol. 62, No. 4 (April 1996): 205–231.

———. "Engendering Eugenics: Feminists and Marriage Restriction Legislation in the 1920s," in Molony, Barbara and Uno, Kathleen, eds., *Gendering Modern Japanese History*. Cambridge, MA: Harvard Asia Center, 2005: 225–256.

———. "The Female Body and Eugenic Thought in Meiji Japan," in Low, ed., *Building a Modern Japan*: 61–81.

———. "Women Scientists and Gender Ideology," in Robertson, Jennifer, ed., *A Companion to the Anthropology of Japan*. Malden, MA and Oxford, GB: Blackwell, 2005: 467–482.

Sitcawitch, Sumiko Otsubo. *Eugenics in Imperial Japan: Some Ironies of Modernity, 1883–1945*, Ph.D. dissertation, Ohio State University, 1998.

Owada, Michiko. "Sanji seigen kara tasan shōrei e: bosei ni miru shizuoka-ken kindai josei-shi no tenkai," *Shizuoka-ken kindai-shi kenkyū* 14 (October 1988): 61–63.

Ozaki, Kōji. "Bankoku eisei kaigi to kindai nihon," *Nihonshi Kenkyū*, No. 439 (March 1999): 120–114.

Ozaki, Norman. *Conceptual Changes in Japanese Medicine during the Tokugawa Period*, Ph.D. dissertation, University of California, San Francisco, 1979.

Poovey, Mary. *Making A Social Body: British Cultural Formation, 1834–1864*. Chicago: The University of Chicago Press, 1995.

Porkert, Manfred. *The Theoretical Foundations of Chinese Medicine: Systems of Correspondence*. Cambridge, MA: MIT Press, 1978, c1974.

Prakash, Gyan. *Another Reason: Science and the Imagination of Modern India*. Princeton, NJ: Princeton University Press, 1999.

Roden, Ronald. "Baseball and the Quest for National Dignity in Meiji Japan," *American Historical Review* 85, No. 3 (June 1980): 511–534.

Rogaski, Ruth. *Hygienic Modernity: Meanings of Health and Disease in Treaty-Port China*. Berkeley: University of California Press, 2004.

Rousseau, Julie. *Enduring Labors: the "New Midwife" and the Modern Culture of Childbearing in Early Twentieth Century Japan*. Ph.D. dissertation, Columbia University, 1998.
———. "A Midwife's Bag," *Impressions*, no. 30 (2009): 92–97.
Rubinger, Richard. *Private Academies of Tokugawa Japan*. Princeton, NJ: Princeton University Press, 1982.
Sakai, Shizu. *Nihon no iryōshi*. Tokyo: Tōkyō Shoseki, 1982.
Sakamaki, Hisa. *Otenba saijiki*. Tokyo: Sōshisha, 1979.
Sakurai, Yuki. "Mabiki to datai," in Hayashi, ed., *Nihon no kinsei*: 97–128.
———. "Kinsei no ninshin shussan gensetsu," *Rekishi hyōron*, no. 600 (April 2000): 27–38.
Sato, Barbara. *The New Japanese Woman: Modernity, Media, and Women in Interwar Japan*. Durham, NC: Duke University Press, 2003.
Sawayama, Mikako. "'Umu' shintai to kaitai shussan torishimari," Edo no Shisō Henshū Iinkai, ed., *Edo no shisō, Vol. 6*. Tokyo: Perikansha, 1997: 7–27.
———. *Shussan to shintai no kinsei*. Tokyo: Keisō Shobō, 1998.
Screech, Timon. *The Shogun's Painted Culture: Fear and Creativity in the Japanese States 1760–1829*. London: Reaktion Press, 2000.
———. *The Lens Within The Heart: The Western Scientific Gaze and Popular Imagery in Later Edo Japan*. Honolulu: University of Hawaii, 2002.
Shirai, Chiaki, ed. *Umi sodate to josan no rekishi*. Tokyo: Igaku Shoin, 2016.
Segi, Mitsuo, "Dai-ikkai: Boshi eisei gyōsei no taisei-ki," *Sanfujin-ka no sekai* 29:4 (April 1977): 519–521.
———. "Dai-nikai: Boshi eisei gyōsei no taisei-ki," *Sanfujin-ka no sekai* 29:5 (May 1977): 661–663.
———. "Dai-sankai: Boshi eisei gyōsei no taisei-ki," *Sanfujin-ka no seikai* 29:6 (June 1977): 785–786.
———. "Ninsanpu techō kitei no seitei," *Jinkō mondai kenkyū* 3:8 (August 1942): 32–34.
Silverberg, Miriam. "The Modern Girl as Militant," in Bernstein, ed., *Recreating Japanese Women, 1600–1945*: 239–266.
———. *Erotic, Grotesque, Nonsense: The Mass Culture of Japanese Modern Times*. Berkeley: University of California Press, 2006.
Smith, Thomas C. *Nakahara: Family Farming and Population in a Japanese Village, 1717–1830*. Stanford, CA: Stanford University Press, 1977.
Sōda, Hajime. *Zusetsu Nihon iryō bunkashi*. Kyoto: Shibunkaku Shuppan, 1989.
Stoler, Ann Laura. *Race and the Education of Desire: Foucault's History of Sexuality and the Colonial Order of Things*. Durham, NC: Duke University Press, 1995.
Sugimoto, Isao. *Kinsei jitsugaku shi no kenkyū*. Tokyo: Yoshikawa Kōbunkan, 1962.
Sugimoto, Masayoshi and Swain, David L. *Science and Culture in Traditional Japan*. Cambridge, MA: MIT Press, 1978.

Sugitachi, Giichi. "Kagawa Gen'etsu to Kagawaryū sanka," in Ōtsuka, Keisetsu and Yakazu Dōmei, eds., *Kinsei kanpō igakusho shūsei, Vol. 106, Kagawa Gen'etsu, Kagawa Genchū.* Tokyo: Meicho Shuppan, 1979: 7–51.

———. "Kagawa Gen'etsu to Kagawa ryū sanka no hatten," in Yamada Kazuo and Sugitachi Giichi, eds., *Sanron, sanron-yoku, dokusan ron.* Tokyo: Shussan Kagaku Sōgō Kenkyūjo, 1977.

Sutō, Mikako. *Kagawa-ryū kaiseijutsu no kenkū: Edo jidai no shussan kan, taiji kan ni kansuru ichi kōsatsu.* M.A. thesis, Ochanomizu University, 1990.

———. "Edo jidai no shussankan, taijikan," in Fujiwara Noboru, ed., *Nihon seikatsu shisō kenkyū, Vol. 2.* Morioka: Seikatu-shisō Kenkyūkai, 1990: 61–94.

———. "'Umu' shintai no kindai," *Gendai shisō,* Vol. 19-3 (March, 1991): 220–229.

Suzuki, Zenji. *Nihon no yūseigaku.* Tokyo: Sankyō Shuppan, 1983.

Takahashi, Bonsen. *Datai mabiki no kenkyū.* Tokyo: Daiichi Shobō, 1936.

Takenoshita, Kyūzo and Kishino Yūzō. *Kindai nihon gakkō taiiku shi.* Tokyo: Tōkōkan Shuppansha, 1959.

Takahashi, Hiroyuki. "Senji 'jinkō seisaku' no sai-kentō," Kawagoe and Tomobe, eds., *Seimei to iu risuku*: 141–176.

Tama, Yasuko. *"Kindai kazoku" to bodī poritikkusu.* Kyoto: Sekai Shisō-sha, 2006.

Takaka, Stefan. *Japan's Orient: Rendering Pasts into History.* Berkeley: University of California Press, 1993.

Tavris, Carol. *The Mismeasure of Woman.* New York: Touchstone, 1992.

Terazawa, Yuki. "The State, Midwives, and Reproductive Surveillance in late nineteenth- and early twentieth-century Japan," *U.S.-Japan Women's Journal. English Supplement,* No. 17 (1999): 59–81.

———. "Racializing Bodies through Science in Meiji Japan: The Rise of Race-Based Research in Gynecology," in Low, ed., *Building a Modern Japan*: 83–102.

Tokyo Rekishi Kagaku Kenkyū-kai Fujin Undō-shi Bukai, ed., *On'na to sensō.* Tokyo: Shōwa Shuppan, 1991.

Tomono, Kiyofumi. "'Umu' koto e no kokkateki manazashi no seiritsu: sensō ga umi otoshita 'boshi techō'," *Nihon fujin mondai konwa-kai kaihō* 52 (1992): 57–61.

Tonomura, Hitomi. "Birth Giving and Avoidance Taboo: Women's Body Versus the Historiography of "Ubuya," *Japan Review,* No. 19 (2007): 3–45.

Tonomura, Hitomi, Walthall, Anne, and Wakita, Haruko, eds. *Women and Class in Japanese History.* Ann Arbor: University of Michigan Center for Japanese Studies, 1999.

Uno, Kathleen. *Passages to Modernity: Motherhood, Childhood, and Social Reform in Early Twentieth Century Japan.* Honolulu: University of Hawaii Press, 1999.

Unschuld, Paul U. *Medicine in China: A History of Ideas.* Berkeley: University of California Press, 1985.

Wakita, Haruko and Hanley, Susan, eds. *Jendā no nihonshi, Vol.1.* Tokyo: Tōkyō Daigaku Shuppankai, 1994.

Walker, Brett L. "The Early Modern Japanese State and Ainu Vaccinations: Redefining the Body Politic, 1799–1868," *Past and Present* 163 (May 1999): 121–160.

———. *The Conquest of Ainu Lands: Ecology and Culture in Japanese Expansion, 1590–1800.* Berkeley: University of California Press, 2001.

———. "Epidemic Disease, Medicine, and the Shifting Ecology of Ezo," in Weiner, Michael, ed., *Race, Ethnicity and Migration in Modern Japan, Vol. 1 Race, Ethnicity and Culture in Modern Japan.* London: RoutledgeCurzon (2004): 397–425.

Yoshinaga, Naoko. "Shōwa zenki ni okeru shussan no henyō 'bosei no kyōka,'" *Tōkyō Daigaku daigakuin kyōikugaku kenkyūka kiyō 37* (1997): 21–29.

———. "1930-nen dai ni okeru nōson no san'iku e no kanshin to shisaku," *Tōkyo daigaku daigakuin kyōikugaku kenkyūka kenkyū-shitsu kiyō* 29 (June 2003): 1–13.

Yanagita, Kunio. *San'iku shūzoku goi.* Tokyo: Onshi Zaidan Aiikukai, 1935.

Yasuda, Muneo, "Meiji-ki no sanba yōsei ni tsuite," *Shikyō,* No. 35 (September 1997): 1–19.

Yatanibe, Mariko. "Akago yoiku shiho ni tsuite," in Watanabe Nobuo, ed., *Miyagi no kenkyū Vol. 4.* Osaka; Seibundō Shuppan, 1983: 231–269.

Yoshimura, Noriko. *Osan to deau.* Tokyo: Keisō Shobō, 1985.

———. *Kodomo wo umu.* Tokyo: Iwanami Shoten, 1992.

Yoo, Theodore Jun. *The Politics of Gender in Colonial Korea: Education, Labor, and Health, 1910–1945.* Berkeley: University of California Press, 2008.

Yumoto, Atsuko. "Nagano-ken ni okeru junkai sanba no seiritsu to fukyū," *Shinano* 53: 7 (July 2001): 489–511.

Zainichi no ianfu saiban wo sasaerukai. *Oreno kokoro wa maketenai.* Tokyo: Kinohana-sha, 2007.

Index[1]

[1] Note: Page numbers followed by 'n' refer to notes.

Y. Terazawa, *Knowledge, Power, and Women's Reproductive Health in Japan, 1690–1945*, Genders and Sexualities in History,
https://doi.org/10.1007/978-3-319-73084-4

P

The manufacturer's authorised representative in the EU is Springer Nature Customer Service Centre GmbH, Europaplatz 3, 69115 Heidelberg, Germany. If you have any concerns regarding our products, please contact ProductSafety@springernature.com

Printed and bound by CPI Group (UK) Ltd, Croydon, CR0 4YY
05/10/2026
02228140-0010